Fine Homebuilding
Construction Techniques

Fine Homebuilding
Construction Techniques

The Taunton Press

International Standard Book Number: 0-918804-23-X
Library of Congress Catalog Card Number: 84-50164
Printed in the United States of America

A FINE HOMEBUILDING Book

FINE HOMEBUILDING® is a trademark of
The Taunton Press, Inc.,
registered in the U.S. Patent and Trademark Office.

The Taunton Press, Inc.
63 South Main Street
Box 355
Newtown, Connecticut 06470

Contents

7   Introduction

General Construction

10   Site Layout
13   Waterproofing Earth-Sheltered Houses
16   The Superinsulated House
20   Truss Frame Construction
24   The Rafter Square
30   Roof Framing Simplified
31   A Glossary of Roofing Terms
32   Putting the Lid On
38   Roof Shingling
44   Sidewall Shingling
47   Installing a Factory-Built Skylight
50   Site-Built, Fixed-Glass Skylights
54   Flashing
59   Tips and Techniques

Masonry

62   Laying Brick Arches
66   Putting Down a Brick Floor
70   A Russian Fireplace
74   Form-Based Stone Masonry
79   The Point of Repointing
82   Rammed Earth
88   Surface-Bonded Block
92   Building a Block Foundation
97   Facing a Block Wall With Stone
100   Tips and Techniques

Tools and Materials

104   On-Site Shop
106   Working with Green Wood
109   On-Site Carpentry with a Circular Saw
112   Drywall
118   Veneer Plaster
121   Nail Guns
126   The Renovator's Tool Kit
128   Portable Power Planes
132   Acrylic Glazing
136   Concrete
142   Small-Job Concrete
144   Wood Foundations
147   Tips and Techniques

Log Construction and Timber Framing

150   An Introduction to Timber Framing
155   Tools for Timber Framing
160   Sizing Roughsawn Joists and Beams
165   Raising Heavy Timber
168   Timbers and Templates
170   Appalachian Axman's Art
176   Round-Log Construction
180   Log-Building to Last
182   An Island Retreat
185   Tips and Techniques

Finish and Woodwork

188   The Kitchen Cabinet
195   Counter Intelligence
198   Greenhouse Shutters
200   Cornice Construction
204   Table-Saw Molding
206   Batten Doors
208   Making An Insulated Door
210   Hardwood Strip Flooring
216   Floor Sanding
218   Connecticut River Valley Entrance
224   Staircase Renovation
229   Making Curvilinear Sash
232   Classical Style in a Porch Addition
236   Tips and Techniques

237   Index

Introduction

CONSTRUCTION TECHNIQUES is an anthology of 55 articles from issues 1 through 15 of *Fine Homebuilding* magazine. Here you'll find information on most aspects of building houses—from laying out and laying up foundation walls to roofing and interior finish. The subjects are divided into five chapters and the book is indexed for quick reference.

If you are a professional builder, an architect or a homeowner who wants to participate seriously in making your own shelter, we think you will find this book interesting and helpful.

— The editors

General Construction

Site Layout

On a flat lot, footings can be oriented with precision using batter boards, string and a water level

by Tom Law

Driving that first stake is always exciting. It doesn't matter whether you've been designing and dreaming about the house for years, or you're beginning the first day of actual work on what you hope will be a profitable contract. Laying out the site in preparation for foundation work is your first chance to visualize the house full scale in its setting. The accuracy of your layout and the foundation it defines will also determine how much you will have to struggle to make your house tight and square.

Unless you are building on a sloped site, all you will need to do the job right is a 100-ft. tape measure, a water level, a ball of nylon string, enough lumber for batter boards, a helper and the application of some practical geometry.

Let's assume that the house is a simple rectangle, and that one of the long walls faces south. If precise solar orientation is important, use a compass or one of the many commercially available siting devices (see *FHB* #5, p. 52). If such precision is unnecessary, stand facing the midday sun. Your outstretched left arm will point to the east and your right arm to the west. Unroll a ball of string along this axis for a distance a few feet greater than one of the long

walls of your building. I use braided nylon string because it will take an awesome amount of tension before it breaks, and because braided string doesn't unravel and can be used indefinitely. It comes in a highly visible yellow as well as in white.

Preliminary layout—Select one end of the layout as a starting corner and drive a small stake into the ground. You can use almost anything for a stake—a timber spike or a tent peg—as long as it holds the string off the ground. Now measure back down the string the length of the wall, drive another stake, stretch the string between the two stakes and tie it off. This lets you adjust the placement of the house on the site, and gives you an idea of where to locate the batter boards. Taut strings and accurate squaring are not necessary at this point, as long as the outside dimensions of the house are accurate.

To lay out the rectangle, pull another string from the corner you just established, at a right angle to the long wall. This is where your helper is needed. One person pulls the string while the other guesses at 90°. Measure along

this second string the width of the house and drive the third corner stake. The fourth corner can be found by measuring. Now step back and study the house placement on the site. Stop and think on this one a while—it's a permanent decision. If you are satisfied, you can begin preliminary squaring.

With a 100-ft. tape, measure the diagonals. If they measure the same, then you have created a rectangle. If not, you have a rhomboid, and you will have to adjust the two corner stakes opposite the long, south wall until the diagonals are about equal. Getting to within one or two inches at this point is close enough. I use a steel tape because cloth tapes stretch. A leather thong tied to the metal loop on the zero end of the tape will help you to pull hard and hold a dimension at the same time.

Crouch with your forearms braced against your thighs, and use your body weight to pull against the person on the other end of the tape. On the zero end, hold the leather strap, not the tape, and when you are squaring strings that are suspended from batter boards later on in the layout, keep the tape from lying on the string and deflecting it. Another method of get-

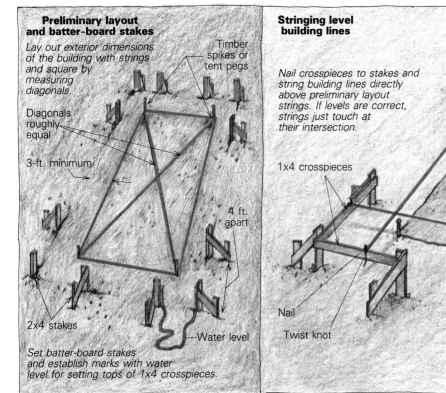

Preliminary layout and batter-board stakes

Lay out exterior dimensions of the building with strings and square by measuring diagonals.

Diagonals roughly equal

3-ft. minimum

Timber spikes or tent pegs

4 ft. apart

2x4 stakes

Water level

Set batter-board stakes and establish marks with water level for setting tops of 1x4 crosspieces.

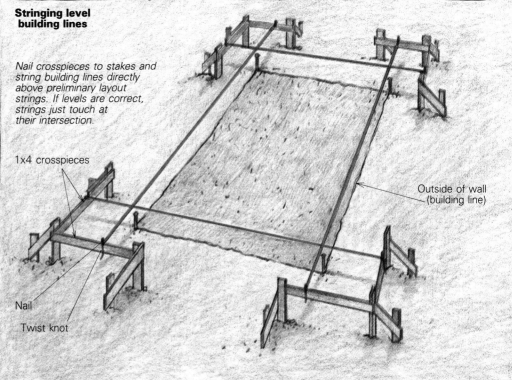

Stringing level building lines

Nail crosspieces to stakes and string building lines directly above preliminary layout strings. If levels are correct, strings just touch at their intersection.

1x4 crosspieces

Nail

Twist knot

Outside of wall (building line)

ting an accurate reading on a tightly stretched tape is to hold the 1-ft. mark rather than the zero. This allows you to grasp the tape with both hands when holding it over a string intersection. With this method, remember to tell the person on the other end of the tape that you are "burning a foot," or "cutting a foot," so the measurement can be adjusted accordingly.

Once this preliminary layout is approximately square and located where you want it on the site, you can set up batter boards. Batter boards are fixed in the ground out beyond the excavation lines. They are temporary wooden corners used to tie the string that accurately defines the perimeter of the building at the outside-of-wall line, or *building line*, and the outside-of-footing line, or *excavation line*. If excavation is required within the perimeter of the structure, for an interior footing or a line of piers, you may want to establish batter boards to hold strings for these lines as well.

I use 2x4 stakes, 3 ft. to 4 ft. long, with 1x4 crosspieces. Usually this is lumber that has been used at least once before. Sharpen the stakes with a circular saw so they will drive easily with a sledgehammer. You'll need three stakes per corner, set about 4 ft. from each other, for a total of 12 for the rectangular house we're using as an example. Drive these stakes about 3 ft. outside the preliminary strings and parallel to them. This placement gives you enough room so that the excavation won't undermine the batter boards, and they can be used until the walls of the house are actually framed. It's a temptation to be exacting in placing your stakes, but you needn't take the time to be too fussy. A good foundation requires precision, but this comes from the strings the batter boards will eventually carry, not from the batter boards themselves. Be sure that the 2x4 stakes are rigid enough to withstand an occa-

sional bumping. If they aren't, nail 1x4 braces near the top of the 2x4 stakes, drive another stake where the brace touches the ground, and nail them together.

Stringing level building lines—I like to nail all the crosspieces at the same level whenever possible. When the strings are in place, this gives me a vertical reference anywhere on the perimeter, which is a real advantage in determining the depth of footings, or the height of foundation walls and concrete block. Since a lot of my foundations are block, I like to set my crosspieces (where the strings will eventually be tied) so that their top edges will be at the same height as the top course of block. To figure this, you must start at the bottom of the footings. In the colder parts of the country, the bottom of footings must always be at or below the frost line to prevent heaving during the winter. If the frost line is 32 in., for instance, and the depth of the footing itself is 8 in., then it will take 24 in., or three courses of 8-in. block to reach grade. Add another three courses, as a convenient and attractive foundation height, and you have a total of 56 in. from the bottom of the footing trench. Keeping this number in mind, measure 24 in. up from the ground on any of the batter-board stakes—the finished height of the block foundation—and make a level line.

I don't own a builder's level or transit, and I've never needed one in my 20 years in the trades. Instead, I use a water level (see *FHB* #6, p. 8). Whatever you use, mark level lines on all the batter boards at 24 in. Accuracy is very important here. Then nail the 1x4 crosspieces with their top edges even with the level marks.

When the crosspieces are up, pull a new string for the south side of the building over the crosspieces of batter boards on each end. Align

it directly above the preliminary layout string by sighting it from above or using a plumb bob. Tack a nail in the top of the cross member and tie one end securely. At the other end pull the string as tight as you can. This establishes a line of elevation, so you don't want it to sag. Use a twist knot to tie it off to another nail. The twist knot (drawing, below right) will keep a nylon string taut, while still allowing it to be released instantly for resetting. This knot doesn't work well with cotton line. Continue stringing until all the lines are up. If everything is level, the strings will just touch as they intersect a few feet in front of the batter boards.

Squaring the corners—The next step is to square the stringed corners, this time using the 3-4-5 check. These numbers refer to the sides and hypotenuse of a right triangle. Since 6-8-10 and 12-16-20 triangles are proportional to a 3-4-5, use the largest one you can for optimum accuracy. The intersection of the strings of each corner defines the 90° angle of the 3-4-5 triangle. You'll need a helper to measure and adjust the strings until the hypotenuse is exactly proportional. With one person holding zero (or 12 in., if you are "burning a foot") on the tape, the other person can mark the legs of the triangle on the string with a pencil, and then knot a short length of string loosely around the mark. Double-check the measurement, and then tighten this knot. Measure from knot to knot to get the hypotenuse, and adjust the strings on the batter boards if necessary. These adjustments will require driving new nails into the top edge of the crosspieces. Pull the previous nail as you correct the position of the string. If you don't, it can get very confusing when the strings come down temporarily for digging the footing trenches. When you take the string down and put it back up, check

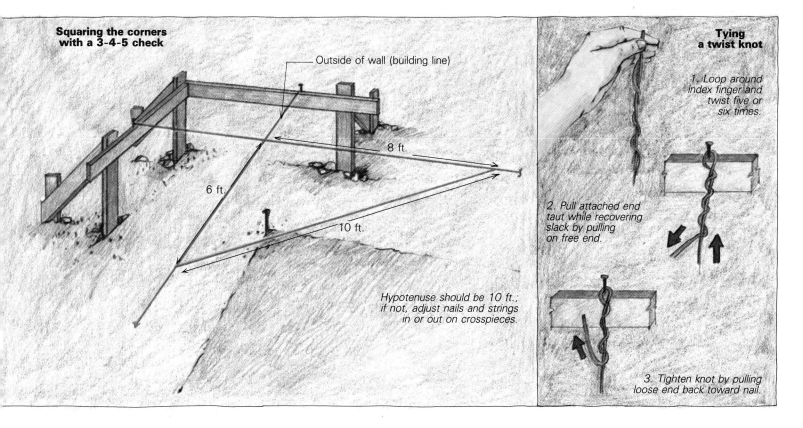

Squaring the corners with a 3-4-5 check

Outside of wall (building line)

8 ft.

6 ft.

10 ft.

Hypotenuse should be 10 ft.; if not, adjust nails and strings in or out on crosspieces.

Tying a twist knot

1. Loop around index finger and twist five or six times.

2. Pull attached end taut while recovering slack by pulling on free end.

3. Tighten knot by pulling loose end back toward nail.

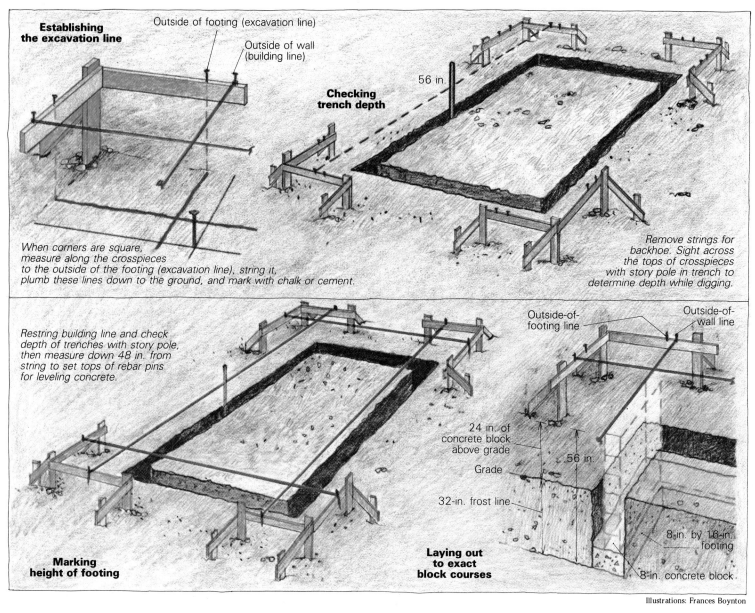

Establishing the excavation line

Outside of footing (excavation line)

Outside of wall (building line)

When corners are square, measure along the crosspieces to the outside of the footing (excavation line), string it, plumb these lines down to the ground, and mark with chalk or cement.

Checking trench depth

56 in.

Remove strings for backhoe. Sight across the tops of crosspieces with story pole in trench to determine depth while digging.

Restring building line and check depth of trenches with story pole, then measure down 48 in. from string to set tops of rebar pins for leveling concrete.

Marking height of footing

Laying out to exact block courses

Outside-of-footing line

Outside-of-wall line

24 in. of concrete block above grade

56 in.

Grade

32-in. frost line

8-in. by 16-in. footing

8-in. concrete block

Illustrations: Frances Boynton

the length to the knots, because nylon string stretches. To finish squaring up, use the 3-4-5 check on another corner, then check the diagonals again.

With the strings squared up to represent the eventual building lines (the outside-of-wall lines), and an elevation established, draw a plumb line down from the strings on the face of each cross-member, and write the wall thickness and the amount that the footing will project beyond the outside of the wall on the batter board. To reduce confusion, I drive nails and hang strings only on the outside-of-wall line and outside-of-footing line.

Excavation lines—Usually a footing is twice as wide as the foundation wall it supports, and as deep as the wall is thick. Footings are contained either by building a wooden form, or by digging a trench and using the undisturbed earth as formwork. I usually use the trench method. To show the backhoe operator where to dig, I plumb down to the ground from the outside-of-footing line and stretch a string at grade. I mark over this string with lime or cement dust as if I were marking out an athletic field. You can also use scouring powder with a shaker top. The backhoe operator should hold

the outside tooth of his bucket to the line, and dig to the inside.

Checking trench and footing depth—The batter boards give a quick vertical reference for determining how deep to trench. In our example, the bottom of the footing is 56 in. down from the top edge of the crosspiece. Instead of strings, which during excavation should be wound around a stick, use a story pole with a 56-in. mark on it. Stand the stick in the trench and sight from the top of one batter board to another. The mark on the story pole should line up with them.

When the machine work is finished, string the lines on the outside-of-wall line (the building line). Pull them very tight. Shape up the sides and corners of the trench with a shovel, maintaining the 56-in. depth you can now check by measuring from the string.

Next, set the depth of pour for the concrete. I use ⅜-in. or ½-in. steel reinforcing rods about 6 ft. to 8 ft. apart to indicate depth during the pour. Cut them about twice the depth of the footing so that you can drive them into the ground. Measuring down from the string to the top of the rebar, carefully tap them with a sledge until you read 48 in. on the tape. This

will give you an 8-in. footing at the 32-in. frost line, and six courses of block on top will bring you up to the string (drawing, above right). Then pour the concrete level with the top of the rebar. I use a garden rake to push the concrete around and for initial screeding. I hold the rake in a vertical position to smooth the top of the concrete and jitterbug the coarse aggregate down into the mix. You also might want to use a 2x4 screed short enough to fit between the rebar depth indicators, but it's not necessary to trowel the surface smooth. If you are pouring a foundation wall on top of the footing instead of laying block, the same techniques can be used, but remember to form a keyway in the footing to receive the next pour, and check with local codes to see if vertical rebar is required to tie the footing to the foundation wall.

The next day the concrete will be hardened sufficiently to begin working on the foundation walls. I usually drop a plumb bob down from the outside of wall lines and snap chalklines on the green footing. If the foundation is to be concrete block, then marking the corners will be enough since the mason will be pulling his own lines from corner to corner on each course. □

Tom Law is a builder in Davidsonville, Md.

Waterproofing Earth-Sheltered Houses

There's a lot more to it than tar on the walls

by Charles A. Lane

Waterproofing an underground house is a one-time, do-it-right job. It requires as much care as choosing the proper structural system, and seems to cause most builders a lot more trouble.

No single waterproofing material could possibly withstand all the environmental forces that will test an earth-sheltered building. The answer is a total waterproofing system, one that is properly designed and carefully installed. Such a system requires: 1. Thoughtful site planning and landscaping, so that as little water as possible will have to be dealt with at all; 2. A good drainage system, including proper backfilling material and techniques, so most of what water there is will be diverted from the building, and 3. A well-chosen and carefully installed waterproofing material that is applied to the structure itself and is the final line of defense.

The site—Choosing the right site is the most important step in building an earth-sheltered house. Avoid floodplains, gullies and high water tables. It is easier and cheaper to avoid water problems than to try to cope with them. Although it is possible to build a watertight structure on virtually any site, the costs and problems inherent in constructing a "landlocked submarine" are enormous. An energy-efficient design above ground level would probably be more appropriate to a wet site.

A simple site inspection often reveals areas with potential groundwater or runoff problems. Surface conditions, vegetation and soil layers exposed to nearby road cuts or excavations can provide clues to seasonal variations in surface and subsurface water levels. Nonetheless, you should always have a soils test for your site. It will add an additional $400 to $800 to construction costs but will yield vital information, not only for determining waterproofing methods, but also for structural calculations.

You may want to build your earth-sheltered house into a hillside. If so, pay special attention to the uphill surface runoff and to subsurface water percolation. You'll have to plan drains so that the water will flow naturally to an exposed surface outlet. If this is impossible, you can route drains to a storm sewer if codes allow, or to a sump that can be pumped out when necessary. If you neglect this sort of planning, your house will

obstruct the normal flow of water and will at least leak a bit. At worst, it will leak badly, begin to "float," or even fail structurally.

Waterproofing materials—Underground houses are vulnerable to dynamic forces created by the earth. They settle, rotate and slide under the soil load. Retaining walls often deflect and crack from house walls. Roof parapet walls can crack from frost action. Cracks are also common at footings, skylights, roof projections and cold joints (joints between structural elements poured in two separate operations). Your waterproofing material will have to be both flexible and strong enough to function under such stresses. Liquid-applied materials like polyurethanes, polysulfides and silicones tear easily. They adhere tightly to the surface of the concrete, but cannot give to compensate for structural movement or cracking.

I've found four generic waterproofing products that are well suited for use in underground houses; isobutylene isoprene elastomer sheets, ethylene propylene diene monomer (terpolymer) elastomer sheets, modified bitumens and sodium montmorillonite.

There are many other products on the market that are excellent for less demanding purposes, but are not suitable for long-term below-ground use. They include polyethylene sheets, polyvinyl chloride sheets (PVC), bituminous coatings (pitch and asphalt), oil-based paints, rubber-based paints, latex paints, portland cement fill coats, iron oxide cement coatings, silicone compound coatings and urethane coatings.

A good cement pour is important if your waterproofing materials are going to be effective. Check the structure carefully, and fill and repair any gaps. Be sure the surface is clean and dry before proceeding.

The elastomers—Isobutylene isoprene (more commonly called Butyl), and ethylene propylene diene monomer (EPDM) have a lot in common. Both materials are soft, strong, flexible and resistant to soil fungi and bacteria. They cannot reseal themselves, so they are sometimes completely glued to the concrete to keep moisture from moving behind the membrane. Any structural shifting, however, can cause an

elastomer to rip if it is applied this way. It is better to glue both Butyl and EPDM down in a grid pattern, which will contain any leakage, while allowing the membrane to give without tearing. Likewise, you should use extra sheet material at flashing points and odd angles so that there is plenty of room for play if the building shifts or cracks.

Both EPDM's and Butyl can soak up heat on sunny days and stretch under their own weight. If they are seamed and glued in place while stretched, they will rip as they cool off in the ground. Using separate sheets for the roof and walls makes it unnecessary to drape material over the roof edge, and grid gluing should leave enough slack to compensate for contraction and structural movement.

Improperly installed seams are probably the membranes' most frequent cause of failure. Most sheet systems include adhesives for seaming overlapped sheets, and are usually installed by professionals rather than owner-builders. Elastomer sheets come in rolls up to 50 ft. wide, so few seams are needed. Both Butyl and EPDM's are available in various thicknesses, but $\frac{1}{16}$ in. is best for waterproofing underground houses.

Price and availability are really the only reasons for choosing one of these materials over the other, since their performances underground are similar. In November 1980 Butyl cost about $1.60 per square foot (installed), or $2.70 with a 10-year guarantee. EPDM's were priced at about $1.50 per sq. ft. (installed).

Modified bitumens—Modified bitumens are a combination of asphalt or tar and synthetic rubber. They resist chemicals, and are soft, flexible and strong. Most of them use a polyethylene-sheet backing that gives the membranes additional strength and adhesion control at the seams. This backing becomes brittle with exposure to ultraviolet rays, so you should backfill quickly to shield it from the sun. Some rubberized bitumens are self-adhesive, while others rely on an adhesive you apply at the seams.

Modified bitumens usually come in narrow rolls (36 in. to 43 in. wide), so more seams are required than with Butyl or EPDM sheets. The surfaces to which they are applied must be smooth, completely dry (including invisible condensa-

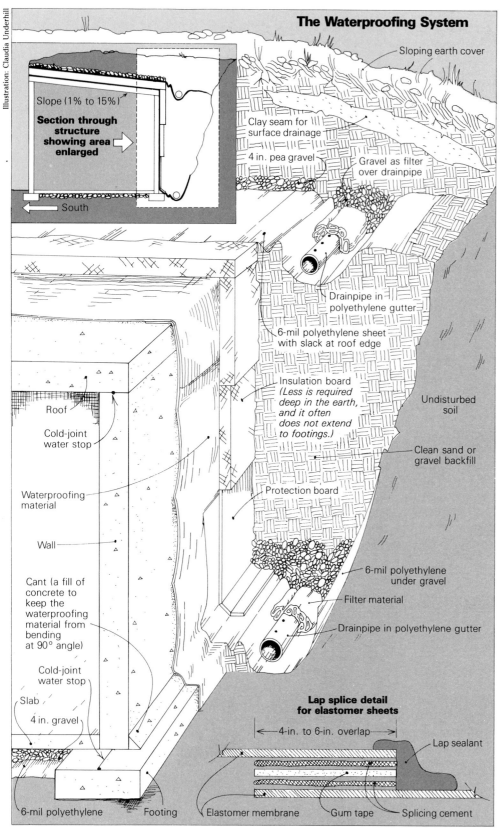

The Waterproofing System

Sloping earth cover

Section through structure showing area enlarged

Slope (1% to 15%)

South

Clay seam for surface drainage

4 in. pea gravel

Gravel as filter over drainpipe

Drainpipe in polyethylene gutter

6-mil polyethylene sheet with slack at roof edge

Insulation board *(Less is required deep in the earth, and it often does not extend to footings.)*

Undisturbed soil

Clean sand or gravel backfill

Protection board

Roof

Cold-joint water stop

Waterproofing material

Wall

Cant (a fill of concrete to keep the waterproofing material from bending at 90° angle)

Cold-joint water stop

Slab

4 in. gravel

6-mil polyethylene under gravel

Filter material

Drainpipe in polyethylene gutter

Lap splice detail for elastomer sheets

4-in. to 6-in. overlap

Lap sealant

6-mil polyethylene   Footing   Elastomer membrane   Gum tape   Splicing cement

Elastomer sheet waterproofing is shown here, but the system would be the same for any of the materials we discuss.

tion) and above 40 °F. Roofs waterproofed with modified bitumens must always slope to keep the material from being subjected to continuous water head from ponding. Seams are joined by overlapping the self-adhesive rolls, or by overlapping and gluing the non-adhesive products with a bitumen mastic.

Unlike Butyl and EPDM membranes, modified bitumens have some inherent resealing ability, and can close off small punctures or cuts. They also are elastic enough to bridge small cracks

without tearing, but installing them in temperatures below 40°F results in wrinkling and cracking. In November 1980 they cost about $.85 per sq. ft, including installation.

Avoid bends of 90° or more in any sheet material, whether elastomer or bitumen. Such sharp folds induce stresses that, in time, may cause the material to crack and leak.

Montmorillonite clays—The montmorillonite clay used for waterproofing is called bentonite.

It was produced by volcanic action, and is now mined in Wyoming and the Black Hills of South Dakota. The loose bentonite can swell, when wet, to 12 or 15 times its dry volume. When it is confined to a definite space, however, its density increases rather than its volume. Moisture doesn't pass through bentonite, but is absorbed by it; so bentonite becomes an excellent waterproofing agent when saturated. It can become saturated, dry out, and become saturated again time after time, without losing its effectiveness. Under the trade name Volclay, bentonite comes in 4-ft. by 4-ft. by $\frac{3}{16}$-in. panels with a minimum of 1 lb. per sq. ft. of clay between two kraft paper sheets. Volclay panels should only be used to waterproof vertical surfaces. Seams between panels are little trouble, because the kraft paper rots and the clay joins to form an effective water barrier. Volclay cost $.60 per sq. ft. installed, in November 1980.

In 1967 a spray-on application of bentonite was developed (Bentonize-R-80-S). The spray-on application ($1.20 per sq. ft installed, in November 1980, including a 5-year guarantee) results in a $\frac{3}{16}$-in. to $\frac{1}{4}$-in. layer of bentonite mixed with a gel-like binder. This amounts to about 1.5 lb. per sq. ft. of clay applied on the structure.

Bentonite can also be troweled onto the structure. In Bentonize R-80-T the bentonite is mixed with a gel and is reinforced with thousands of lint-like synthetic fibers. This should also be troweled to a thickness of $\frac{3}{16}$ in. to $\frac{1}{4}$ in. This can be a do-it-yourself operation, and as such, it cost $.70 per sq. ft. in November 1980. Both Bentonize products can be used on roofs, as well as walls. Bentonize R-80-T should be covered with a sheet of 6-mil polyethylene on the roof and over all uninsulated portions of the wall. It cures in four hours. Neither panel Volclay nor spray-on Bentonize needs time to cure.

Applying the proper thickness of trowel-on and spray-on bentonite is important. Areas prone to structural movement should get a thicker application. Both types adhere well to most surfaces, and have excellent resealing and crack bridging characteristics because of their monolithic application. Bentonite clays should not be used in high-brine soils, because salt affects the absorption of water by the clay. All clays should be backfilled immediately, because rain can wash them away.

Another bentonite product can be used as a water stop at cold joints. Waterstop-Plus is a clay gel that is applied in a bead between concrete pours. It is cheaper and more convenient than traditional vinyl or copper water stops, which must be placed during the pour.

Making your choice—All of these waterproofing materials will do the job if they are installed properly. Consider the specific strengths and weaknesses of each in relation to the requirements of your project. If your design, for example, calls for a lot of angles and projections,

Illustration: Claudia Underhill

bentonite might be your best bet, because sheet materials would require too many seams. On the other hand, clay must be backfilled as soon as it's installed. If you can't schedule things this tightly, go with one of the other materials. Consider the weather, too. Modified bitumens, for example, aren't at their best when installed in extreme cold. You'll probably want to use something else if you're building during the winter.

The idea is to cut down on the chances for failure. Cost and availability aside, choose the material you think will give you the least trouble during installation on your site, with your design, at your schedule.

Liability—Most waterproofing manufacturers offer a limited guarantee, sometimes at extra cost, stating that the manufacturer will replace or repair a defective product if it was applied in accordance with their specifications. The owner pays to find the leak, remove the backfill, repair any interior damage caused by the leak, and replace the backfill. If the owner can prove that the material has been properly applied, the manufacturer will pay to patch or replace the defective area, but manufacturers usually deny liability for consequential damage.

Most building insurance policies do not cover damage to the building or its contents caused by groundwater leakage; water seeping through man-made soil berms and roof cover would undoubtedly be classified as groundwater. Simply put, the owner, architect or contractor really carries the risk of waterproofing failure.

Read the material specification manual from cover to cover, and follow it to the letter. If you hire a crew to do the work, strict supervision by architect or general contractor is a must.

Drainage and backfilling—When the waterproofing material has been installed, drainage ducts and construction-placed earth will complete the total waterproofing system. They capture and remove groundwater from contact with the structure.

Place perforated plastic or clay drain tile around the perimeter of the footings as in standard building practice. To prevent drains from clogging, wrap porous filter fabric (fiberglass insulation batts will also work) around the top of the drain tile. Except for a capping layer of less permeable soil, such as clay, you should use a porous, granular material (coarse sand, pea rock or gravel) for backfill. Besides providing good drainage, these materials prevent the surrounding soil from settling and clogging foundation drain tiles. Bring in sand and gravel, if necessary. Commercial wall-draining aids are available. They are placed over the wall surface and provide a space for water to drain between the wall and the inadequate soil backfill. These commercial draining aids work, but shouldn't be used on an underground house because they result in reduced thermal performance. The water

flowing along the wall surface removes large quantities of heat which should be transferred into the adjacent earth berm.

One inch of rain deposits a thousand gallons (8,000 lbs.) of water on the roof of a 1600-sq. ft. structure. A porous backfill material above the roof's final waterproofing layers and beneath the topsoil is critical to hasten drainage. Filter fabric placed over the backfill and beneath the topsoil helps prevent clogging of the porous granular fill. When placing soil on the roof, dump the backfill and pull it back rather than pushing it forward. The weight of the Bobcat or spreader can then compact the previously spread soil, resulting in less soil sliding and friction. To discourage ponding, design your roof to have a shallow slope of between 1% and 15% for drainage. Too steep a slope can cause soil to slip and waterproofing to shift.

Roofs should allow water to flow unobstructed to earth-bermed walls, from which it should drain down and away from the wall surface. One way to do this is to lay a piece of 6-mil polyethylene over the draining roof edge (usually, but not always, at the rear of the house) to form a gutter under a drainage tile in the bermed earth, as shown in the drawing at left. This gutter channels roof water to the drainage pipe, which is set an an angle to carry it away. A fold in the polyethylene will allow the sheet to give as the soil settles. Assemble the same sort of system at all footings as well, so that water won't bypass the drain tile and cause problems at the footing's joint with the floor slab. Don't drain water to normal drain gutters and spouts; they clog and freeze up.

Avoid designs that encourage surface or soil water to drain to any flashing details such as skylights or mechanical projections (plumbing vents, flues, air intakes). Water trapped at these locations causes backup and leakage problems. Because floors are also subject to leaking and moisture migration problems, you should lay a 6-mil polyethylene sheet under 4 in. of rock fill or sand before pouring the slab. This underlying fill should drain to the perimeter tiles, to remove any accumulated soil moisture. In damp conditions, you may need to lay additional drain tiles and up to a foot of fill under the floor.

Protect all waterproofing materials from the soil during backfilling, because friction, sharp aggregates and debris can tear the material. In many cases, external insulation placed over the waterproofing will offer protection, but insulation on an earth-sheltered house varies in thickness, and it often doesn't extend all the way to the footings. Use a separate, unfastened protective covering where there is no insulation. Some materials that will work are: 15-lb. building felt below grade; heavy-gauge (45-mil) PVC or polyethylene plastic sheeting below grade; $\frac{1}{4}$-in. fiberboard above and below grade; a 4-in. layer of sand over the roof waterproofing material; bead board, and asphalt board. If your backfill

consists of especially large, heavy or sharp stones, use cement-asbestos board.

No waterproofing material should ever terminate below the ground. Ask the manufacturer how to blend the below-grade waterproofing with the above-ground wall sections. Some of the materials listed above can be used above grade to protect this area of flashing. You can also install a strip of either of the elastomers at this point, or extend rigid insulation board above the surface and coat it with a fiberglass-reinforced cement coating.

You should compact wall backfill in 9-in. to 12-in. layers. This requires less force than trying to compact all of it at once, and offers superior results. Insufficient compaction allows the earth to settle on its own, sometimes displacing insulation and waterproofing materials. If you backfill before the slab, floors and roof are in place, you'll risk structural damage; but if that's your only choice, be sure to shore up the walls with sturdy bracing.

Once everything else is in place, landscape your site to control ponding and erosion. Pay attention to your site's contours and soil types, and plant hardy shrubs and grasses on inclines and where water might pool. Be careful, though, not to locate plants with deep root systems in spots where they could damage the waterproofing material or interfere with drainage pipes.

An earth-sheltered house that is well-sited and landscaped, protected by a suitable waterproofing material, and properly drained and backfilled, will be as free of unwanted moisture as any conventional above-ground building—and snugger than most. □

Charles A. Lane is the assistant director of the Underground Space Center at the University of Minnesota in Minneapolis.

Sources of Supply

Isobutylene isoprene (Butyl)
American Hydrotech, Inc. (Uniroyal)
   541 N. Fairbanks Ct., Chicago IL 60611
Carlisle Tire and Rubber Co.
   Construction Materials Department, P.O. Box 99, Carlisle PA 17013
B.F. Goodrich
   Construction Products Division, 500 S. Main St., Akron OH 44318
Ethylene propylene diene monomer (EPDM)
American Hydrotech, Inc. (above)
Carlisle Tire and Rubber Co. (above)
Firestone Tire and Rubber Co.
   Industrial Products Div., 1700 Firestone Blvd., Noblesville IN 46060
Gates Engineering Co.
   P.O. Box 1711, Wilmington DE 19899
B.F. Goodrich (above)
Modified bitumens
W.R. Grace
   Construction Products Division, 62 Whitemore Ave., Cambridge MA 02140
Koppers Co., Inc.
   Koppers Bldg., Pittsburgh PA 15219
Sodium montmorillonite (clays)
American Colloid Co. (Volclay)
   Building Material Division, 5100 Suffield Ct., Skokie IL 60076
Effective Building Products (Bentonize, Waterstop-Plus)
   1724 Concordia Ave., St. Paul MN 55104

The Superinsulated House

Thick walls and airtight construction keep the cold outside

by John R. Hughes

Superinsulation is usually thought of as a recent development in building design; it isn't. The first superinsulated buildings in North America were built in New England by early settlers. They were ice houses, double-walled log structures with about 2 ft. between their inner and outer walls. This cavity was filled with sawdust, yielding an insulation value of about R-50 (or $R_{SI}8.8$; a metric designation for insulative value; $R_{SI}1 = R\text{-}5.68$). Today's high heating costs have brought about new interest in superinsulated construction. In the Canadian prairies where I live, we've found that superinsulation is the best way to cut heating bills during the long, cold winter.

The first modern superinsulated house, called the Lo-Cal house, was designed in 1974 by the Small Homes Building Council at the University of Illinois. It features double stud walls with a cavity between them for extra insulation. Superinsulation accounts for 80% of its heat retention; window placement and proper siting for the rest. In 1977 the Canadian federal and Saskatchewan provincial governments collaborated on the design and construction of an energy-efficient house suitable for a northern prairie climate. It was originally to have been primarily an active solar design with a bit of extra insulation thrown in for good measure. Luckily the priorities were reversed, and the Saskatchewan Conservation House (SCH) was born. Since then we have been building superinsulated houses in this part of the country based on the SCH design, though a few modifications have been introduced along the way. So far the designs have worked as well as or better than we expected. For many of these houses, a

woodstove is the primary source of heat. Typically, it consumes just one or two cords of wood per winter.

A tight vapor barrier—The word superinsulation probably makes most people think of thick layers of fiberglass. Actually, a tight vapor barrier is just as important as the thickness of the insulation. Builders in Canada usually refer to it as the air/vapor barrier (a/vb) because its job is to stop air infiltration as well as to retard the movement of water vapor through the wall. If you keep drafts of frigid air out of the house, the extra insulation will be able to perform the way it should.

Of course, the key to all this is ensuring an airtight a/vb. We use 6-mil polyethylene because it's resistant to tearing. Solid backing is necessary where two sheets of a/vb are joined together, and all sheets must be overlapped at least 4 in. (100 mm). We like to overlap the a/vb one stud space or 16 in. (400 mm). Sealant, in the form of a continuous bead running the full length of each joint between two sheets, is critical to air-tightness. Since the sealant acts as a form of glue, it must be a type that never dries out or becomes brittle. We have found that acoustic sealant is best. It never sets up or hardens, and changes in temperature or humidity don't affect it. Brands commonly available in Western Canada include Chemtron Metaseal, PRC and Tremco.

Once installed, the a/vb has to be protected from punctures. The tearing that can easily occur during construction will ruin its effectiveness. The modified SCH wall design calls for plywood sheathing right over the 6-mil a/vb,

and this protects it very well. We keep the a/vb continuous from crawl space or basement to attic by carefully sealing adjoining sections of a/vb. Windows and doors, plumbing vents, metal chimneys, and wall-to-ceiling junctures require special a/vb joints, which I'll discuss when we get to construction details.

The location of the a/vb varies in superinsulation designs. In the Lo-Cal wall and in the original SCH wall, the a/vb is installed on the inside stud face of the interior wall, just under the finished interior wall surface. There are several drawbacks to this placement. The installation job must be done after the wall is up, and working on a vertical wall with the thin sheeting is very tough. Secondly, in this location the a/vb is extremely vulnerable to construction damage. And once installed, it must be broken for electrical wall outlets. We prefer to install it on the outside stud face of the interior wall (drawing, facing page). This way, we can work horizontally, just laying the sheeting over the finished frame of the interior wall after we've built it on the floor. In our design, ¼-in. (6-mm) or ½-in. (12-mm) plywood sheathing goes over the a/vb. This solid backing, along with the recessed location of the sheeting, protects the a/vb from wear and tear.

Building a superinsulated wall—We construct our walls on the completed subfloor of the house and then tilt them up. All that's required after tilt-up is the finished interior surface (usually gypsum board or paneling) and exterior siding. We frame the inner wall first, using 2x4 studs on 16-in. (400-mm) centers and either 2x10s or 2x12s for the window and door

---

Lo-Cal	Saskatchewan Conservation House	Modified SCH	Three systems for superinsulation using double-wall construction
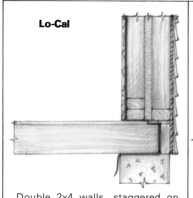		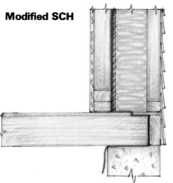	*The earliest modern system for a superinsulated building was the Lo-Cal house, which was designed in 1974 by the Small Homes Building Council at the University of Illinois. In 1977, the Canadian government collaborated with the province of Saskatchewan to design the Saskatchewan Conservation House, or SCH. This design was later modified by Canadian researchers in an effort to cut construction costs and to protect the air/vapor barrier (■) from damage during the building process.*
Double 2x4 walls, staggered on 24-in. centers. 8½-in. wall thickness (R-32); R-19 floor insul., R-39 attic.	2x4 inner wall, aligned with 2x6 outer wall 24 in. o.c. 12½ in. thick (R-44); R-36 floor insul., R-60 attic.	2x4 inner wall, 16 in. o.c., 2x3 outer wall, 24 in. o.c. 12 in. thick (R-42); R-36 floor insul., R-60 attic.	

headers. This is a conventional stud wall, with the usual double top and bottom plates.

The complete frame of the inner wall rests on the subfloor with its outside face exposed, ready to be covered with the poly a/vb. Window and door openings are also covered at this stage. Always leave extra sheeting—about 18 in. (450 mm) of it if possible—hanging over all edges of the wall. You'll need these overhangs to tie into adjoining sections of poly.

Nailing plywood sheathing over the a/vb is the next step. The inner wall is load-bearing, so the sheathing is a structural necessity, as well as being good protection for the poly. Cut the plywood to fit around rough window and door openings before you nail it down.

Once the inner wall is sheathed, you can start building the outer wall right on top of it. This is the easiest way to ensure that the rough openings will be identical in both walls. This wall isn't structural, so we usually use 2x3 studs to save money and reduce weight. We also frame to 24-in. (600-mm) centers rather than 16, and instead of using conventional headers, we span all openings with 2x3s. The outer wall has to be set away from the inner one. For this we use 2x6 spacer blocks, making sure the wall sections are in vertical alignment. To tie them together we nail plywood plates along the top and bottom of the wall, and later we sheathe the insides of door and window openings with plywood strips of the same width. Plate width is almost 12 in. (300 mm), with the actual width of the framing lumber taken into account ($3\frac{1}{2}$ in. for the 2x4; $\frac{1}{2}$-in. plywood; $5\frac{1}{2}$ in. for the 2x6 spacer; $2\frac{1}{2}$ in. for the 2x3). We get four strips from a sheet of plywood, with no waste.

Once the two walls are connected, remove the 2x6 spacer and start stuffing in insulation. A

Stuffing the double wall. **Assembled flat on the subfloor, the exterior wall gets its unfaced fiberglass insulation. The air/vapor barrier, here installed against the inside stud face, is left intact whenever possible over all rough openings for doors and windows. It will be cut and folded back against the inner studs once the wall is up.**

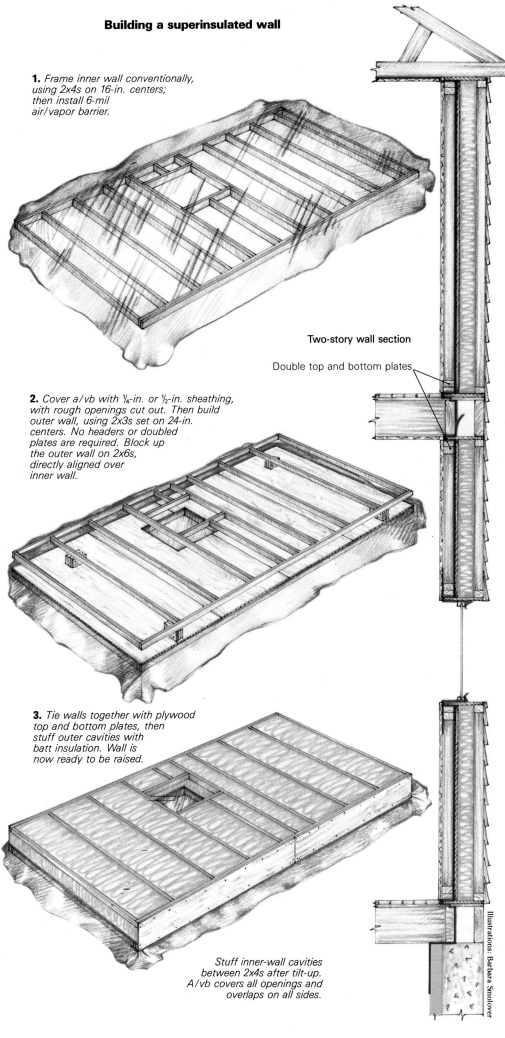

Building a superinsulated wall

1. *Frame inner wall conventionally, using 2x4s on 16-in. centers; then install 6-mil air/vapor barrier.*

2. *Cover a/vb with $\frac{1}{4}$-in. or $\frac{1}{2}$-in. sheathing, with rough openings cut out. Then build outer wall, using 2x3s set on 24-in. centers. No headers or doubled plates are required. Block up the outer wall on 2x6s, directly aligned over inner wall.*

3. *Tie walls together with plywood top and bottom plates, then stuff outer cavities with batt insulation. Wall is now ready to be raised.*

Two-story wall section

Double top and bottom plates

Stuff inner-wall cavities between 2x4s after tilt-up. A/vb covers all openings and overlaps on all sides.

Preventing infiltration around window and door openings

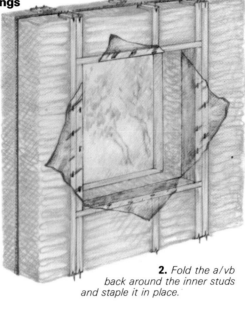

1. With the a/vb installed over the rough window or door opening, cut an X in the poly sheet from corner to corner.

2. Fold the a/vb back around the inner studs and staple it in place.

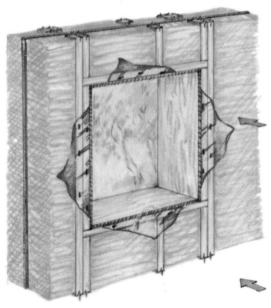

3. Sheath the rough opening with plywood.

4. Enclose the window frame with the poly a/vb before shimming it in position. Use a bead of acoustic sealant and staples to seal wall and window a/vb sections together. Adding the finish trim covers the a/vb and completes the job.

wall of this design can hold 8 in. (200 mm) in the outer cavity, which is accessible at this stage of construction. The 3½-in. (88-mm) inner cavity can be filled only after tilting the wall up, which is the next step. Double walls are heavy, so build them close to where they'll be placed. Plywood is applied to the outer wall only if the building is to be stuccoed, as was the case here. Otherwise, siding can be applied directly to the 2x3 outer wall.

Windows and doors—With the wall in place, the poly sheeting should still be intact over all openings. In ordinary construction the a/vb is simply cut off at the edge of the rough opening, but we follow a different procedure. Cut a giant X in the poly, with the legs of the X running to the corners of the opening. Then fold the poly back around the inner wall studs framing the opening and staple it to their inside faces before trimming off the excess.

The next step is to cover the double-walled sides of the opening with plywood. The plywood strips will be the same width as the plywood top and bottom plates. This gives you an enclosed rough opening into which the door or window is shim-fit. As you install each window unit, you should enclose the window frame with the a/vb. To do this, attach one edge of a strip of poly to the outside face of the window frame. The poly strip should be wide enough to overlap the a/vb stapled to the inner wall studs that frame the opening (drawing, left). Run a bead of acoustic sealant along the joint between wall and window a/vb sections. Once this is done, the casings and finished wall can go up conventionally. This detail, which has to be repeated for all windows and doors, requires extra time and effort, especially if you haven't done it before. But it keeps the a/vb continuous and solves the age-old problem of infiltration around door and window frames.

As far as window or door position in the opening is concerned, you have a number of choices. If a window is positioned flush with the exterior wall, you will have a deep interior sill, which makes a nice window seat or shelf for a planter box. If you set the window back flush with the inside wall, the exterior depth provides protection from heat-robbing winds. I have also seen windows positioned halfway between. Doors are generally framed against the inner wall studs to take advantage of the stronger framing.

Roof and ceiling—Pre-manufactured roof trusses are used on most of the superinsulated houses we build. We recommend from 10 in. to 15 in. (250 mm to 400 mm) of attic insulation, which is dropped or blown into place after the a/vb and finished ceiling have been installed. In most instances, the roof trusses span the entire width of the house, so no interior bearing

A superinsulated wall section 1 ft. thick is tilted up on the subfloor. The poly air/vapor barrier is visible behind the inner wall studs, which must still be packed with 3½-in. thick fiberglass batts. Outer walls contain 8 in. of insulation, for a total R-value of 42.

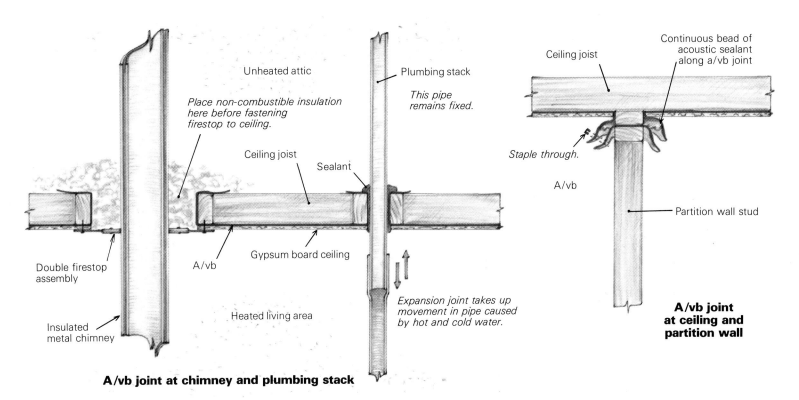

Unheated attic

Place non-combustible insulation here before fastening firestop to ceiling.

Plumbing stack

This pipe remains fixed.

Ceiling joist

Sealant

Ceiling joist

Continuous bead of acoustic sealant along a/vb joint

Double firestop assembly

A/vb

Gypsum board ceiling

Staple through.

A/vb

Partition wall stud

Insulated metal chimney

Heated living area

Expansion joint takes up movement in pipe caused by hot and cold water.

A/vb joint at ceiling and partition wall

A/vb joint at chimney and plumbing stack

walls are needed. Having the ceiling uninterrupted by partition walls is an advantage at this stage of construction because it's easy to install an unbroken vapor barrier across the ceiling joists. Once the a/vb has been stapled up, you can do all the ceiling drywall, then build and finish the interior walls last. This sequence means the drywall crew will have to come in twice, but they'll find the ceiling job much easier to do, since no fitting around closets or interior walls will be required.

If you have to frame up interior walls first and install the ceiling a/vb room by room, keep the barrier continuous by sandwiching a 10-in. (250-mm) wide strip of poly between the top and double plates. Then as you install the ceiling a/vb, join the excess poly along each ceiling edge to the protruding edge of the sandwiched strip. Use acoustic sealant to make the joint, and staple through both pieces into the frame (drawing, above right). This work has to be done conscientiously, even though it will be totally hidden by the finished interior wall.

Chimneys and plumbing stacks—Ordinary fireplaces with brick chimneys aren't recommended in superinsulated houses because of the amount of heat energy they waste, the large drafts of air they require, and the difficulty of sealing the a/vb around the masonry. Woodstoves with a controlled draft and insulated metal chimneys work far better. Still, metal chimneys and plumbing stacks have to go through the roof at some point, and sealing the opening between unheated attic and warm living space is a problem. In both instances, the hole in the ceiling should be cut carefully. Where plumbing stacks break through top plates, use a hole saw and make the opening not more than ⅜ in. (10 mm) greater than the stack diameter. This way, you can brace and caulk the upper section of pipe effectively. The detail that we recommend (drawing, above left) will keep the seal between cold and warm

space intact. The upper pipe remains fixed, while the expansion and contraction of the plumbing stack is taken up inside the house by the expansion joint.

For chimneys, use a double firestop assembly to seal the rough opening in the ceiling, as shown. You should also place some noncombustible insulation along the joint where the firestop meets the ceiling opening before screwing the firestop in place.

Attic hatches can be critical heat-loss areas, and in a superinsulated house you're better off with no hatch whatsoever. If one is required, it should be located outside the house so that no elaborate sealing or insulation is needed. If the hatch must be inside, then it should be fitted with high-quality, compression-type weatherstripping and latches. It should also be insulated as heavily as the rest of the ceiling.

Heat exhangers—In a conventional house, infiltration through breaks in the a/vb usually brings an adequate supply of fresh air inside the house. This doesn't happen in a superinsulated house, and the tight a/vb can create several problems. First of all, humidity can rapidly build up to an uncomfortable level, causing condensation on windows, mold on food, mildew on cloth and even wood rot. This moisture comes from nothing more dramatic than respiration, cooking, washing clothes, bathing and watering plants. Cooking odors can also linger longer than they normally do.

Another problem caused by the right a/vb has to do with the retention of gases emitted from construction materials and plastic or foam products inside the finished house. Recent research indicates that high levels of formaldehyde, radon 222, carbon monoxide and nitrogen oxide can build up in tight houses.

Keeping a window open is not enough to alleviate these problems, and in winter this would defeat the purpose of superinsulating anyway. Fortunately there's a device that can provide

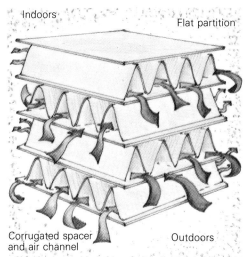

Indoors

Flat partition

Corrugated spacer and air channel

Outdoors

How an air-to-air heat exchanger works. **Cold, fresh outside air is drawn through channels in the heat exchanger, which have been warmed by stale air vented from inside the house. About 75% of the total heat is reclaimed.**

safe ventilation without significant heat loss. It's called an air-to-air heat exchanger and operates on the principle shown in the drawing above. Driven by a small fan, stale interior air is blown outside, while fresh air is drawn into the house from outside. Because the separate air passageways are adjacent, the incoming air is warmed by the outgoing air. Residential heat exchangers are usually small enough to fit into a closet or basement corner, and more and more of them are being sold in Canada and the U.S. The most popular brands around here are produced by vanEE Air Exchangers (Conservation Energy Systems, Inc., Box 8280, Saskatoon, Sask. Canada S7K 6C6), Enercon Industries Ltd. (2073 Cornwall St., Regina, Sask. S4P 2K6) and The Air Changer Co. (334 King St. E., Toronto, Ont. M5A 1KB). ☐

John Hughes lives in Edmonton, Alberta, and designs energy-efficient homes.

Truss Frame Construction

A simple building method especially suited to the owner-builder

by Mark White

Standard frame construction is complex and can be baffling to the first-time owner-builder. When the purposes and natures of foundations, sills, sill plates, floor joists, partition walls, studs, cripples, headers, ceiling joists, top plates, rafters and sheathing are taken one at a time they can be understood. But novices have a hard time handling the complexity once they are staring at all the pieces on their sites.

I have been teaching building on the college level for the past six years, trying to find a method that would reduce house construction to simple elements. Having tried balloon framing, platform framing, tilt-up walls, post and beam, and variations of them all, I now think I've found an answer: truss frame construction.

The truss frame is not new. Contractors, and individuals all over the world, have fiddled with the idea for years. It began to attract more attention in the United States when the Department of Agriculture's Forest Products Laboratory in Madison, Wis., erected an experimental building that combined floor, walls and roof in single truss sections. I was immediately taken with the simplicity of the concept. It looked like a system that would enable the owner-builder to come up with a sound, useful structure on the first attempt, given some basic training and guidelines.

To test this building concept, I sketched up a set of plans and ordered the appropriate lumber from one of our local sawmills. The lumber was ready in January. In February, I began work on the foundation, and I set the sill timbers in early March. The building, which I planned to build alone to see how well the system would work, was to be used as a rental unit. It would have 12-in. floors, 10-in. walls and a 12-in. roof, all stuffed with a nominal 12 in. of insulation, for an insulation value of R-45.

Our climate is quite mild, as Kodiak, an Alaskan island in the north Pacific, is warmed by the Japanese current. Winter temperatures rarely drop to 20°F, and average between 30°F and 40°F most of the time. Still, the winters are long and our primary heating fuel is oil, which is delivered by tanker from the lower forty-eight. The price of oil hasn't gone down in years, which leads us to think hard about proper insulation.

Foundations—Concrete costs a lot up here (about $165 a cubic yard), so many foundations are either creosote posts or treated wood. I opted for posts, because this is the fastest method and disturbs the soil the least.

The frost line here is 6 in., and bedrock is usually between 18 in. and 36 in. below the surface of the soil. I dug into glacial till—a mixture of hard clay and shale gravel a few inches above the actual bedrock.

The posts are Douglas fir, pressure-treated with creosote and about 12 in. in diameter. The ends that go into the holes are cut off squarely, then covered with another coat or two of creo-

The author built his first truss frame house alone to test the simplicity of the technique. At left, he winches the completed trusses onto sills set atop a post foundation. Center, a framed and sheathed partial truss is tilted into place as an end wall. It will be toenailed and temporarily braced. Right, most of the trusses are

sote to protect the center where the pressure treatment has failed to reach. I then nail a few pieces of heavy asphalt shingles (smooth side in) over that end to keep water from wicking up through the center of the post. When the post is in the hole, an eventual burden of 6,000 lb. to 10,000 lb. of house forces the asphalt into the wood fibers of the end grain and pretty much seals the pores. It takes 20 or 30 years to rot out the untreated center of a Douglas fir post, but I've seen it happen. The houses we build should last at least 200 years—for this reason I'm interested in having the posts last that long as well.

We carefully dig holes by hand into the 6-in. layer of glacial till and pour about a gallon of clean dry sand into the bottom of the hole instead of using a concrete pad. The sand is easier to work with if the posts need to be shifted to line up properly, and we have had little evidence of settling. We wrap the part of each post that is going to be in the ground with a few layers of 6-mil polyethylene to reduce the leaching of poisonous creosote into the groundwater. The plastic wrapping would deteriorate rapidly in sunlight, but it lasts a long time underground.

After the plastic goes on, the post is dumped into its hole, rotated a half turn in the sand and then propped into correct alignment with a few wedges jammed into the hole on one side or another. Once all the posts are in position, they are aligned with a transit, and everything is tied together with rough-cut 2x6s fastened with hot-dipped, galvanized, 20d nails. Extensive cross-bracing is installed before any weight is placed on the piles. The bracing is extremely important, because the soil is so shallow that it lends little racking resistance to the system. Once the bracing is in, the holes can be filled and tamped around each post. Sand makes the best fill, but we usually wind up using the dirt that came out of the hole.

We usually space posts 6 ft. o.c., forming "strings" of them to support two 8x12 sill timbers along the length of the building. Near each end, we reduce the spacing to 3 ft. or 4 ft. o.c. to support the greater weight of the end walls and the extra load transmitted to them by the roof overhangs. We space the parallel strings of posts between 14 ft. and 20 ft. apart, depending on the carrying capacity of the floor joists the sills are supporting. Post foundations on sand and bedrock work well as long as the quality and spacing of the individual members is kept within reason, and the cross-bracing is adequate. We tend to be conservative, planning shorter spacing than the maximum indicated by charts and tables. A foundation is not worth skimping on. Besides, our cost only runs between $200 and $300 per structure—dirt cheap compared to the cost and labor associated with concrete.

The first house—Teaching duties and a building project in a remote village kept me from further work on the truss house until May. Then I cut out and assembled a single truss on the sill timbers. I used it to pull master patterns, from which I then traced the necessary shapes on 10-in. and 12-in. rough-cut spruce planks. I used a portable circular saw and a small chainsaw to cut out the pieces for the rest of the trusses.

The Forest Products Lab's original truss frame design called for the use of standard 2x4 material for all chords and webs and results in many more pieces in each truss. I stayed with 1½-in. by 10-in. and 1½-in. by 12-in. material in the interest of simplicity. My trial structure was to be 20 ft. wide by 24 ft. long, with outward-sloping walls and generous porches all around. The outward-sloping walls were an experiment aimed at providing more visual interior room for a given area of floor space. They did provide the room, but for a few dollars more the side walls near the floor could have been kicked out a bit under the same roof and I would have had even more room. In a word, the experiment was successful, but I wouldn't repeat it. A frame spacing of 24 in. called for 11 full trusses, two partials for the end walls, and a total of four roof trusses to support the porch overhang at the ends.

A partial truss is a truss that has gussets on one side and studs on the other side. Trusses are used on the end walls only to define the shape and outline of those walls and to hold the studs in that configuration before the walls are tilted up into position.

Roof trusses are made up of rafters and collar ties, without vertical members. If the overhang at the gable end is less than 3 ft., it is possible to use roof trusses supported only by a sturdy 2x12 fascia board nailed to the other rafters along the eaves. If the overhang is greater than 3 ft., some arrangement of conventional headers and cripples is necessary. This often makes even the cross tie unnecessary.

Construction of the trusses went according to plan, but including a floor joist as a part of each unit turned out to cause more trouble than it was worth. It meant that there was no floor to work

erected, toenailed to the sills, and stabilized by the plywood nailed along their sides. In this first house. floor joists were a part of each truss. This resulted in great strength and stability, but made construction a bit awkward, since there could be no platform to work on until all the trusses were in place.

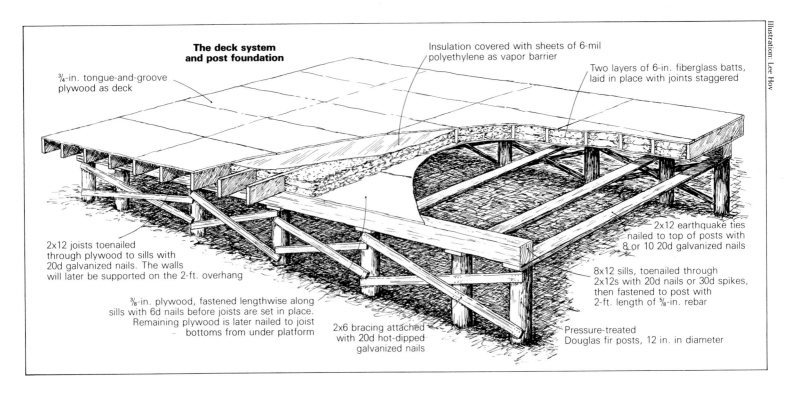

The deck system and post foundation

¾-in. tongue-and-groove plywood as deck

Insulation covered with sheets of 6-mil polyethylene as vapor barrier

Two layers of 6-in. fiberglass batts, laid in place with joints staggered

2x12 joists toenailed through plywood to sills with 20d galvanized nails. The walls will later be supported on the 2-ft. overhang

⅜-in. plywood, fastened lengthwise along sills with 6d nails before joists are set in place. Remaining plywood is later nailed to joist bottoms from under platform

2x6 bracing attached with 20d hot-dipped galvanized nails

2x12 earthquake ties nailed to top of posts with 8 or 10 20d galvanized nails

8x12 sills, toenailed through 2x12s with 20d nails or 30d spikes, then fastened to post with 2-ft. length of ⅝-in. rebar

Pressure-treated Douglas fir posts, 12 in. in diameter

on, so moving materials, assembling them and erecting them was awkward work.

The building did go together with less effort than one built with either the standing stick or the tilt-up wall method. And the truss assembly with its plywood gussets made an extremely strong and rigid structure. If anything is wind and earthquake-proof, it is this building. I was heartened by the progress and by the rigidity of the building, but felt the system could be simplified even further.

An improved design—The following fall, I designed a 20-ft. by 26-ft. house with straight walls 6¼ ft. high. This time the floor would be built and insulated as a separate unit (drawing, above), with the trusses and end walls erected on top of it. This is much easier, and almost as strong.

It took an inexperienced, seven-person crew (my class) six hours to build and insulate the floor. First, we laid our joists (in this case 20-ft. long 2x12s) over two sill timbers, spacing them 24 in. on center. Then we nailed ⅜-in. CDX ply-

wood to the joists from underneath and dropped in two layers of 6-in. fiberglass insulation, making sure to stagger their joints. On top of this went a 6-mil vapor barrier, then either a plank or a plywood floor. Dropping insulation in from above is an easy 10-minute task. If we had had to install it from below, it would have become a frustrating, eye-irritating chore.

Each truss designed for this building consists of five main pieces: two wall studs, two rafters and one collar tie. The class cut and assembled the required twelve full trusses, two end walls and three roof trusses in exactly 14 hours of hard labor. To eliminate inaccuracies, each truss was assembled on top of the master truss. All of the ⅝-in. plywood gussets were nailed with great quantities of #6 galvanized nails.

I have used gussets of ½-in. plywood, but they sometimes break when they're fastened to only one side and the truss is being flipped over to have the rest nailed onto the other. Using ⅝-in. plywood has eliminated the problem. I would recommend ¾-in. plywood for gussets on larger trusses and in two-story houses. Once gussets are nailed onto both sides of a truss, breakage is extremely unlikely, even with ½-in. plywood.

Truss members are very thick to provide room for lots of insulation. The trusses are thus massively overbuilt, so a builder need not worry much about a structural failure. A neophyte designing a truss pattern need only include a substantial enough collar tie to keep the rafters from spreading out near their bottoms. The collar ties should equal about half the span of the rafters. We usually use 2x12 ties, which can safely span 16 to 18 ft. On longer spans, it's safer and cheaper to use smaller dimensioned lumber, with a center tie fastened between the center of the collar tie and the peak of the roof. The center tie adds strength and keeps the ceiling from sagging and the long collar ties from warping.

We scheduled raising the frame for a Saturday so the students could work all day. Moving the sections and clearing ice from the floor's plat-

The completed first house. Sloping side walls increase the volume with limited floor space.

form took about two hours. Putting up the end walls and the trusses themselves took 17 minutes. Passers-by were amazed at the speed with which the building took shape.

Initially, the bottoms of the studs were toe-nailed to the floor, with each stud positioned over a floor joist. The first course of plywood sheathing was nailed to the floor perimeter and to the studs as the trusses were raised. This course secures the studs to the deck and keeps the trusses erect. More plywood sheathing on the walls and 1x6 decking on the roof locks everything together.

Alignment of the walls and roof was perfect. This turned out to be the straightest and squarest building I've ever seen assembled—thanks to a carefully leveled floor and to the uniformity of the trusses.

Insulation—The class met for four hours one final Saturday to get the building weathered in. To insulate the roof, we strung nylon twine under the rafters and stapled it in place to hold the fiberglass. In about 15 minutes, two layers of 6-in. insulating batts were laid in from above. We then decked over the roof with 1x6 spruce, trimmed and covered it with a layer of 55-lb. roofer's base felt. This layer will eventually be covered with asphalt shingles, but it will protect the building until good weather returns.

We have built a number of buildings with floor, wall and roof sections insulated to R-45. They are effective, allowing us to heat the average home with 15 to 20 gallons of fuel a month during the winter. A conventionally insulated house of equivalent size may gobble up to 350 gallons of fuel a month during a similar heating season.

In most of these buildings, I used Owens Corning 6-in. fiberglass in the roof. It seems to come 5½ in. thick, so two layers gave me 11 in. of insulation and a 1-in. ventilation space. I used only friction-fit batts or rolls, no foil face. In the floors and walls, I used Johns-Manville 6-in. fiberglass, two layers of which usually measure out to 13 in. (Owens Corning and Johns-Manville claim the same R-value, and they cost the same.) Recently I have switched to 12-in. wall cavities to make better use of the Johns-Manville insulation.

Plumbing, electrical and heating—The structures my classes and I have built appear deceptively simple in shape, form and function. They are not. Their cross-sectional designs have been carefully worked out to provide maximum floor space and volume with minimum exposed surface area. There are a number of deviations from typical construction practice that could be imposed upon any building method, but seldom are. The very nature of the truss frame structure, along with our floor-building technique, simplifies and encourages their use.

First, the floor is an almost totally sealed unit. Its vapor barrier is penetrated only by a carefully installed waste pipe and a water supply line. We use 6-in. or 8-in. interior plumbing walls, and try to back the kitchen up to the bathroom, utility room and wash room to get all the plumbing into one area.

Under the house, I make no attempt to insulate the 3-in. waste drain that flows to the sep-

Trusses for the second house did not include floor joists. A deck was built first so that work could progress easily, then—in just 17 minutes—trusses were tilted up, toenailed and temporarily braced, before being firmly connected to the deck by the first course of horizontal plywood sheathing. These workers are stringing twine beneath the rafters to hold 12-in. batts of fiberglass insulation.

tic system. It is merely angled properly and enters the ground quickly. None of the many that have been installed this way in our location has ever frozen. I usually wrap the 1-in. PVC water supply line with a short length of thermostatically controlled heat tape extending just below the frost line, and then insulate it. A neon indicator light on the upper end of the tape tells you whether it's working or not.

In Alaska, we install this 1-in. pipe inside a 2-in. pipe, which in turn should be insulated to beneath the frost line with heavy, black neoprene foam made for the purpose.

Electrical power enters the house by way of a single piece of 1½-in. conduit passing from a meter base on an outside wall to a single service panel on the wall inside the vapor barrier. All circuits emanate from this service panel, passing throughout the building in a single channel cut into the interior face of the outer wall studs. (There are two such channels cut into the faces of the studs in a two-story dwelling.) We use either Romex cable or conduit, ½ in. or ¾ in. in diameter. All cables and conduits are inside the vapor barrier, so there are no leaks through the membrane. Switches and electrical outlets don't let cold air get into the room.

The heavy insulation and complete vapor barrier eliminate drafts, convective air currents, and excessive heat losses, so there is little need in our houses for complex heat distribution systems. For heat sources we have tried woodstoves but they typically put out too much heat. We have settled on either a standard oil-fired, hot-air furnace, or an oil hot-water heater with a short loop of pipe run from the water tank to heat the air. In fact, we have a real problem finding an appropriately sized heat source. Right now the smallest furnace available is in the neighborhood of 85,000 Btu. What we really need is one that kicks out 12,000 Btu or less.

Potential uses—The frame truss system lends itself to simple buildings with repetitive sections, and I've used it to build structures with conventional shapes. But different applications and shapes are possible, because the basic truss is highly adaptable. I once built a strong and very lightweight building out of 1½-in. by 1½-in. stock and plywood glued and nailed in place. If you live in an area where labor is considerably cheaper than material you might try a truss composed of 2x2s or 2x3s and light plywood gussets glued and stapled in place to form an intricate webwork. This would eliminate the problem of direct heat transfer through solid joists, studs and rafters. It would, however, introduce the new problems of insulating between the webwork and of sealing off passages to fire and rodents. In our area rough-cut lumber is available at a reasonable price in lengths of up to 24 ft. so trusses of solid lumber are more cost effective than lighter, more intricate designs.

The use of a well-ventilated, heavy truss system instead of concrete in an underground structure merits consideration. The strength of a properly designed truss makes it a good choice in the high load conditions found beneath the earth. Another area of application would be in passive solar designs where the shape would be in one of the many asymmetric configurations, with the tall open side facing south.

The beauty of the truss system is that only one of the many frames that go into the building needs to be laid out with great care. Once that first unit is formed it's an easy and repetitive task to construct the rest of the units, using the first as a pattern to get the rest right. Raising the frame is then a simple matter involving a minimum of fiddling and measurement. □

Mark White teaches at the University of Alaska at Kodiak.

The Rafter Square

Laying out a roof with this basic tool and a new generation of accessories

by Jud Peake

Few carpenters would neglect to include a steel square when packing their toolboxes for a job. Yet, when pressed, quite a few good builders will abashedly admit that they generally use the square just for scribing a cut-off line on stock too large for their combination square. The steel square can serve a variety of functions, from stairbuilding to making simple checks for right angles, but it's especially useful for laying out rafters and other roof-frame members. With a little instruction, anyone can lay out cuts for common rafters, valleys, hips, jacks and gable ends. This doesn't require a knowledge of trigonometry, just a simple understanding of the geometry involved.

The rafter square—This versatile tool consists of two parts—the body, or blade, and the tongue (drawing, facing page). These two meet at the heel. The body is 24 in. long and 2 in. wide, and usually represents the level line, or run, in laying out rafters. Plumb, or rise, is represented by the tongue, which is 16 in. long and 1½ in. wide. The face and back of the square are usually imprinted with edge scales and math tables. The latter distinguish a rafter square from a framing square.

Squares are made of steel, usually painted black, or aluminum. Aluminum squares are typically more expensive, but lighter and less liable to bring second-degree burns to your palms on hot summer days. The best squares have their numbers stamped deeply into the metal rather than painted on. Most of them come with instruction books that are a handy toolbox reference.

The use of the rafter square is based on the geometry of right triangles. All right triangles have a 90° angle, which can be used to describe the intersection of a plumb and level line—the intersection of the tongue and the body of a framing square. Right triangles also share another quality: when the rise and run of a right triangle are increased proportionally, the hypotenuse lengthens proportionally, too, although its slope or pitch remains constant. A stair with a rise of 6¹¹⁄₁₆ in. and a 10-in. tread has the same pitch as an 8-in-12 roof because the proportion between rise and run for each of them is the same. Since rafters are nothing more than the sloping side of the triangle, the rafter square acts as an infinitely expanding intersection of plumb and level, allowing you to use the rafter stock as the hypotenuse. More information on how the proportional nature of

the rafter square can come in handy will be presented later.

A useful quality of right triangles, discovered by Pythagoras, is that the sum of the square of the sides equals the square of the hypotenuse ($a^2 + b^2 = c^2$). To the roof framer this means that if the rise and run are known, then the length of the rafter can be easily calculated. (The glossary on p. 31 defines common roofing terms.)

Scales and tables—On the face side of the square, the side with the maker's stamp on it, inches are broken into eighths and sixteenths. On the back, the outside edge of both the body and the tongue shows twelfths of an inch. This is useful for scaling inches to feet. The inside edge of the tongue is laid out in tenths, and the face of the heel usually has a hundredths scale. By holding your tape measure against these last two scales, you can easily convert back and forth from decimal inches to sixteenths.

Different manufacturers of squares give slightly different information in the tables on the face of the square. Usually, the first line gives in decimal inches the lengths of common rafters per foot of run; the second line does the same for regular hip and valley rafters. These figures are listed in the tables by their unit rise. Units of run are always 12 in. The inch markings on the outside face of the body double as unit-rise headings under which unit rafter lengths can be found. None of the tables that give rafter lengths makes allowance for the thickness of the framing members that they butt. This allowance, referred to as reduction or shortening, is a factor I'll talk about later.

The next two lines in the tables give the actual difference in the lengths of jack rafters in inches and fractions. The first of these lines gives the common difference for jacks on 16-in. centers, and the second line for those on 24-in. centers. Most squares then show two lines of side cuts for hips, valleys and their jacks. Some squares have a seventh line, which gives the angle at which sheathing should be cut where it meets hips and valleys.

Unit measurement—The square can be used in two ways to determine the length of rafters. The first method is unit measurement. As shown in the drawing, top of facing page, this technique uses the proportional qualities of the right triangle, and expresses rise and run as a ratio. For common rafters, unit run is always 12 in., and the unit rise is the rise per foot of

run. The pitch triangle seen in most plans is a representation of unit measurement; so are the rafter tables on the square. For example, if you need the unit length of a common rafter on a 4-in-12 roof, look in the first line of the tables, length of common rafters per foot of run, under the 4-in. mark, the unit rise, on the outside of the body. The number given is 12 65. This means that for every foot of run with a rise of 4 in., the rafter will have to be 12.65 in. long. Check this figure by measuring diagonally with your tape between the 4 and the 12 on the square. You should get slightly more than 12⅝ in., or 12.65 in. The drawing, right, shows how holding the rise and run of the square determines the unit length of the hypotenuse, which, for the purposes of framing a roof, is the edge of a rafter.

Once you have found the unit length of the rafter in the tables for a given rise and run, multiply this figure by the actual run (usually half the width of the building) to get the theoretical, or unadjusted, length. This measurement begins with a plumb cut at the top of the rafter and ends with the plumb cut of the bird's mouth, the 90° cutout where the rafter sits on the top plate of the exterior wall. You will have to add the length of the rafter tail, and subtract half the thickness of the ridgeboard.

Stepping off—The second method for finding an unadjusted rafter length is called stepping off. To step off, lay the square on the rafter with the tongue and body reading the pitch, and repeat this procedure as many times as there are feet in the total run (drawing, bottom of facing page). For example, if the actual run of a rafter on a 4-in-12 roof were 13 ft., you would have to mark the rafter 13 consecutive times to lay out the length of the body of the rafter. You would do this by setting the square on the rafter stock, crown side away from you, so that the heel of the square is toward you and the tongue is on your right. Align the 4-in. mark on the outside of the tongue, representing the rise, and the 12-in. mark on the body, representing the run, on the edge of the rafter. Now scribe a line on the outside of the body at the edge of the rafter, and slide the square along until the outside of the tongue, held on the edge of the rafter at the 4-in. mark, lines up with the scribed line, and scribe again.

Move the square and scribe in the same attitude twelve times from the original position. The plumb cut of the bird's mouth will intersect

Illustrations: Frances Boynton

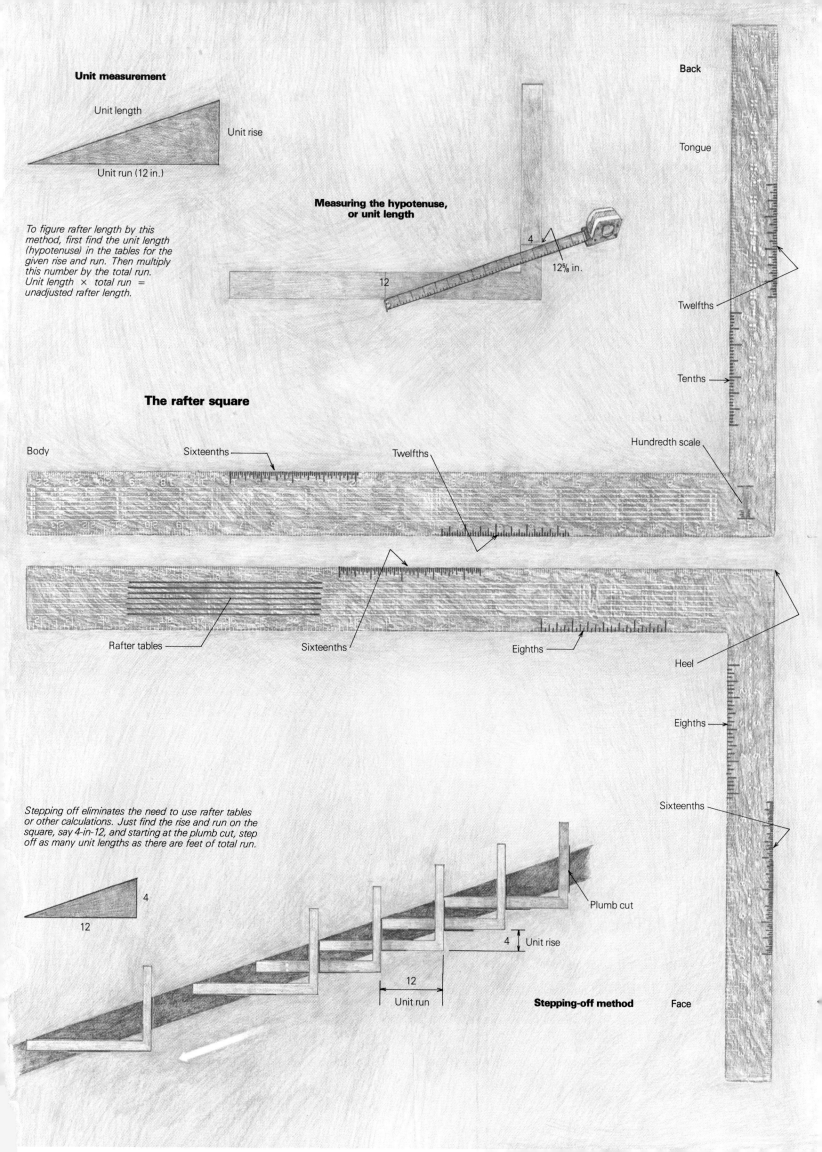

Unit measurement

Unit length

Unit rise

Unit run (12 in.)

To figure rafter length by this method, first find the unit length (hypotenuse) in the tables for the given rise and run. Then multiply this number by the total run. Unit length × total run = unadjusted rafter length.

Measuring the hypotenuse, or unit length

4

12⅝ in.

12

Back

Tongue

Twelfths

Tenths

Hundredth scale

The rafter square

Body

Sixteenths

Twelfths

Rafter tables

Sixteenths

Eighths

Heel

Eighths

Sixteenths

Stepping off eliminates the need to use rafter tables or other calculations. Just find the rise and run on the square, say 4-in-12, and starting at the plumb cut, step off as many unit lengths as there are feet of total run.

4

12

Plumb cut

4   Unit rise

12

Unit run

Stepping-off method

Face

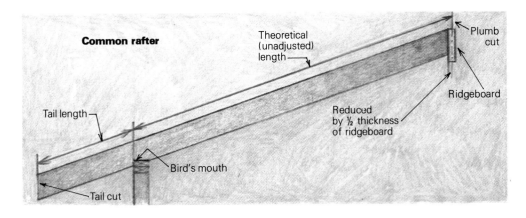

Common rafter

Theoretical (unadjusted) length

Plumb cut

Ridgeboard

Reduced by ½ thickness of ridgeboard

Tail length

Bird's mouth

Tail cut

the edge of the rafter at the thirteenth scribed line. This is the unadjusted, or theoretical, length of the rafter. Stepping off won't tell you its length in inches and feet. You'll have to measure it later.

When the run is not in whole feet, the remaining inches are measured along the level line of the body. This mark is brought back to the edge of the rafter by lining up the tongue with this mark and scribing a plumb line, while still holding the rise and run on the square along the edge of the material. The stepping-off method also requires reduction. As with unit measurement, you'll have to subtract half the thickness of the ridge, and add the length of any overhang. Stepping off must be done carefully because of the danger of accumulated error. It also doesn't give you the precise length of rafter stock at the outset.

Laying out a common rafter—The object of all of these tedious calculations is to cut a rafter pattern, which can then be used to lay out the remainder of the roof without any further headscratching. If you think of a rafter pattern as the only obstacle between you and the goal of calculating an entire roof plane, it will lighten the burden a little.

The first task is to lay out the plumb (top) cut of the rafter, as shown above. This cut will rest against the ridgeboard when installed. Lay it out by setting the square on the rafter stock just as you would for stepping off. Align the inch

Laying out a plumb cut

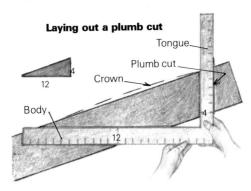

Tongue

Plumb cut

Crown

Body

marks that correspond with rise and run on the outside of the tongue and body on the side of the stock nearest you (drawing, above). Remember to use the tongue for the unit rise, and the body for the unit run. Set your pencil against the outside of the tongue and draw the plumb line. This line represents the very center of the span, as if the rafters from each side were butting together without a ridgeboard. Because

this line doesn't take the ridgeboard into account, you must measure along the body of the square (a level line) half the thickness of the ridgeboard and draw a new line parallel to the original plumb line for the actual cut. For a 2x ridgeboard, you would measure ¾ in. perpendicular to the plumb line—not along the edge of the rafter—to get the shortening line.

Determine the unadjusted length of the rafter either by stepping off or by unit measurement. Measure from the plumb line (not the reduced cut-line) to the plumb line of the bird's mouth, known as the heel cut. This is the part of the bird's mouth that hooks over the outside of the wall (drawing, below).

To lay out the bird's mouth, make a mark 1½ in. up from the bottom of the heel cut. Begin measuring from the rafter edge nearest

Bird's-mouth layout

Heel cut (plumb cut)

Seat cut (level cut)

Seat cut-line

Heel cut-line

Tail

Bird's mouth

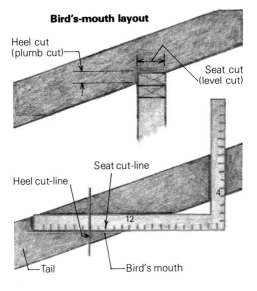

you. Now slide the square, still holding it at the correct rise and run, along the rafter toward the top plumb cut. When the body of the square intersects the 1½-in. mark on the heel cut, scribe along the body from the 1½-in. mark across to the edge of the rafter. This is a level mark; the heel (plumb) cut and the seat (level) cut make the bird's mouth. The 1½-in. depth of the bird's mouth is arbitrary, but it shouldn't be cut so deep that it weakens the rafter tails. The depth of this cut doesn't affect the roof's slope, but does affect the absolute height of the ridge. If you deepen the seat cut by 1 in. the ridge will be lowered by 1 in. This usually doesn't matter unless you already have high walls, purlins or a ridge beam in place.

You can now add the length of the tail, or

overhang. Beginning with the heel cut of the bird's mouth, measure down along the rafter (or step it off) and mark the tail cut. Standard tails can be cut square (perpendicular to the line of the rafter), cut level or cut plumb.

Laying out a regular hip or valley—Hips and valleys are different from common rafters because they take a diagonal path across the building. The run of regular hips and valleys angles across the plan at 45°, completing an isosceles right triangle with the run of the last common rafter and the top plate. This means that the run is longer than that of the common rafter, although the rise remains the same. The drawing below shows the relationship of the run of common rafters and hips. By applying the Pythagorean theorem, or by measuring diagonally between the 12-in. marks on the square, the unit run of the hip figures out to be about 17 in. for every 12 in. of common run. This means that each time you would use the

Relationship of hip to common rafter

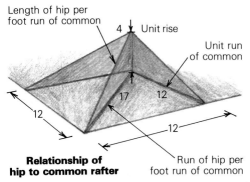

Length of hip per foot run of common

Unit rise

Unit run of common

Run of hip per foot run of common

12-in. mark on the body of your square to find the cuts and lengths of common rafters, you use the 17-in. mark to work with hips and valleys. However, you still use the same rise figure and, in stepping off, take the same number of steps as you would with a common.

The adjustments for hip and valley rafters are more complicated than they are for commons. Because a hip intersects both the ridge and the common rafter at the top, each side requires a vertical 45° bevel to form the plumb cut. This is known as a double cheek or double side cut. The drawing below shows the reductions that

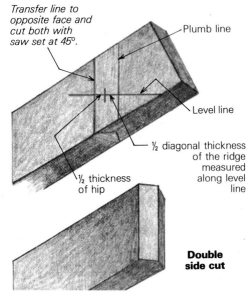

Transfer line to opposite face and cut both with saw set at 45°.

Plumb line

Level line

½ diagonal thickness of the ridge measured along level line

½ thickness of hip

Double side cut

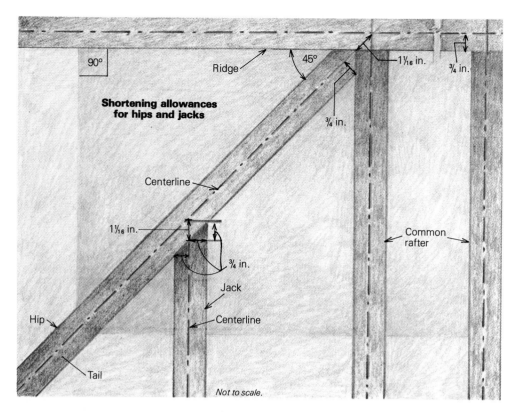

Shortening allowances for hips and jacks

90°

Ridge

45°

1¹⁄₁₆ in.

¾ in.

¾ in.

Centerline

1¹⁄₁₆ in.

¾ in.

Jack

Centerline

Common rafter

Hip

Tail

Not to scale.

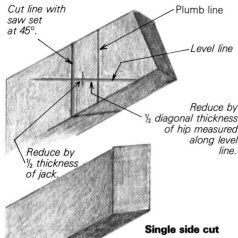

Cut line with saw set at 45°.

Plumb line

Level line

Reduce by ½ diagonal thickness of hip measured along level line.

Reduce by ½ thickness of jack.

Single side cut

are necessary for double side cuts. With the common rafters we deduct one-half the ridge thickness, measured level, on the side of the rafter. With a hip we have to deduct one-half the diagonal measure of the ridge thickness and again lay this out level on the side of the rafter. To move to the side of the rafter, you have to deduct another ¾ in. (for a 2x hip), again measured level, to account for the difference between the long point of the double bevel at the center of the hip and the short point on the side, as shown above.

These cuts can be made easily and accurately with a circular saw set at 45° because at any pitch, regular hips and valleys intersect common rafters at 45°. If you draw the correct double side cut on the top edge of the rafter, it won't appear as a 45° angle because the edge of the rafter is not a level line when installed. However, for a good fit using a circular saw on regular hips and valleys, the only mark you'll need is the plumb line.

In the tables, determine the lengths of regular hips and valleys the same way you find them for commons. Just make sure to look under the heading "length of hips and valley per foot of run." The bird's mouths are also similar except you have to use the 17-in. mark on the body for the level line, and the depth of the heel cut will be different. Hip and valley rafters are usually of wider stock than commons; so to make sure the tails are level, hip and valley bird's mouths have to be deeper, leaving the same amount of uncut rafter above them as you did on the commons. After adding length for the overhang you can lay out and cut the tail. If you are planning on a surrounding fascia, the tail cut has to be a double side cut.

The rafter tail is a good place to practice stepping off inches. For an example, assume that a 4-in-12 roof requires a 1-ft. 6½-in. tail. Step off the first foot of run as you did with the com-

mon, except with the hip, use 4 and 17. Draw a level line. Just as 17 (the diagonal measure of 12 and 12) gives you the first foot of run, the diagonal measure of 6½ and 6½ will give you the remainder of the run. Stretching a tape measure between these marks on the square gives a measurement of 9³⁄₁₆ in. Measure along the level line on the rafter tail 9³⁄₁₆ in., and the overhang will be correct.

Laying out a jack—A jack rafter is a common rafter that intersects a hip or valley before it reaches the ridge or plate. The only way that it differs from a common is its length and the bevel of its plumb cut. This is a single side cut, and in 2x material I make it with a circular saw set at 45°.

Once you have established the plumb line on the jack rafter, make the hip or valley reduction (one-half the diagonal thickness) along a

level line as shown in the illustration, above right. With a 2x hip or valley, this measures 1¹⁄₁₆ in. To reach the short point of the side cut, make a further reduction along the level line for one-half the thickness of the jack itself. This measures ¾ in. The jack reduction is necessary because you are laying out the side cut on the side of the rafter, rather than on the centerline of the top edge.

As with hip or valley rafters, using a saw with a shoe that pivots allows you to cut any regular hip or valley jack correctly by following the reduced plumb line with the saw set at 45°.

Big beams require the angle of the side cut to be laid out on the top edge of the rafter because a circular saw won't handle the depth of cut. Use a rafter book or the rafter tables for the co-ordinates on the square. If two figures appear under each rise, find the first figure on the body of the square, and set it on the edge of the rafter stock. Next find the second number on the tongue and place it on the same edge of the rafter. Check instructions for your square or book for which leg of the square to scribe against. If there is only one figure listed, then use it on the body, use 12 on the tongue, and make your mark along the tongue.

Once you have determined the length of the first jack, the rest are merely multiples. This is called the common difference, and can be seen in the framing diagram (drawing, below). Find

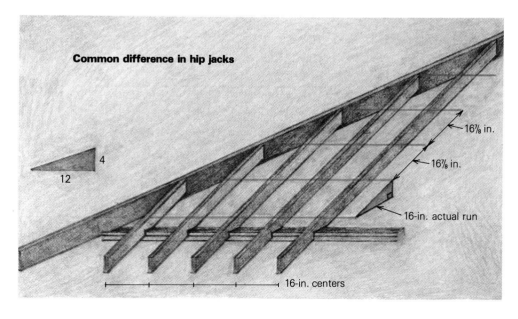

Common difference in hip jacks

4

12

16⅞ in.

16⅞ in.

16-in. actual run

16-in. centers

the appropriate table on your square (16-in. spacing or 24-in. spacing) and look up the pitch of your roof. The figure you see listed is the length of the first jack, excluding the length of the tail, and also the increase in the length of each subsequent jack.

Notice that the relationship that regular hips and valleys have with common rafters and plates is the same one that exists between jacks and their hip or valley rafter. This is an isosceles right triangle in plan, and means that the actual run of the first jack will be the same as its spacing. If the first jack is 16 in. away from the seat cut of the hip, it will have an actual run of 16 in.; if it's 24 in. along the plate, the actual run will be 24 in. This is an actual run, not a unit run.

Solving proportion problems—With a rafter square, you can easily determine the wall height under a shed roof. For example, if a 4-in-12 roof is supported by an 8-ft. wall on the low side, then what is the height of the supporting wall 13 ft. away on the high side? Use the twelfths scale. Lay the 4-in. mark of the tongue and the 12-in. mark of the body on a straightedge, as shown below. Draw a line against the

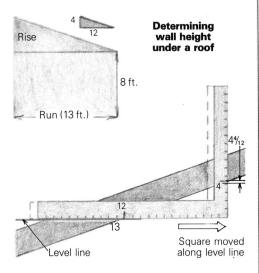

Determining wall height under a roof

Rise

4

12

8 ft.

Run (13 ft.)

$4\frac{4}{12}$

4

12

13

Level line

Square moved along level line

body representing a level line. Then move the square along this line until 13, representing the actual run, lines up with the edge of the material. The answer, $4\frac{4}{12}$, reads on the tongue of the square where it first comes in contact with the wood. This figure (4 ft. 4 in., when multiplied from unit dimensions to actual dimensions) represents the rise of the wall above the established 8-ft. mark. The height of the other wall is then 12 ft. 4 in.

Determining the length of gable-end studs is another proportion problem, and can be solved in the same way as determining wall heights under a roof. Like jack rafters, the length of the first gable stud is equal to the difference in the lengths of the other studs. The length of the first stud can be determined by the proportional method, with a calculator, or by measurement. To use the square, follow the directions for finding wall heights. In this case the spacing of your studs, typically 16 in. or 24 in., is the actual run. ☐

Jud Peake is a carpenter and contractor in Oakland Calif.

Other tools that help

The rafter square is still king. Despite challenges from quite a number of patented devices, few roofs get framed without the use of a rafter square. It remains the best choice because of its accuracy, durability and variety of necessary data. Still, there are several other useful tools that can simplify the job. Most of these are used in conjunction with a rafter square.

Rafter books. These guides give listings of rafter lengths for common, hip, valley and jack rafters referenced by span, and are typically organized by pitch from $\frac{1}{2}$-in-12 to 24-in-12. The pitch angle is given, as well as the corresponding layout numbers on the steel square for all cuts and bevels. Although quick and accurate, these guides will leave you in the dark if you're working on an irregular or polygonal roof. One widely used rafter book is *The Full Length Roof Framer* (A.F. Riechers,

Box 405, Palo Alto, Calif. 94302). It lists 48 different pitches and includes the cuts and bevels for gable and cornice moldings. It is pocket size, bound, and sells for about $6.

Squangle (Mayes Bros. Tool Mfg. Co., Box 1018, Johnson City, Tenn. 37601). This device is small enough to fit into your nail bag and can be used with one hand. It has some limited rafter tables on its tongue, although it's best used with a rafter book or the rafter tables on a square. Unlike a square, the Squangle can convert a unit rise and run into degrees. But once the tool is dropped from a roof, it may not work accurately. It sells for about $9.

Speed Square (The Swanson Tool Co., Box 434, Oak Lawn, Ill., 60453). This tool is also small enough to be carried on your tool belt. It comes with an instruction book and extensive, full-length rafter tables. A solid

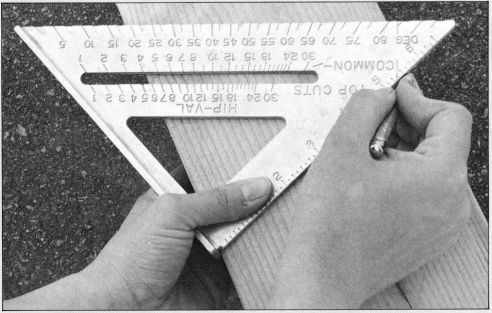

The Speed Square is an aluminum casting small enough to fit into nailbags. Angles in degrees are given on the face of the tool, and it comes with an instruction book and rafter tables.

The Squangle is one of the relatively new tools based on the rafter square. It has limited tables, but is smaller than the square and gives equivalent angles for roof pitches.

aluminum casting, the Speed Square won't bend or break, but because of its sharp points, you wouldn't want it hanging from your belt if you lost your balance and took a fall. Angles in degrees are also given on the face. It takes two hands to use it, and costs $8.35.

Two other squares that are nearly identical to the Speed Square are the Carpenter Handy Square for $7.49 (Macklanburg-Duncan Co., Box 25188, Oklahoma City, Okla. 73125), and the Angle Square at $7.67 (Johnson Level and Tool Mfg. Co., 2072 North Commerce St., Milwaukee, Wis. 53212).

Stair-gauge fixtures. These really help if you use the framing square. They are purchased in pairs, usually made of milled brass, and cost about $8. You fix them to the square by tightening a setscrew on each one. Attach these stops at the points on the square that define the pitch you are using (such as 4 and 12), and you can repeatedly set the square accurately on the rafter stock without having to read the numbers each time. If you are careful when you attach them to the square, they will improve your accuracy greatly, particularly if you are using the stepping-off method. As the name says, they are also very helpful in laying out stair stringers.

Layout tee. This is a roof-framing aid that you make on the job site. Tees are patterns, or templates, of the tail and bird's mouth, or the ridge cut of a rafter that is used to transfer the layout onto rafter stock. They are made of a short length of rafter with a 1x4 nailed to the edge as a fence to reference the tee to the rafter being marked.

Pitch board. This is simply a piece of plywood cut in the shape of a right triangle defined by the rise and run of the roof to be framed. It is used for stepping off rafters in the same manner as a framing square. It can also help mark cuts for gable-end studs and bird's mouths.

Calculator. One of the items you are likely to find in the toolbox of a canny roof framer these days is an extra battery. With the advent of the $10 pocket calculator, the framing square has a new companion. Because roof-framing calculations are based largely on the Pythagorean theorem, any calculator with a square-root key will do. Using a calculator is more accurate and less tedious than reading the tables on a framing square because you are able to deal directly with the equation and the variables involved.

It's important to decide what units you are using—inches, feet or decimal inches/feet—and remain consistent throughout the calculation. I find it easiest to enter measurements and take the answers in inches to save conversion steps, which are a common source of error. I use my calculator to determine pitch angles, rafter lengths, common difference in jacks, gable-end stud heights and other proportional problems, as explained below. —*Jud Peake*

Using a pocket calculator

Proportion. This is a useful calculation in determining wall heights, as in the example given (drawing, facing page) using the framing square. This is a matter of converting a unit rise and run to an actual rise, given an actual run. My calculator uses the algebraic operating system, so your keystrokes may be different; but the logic goes like this:

$$\text{If } \frac{\text{unit rise}}{\text{unit run}} = \frac{\text{actual rise}}{\text{actual run}},$$

$$\text{then actual rise} = \frac{\text{unit rise}}{\text{unit run}} \times \text{actual run}.$$

To find the actual rise of a 4-in-12 roof over a 13-ft. run:

$$\frac{4}{12} = \frac{\text{actual rise}}{13 \text{ ft.}}, \text{ and actual rise} = \frac{4}{12} \times 13 \text{ ft.}$$

This answer is in feet. On the calculator, the keystrokes are:

rise ÷ run ✕ actual run (ft.) = actual rise (ft.).

Difference in the lengths of gable studs. The unadjusted length of the first gable stud is also the difference in the lengths of the gable studs. This is just another proportion problem:

$$\frac{\text{difference in length of studs}}{\text{spacing of stud centers}} = \frac{\text{rise}}{\text{run}}.$$

The keystrokes are:

rise ÷ run ✕ spacing (in.) = difference in length (in.).

The formula is simplified for common stud centers. For studs 24 in. o.c.,

difference in length of studs = 2 × unit rise.

For 16-in. centers,

difference in length of studs = 1.333 × unit rise.

Lengths of common rafters. These are given in a general way by the Pythagorean theorem:

$$\text{length of common} = \sqrt{\text{rise}^2 + \text{run}^2}.$$

However, you need to be clear about which units of measure you are using. If you use the actual rise and run in feet, the answer will be the actual length of the common, expressed in feet. If you use the unit rise and run, the answer will be the unit length of the rafter. This is a length in inches for every 12 inches of run. The simplest formula uses the unit rise and run:

$$\text{length of common rafter (in.)} = \sqrt{\text{rise}^2 + 144} \times \text{actual run (ft.)}.$$

The keystrokes are:

rise x^2 + 144 = √ ✕ actual run (ft.) = length of common (in.).

This answer doesn't account for the thickness of the ridge. There is less chance of error if you make this allowance during the layout of the rafter.

Lengths of regular hips and valleys. Once you determine the run, you can find the length of a hip or valley. Because the run of a regular hip or valley is the diagonal of a square whose sides are runs of the common rafters, the run is the square root of twice the square of the common run.

$$\text{Run of hip} = \sqrt{2 \times \text{run of common}^2}, \text{ and}$$

$$\text{length of hip} = \sqrt{\text{run of hip}^2 + \text{rise}^2}$$

$$= \sqrt{2 \times \text{run of common}^2 + \text{rise}^2}.$$

Or you can use unit rise, unit run, and the run of the common rafter:

$$\text{hip length} = \sqrt{2 \times \text{unit run}^2 + \text{unit rise}^2} \times \text{actual run of common.}$$

Since the unit run is always 12, this becomes:

$$\text{length of hip} = \sqrt{288 + \text{unit rise}^2} \times \text{actual run of common.}$$

The square root of 288 is $16^{31}/_{32}$, or almost 17, the same 17 that you use on the framing square for the run. The keystrokes for finding a regular hip or valley length with a calculator are:

rise x^2 + 288 = √ ✕ actual run (ft.) = unadjusted hip length (in.).

Again, it's useful to make the ridge reduction on the actual rafter. This time, though, the ridge thickness is a diagonal measure:

$$\frac{\text{ridge thickness in inches}}{2} \times \sqrt{2} = \text{diagonal of } \frac{1}{2} \text{ ridge thickness.}$$

In an isosceles right triangle, the square root of 2 multiplied by the side equals the hypotenuse.

Difference in the lengths of hip and valley jacks. The first jack has an actual run equal to the spacing of the jacks. The length of this jack, unadjusted, will be equal to the difference in the lengths of the jacks:

$$\text{difference in length of jacks (in.)} = \sqrt{\left(\frac{\text{rise}}{\text{run}} \times \text{spacing}\right)^2 + \text{spacing}^2}.$$

For jacks 16 in. o.c.:

$$\text{difference} = \sqrt{(\text{rise} \times 1.333)^2 + 16^2}.$$

For jacks 24 in. o.c.:

$$\text{difference} = \sqrt{(\text{rise} \times 2)^2 + 24^2}.$$

To adjust for thickness on the long point side, deduct the following:

$$\text{thickness of hip} \times \sqrt{2} - \frac{\text{thickness of jack}}{2}$$

Bevels and angles. Plumb, level and bevel cuts can be converted to degrees if your calculator has trig functions. The keystrokes are:

for angle A, rise ÷ run = TAN⁻¹
for angle B, run ÷ rise = TAN⁻¹

Roof Framing Simplified

This direct approach involves full-size layouts and stringing rafter lines

by Tom Law

There isn't a cut in roof framing that can't be calculated given a sharp pencil, a framing square and a head for math. But my 20 years in the trade have taught me that in some cases, the theoretical calculation of rafter angles and lengths is slower and leaves more room for error. While I think that it's important to understand the geometry of roof framing, the empirical method can save time and frustration, and contribute to your understanding of the process. I'm better at solving problems when I can grasp them—literally.

When I'm cutting a complicated roof and things get foggy, I use two techniques to help me produce rafters that fit the first time. I chalk lines on the plywood subfloor to represent a rafter pair in relation to its plates and ridge. This two-dimensional diagram is laid out full size. Pattern rafters can be tested right there on the job site. The other method I use is to deal directly with the components involved by getting up on the roof and measuring the relationships between the rafter to be cut and the existing plate and ridge with string and sliding bevel.

I used this method several years ago when I built a Y-shaped house. One wing was for the bedrooms and the other contained the kitchen, dining room and family room. The stem of the Y was the living room. The roof over the wings called for trusses, but the living-room rafters were exposed. The problem was in framing the intersection of the three roofs. These beams were big, long and expensive, and all the cuts would show. Had the house been a T shape, the valley rafters would have been a textbook case, and I could have found the information I needed in the rafter tables on my framing square. But since the intersection was 120° and not 90°, I had to find the angles by calculation or direct observation. I chose the latter.

Full-size layout—After setting the trusses, I chalked a full-size layout of the living-room common rafters and ridge on the subfloor below. The ridge beam was a 4x14, and the common rafters were 4x8s on 4-ft. centers. I decided to tackle the easiest steps first. This gave me time to think about the problem while reducing the parts in the puzzle.

The ridge beam went up first. I found its height by measuring on the full-size layout. The common rafters were then cut using patterns made from the layout. I nailed these in place

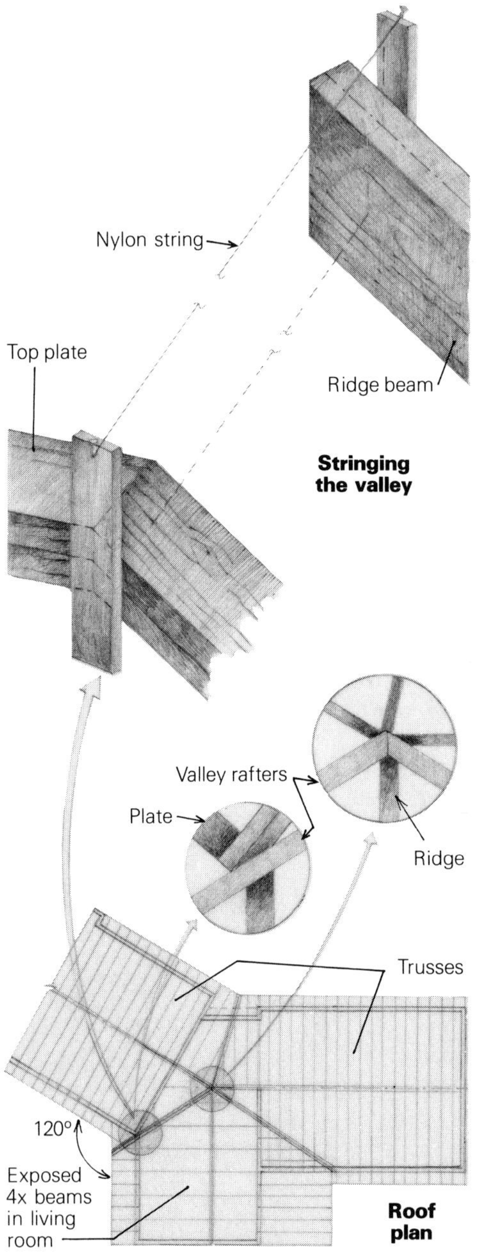

Tom Law is a builder in Davidsonville, Md.

Nylon string

Top plate

Ridge beam

Stringing the valley

Valley rafters

Plate

Ridge

Trusses

120°

Exposed 4x beams in living room

Roof plan

starting at the outside wall, working toward the junction of the Y.

The valley rafter was next. Its location is shown above. Valley rafters are usually heftier than common rafters because they have a longer span, and have to carry the additional weight of the valley jacks. In this case, the valley rafter was a 4x10. Because it's deeper, to achieve the same height above the plate and ridge, its seat and ridge cuts also differed from those of the common rafters. I found it faster and easier to measure the actual distances than

to calculate imaginary ones. This way I could find the length of the rafter and the angles of the cuts without guesswork or error. This took me back up on the roof armed with a ball of nylon string, a sliding bevel and a level.

String lines—First, I tacked a scrap piece of wood vertically on the opposite side of the ridge beam from its intersection with the valley rafter. Then I tacked another block on the outside of the top plate where the valley rafter would sit. Between these sticks, I stretched nylon string at the height of the top of the rafter and along its imaginary center line. The string made it easy to visualize the actual rafter in place. For reassurance, I sighted across the rafter tops from the outside wall to check the alignment. I used the sliding bevel and level to find the angle of all the cuts, being careful not to distort the string.

Before I transferred the angles to the rafter stock, I made two templates (called layout tees) for marking out the ridge cut and rafter seats. I made one for the bottom and one for the top of the valley rafter. With some adjustments, they fit when the center line drawn on the pattern was in line with the string representing the center of the rafter. The tees also allow you to test-fit the bird's mouth to the plate before you carve up costly rafter stock. With these pattern pieces tacked in place, I measured the length of the valley directly. Then I transferred this length and the angles on the tees to the valley-rafter stock, and cut it to its finished dimensions. It fit perfectly the first time.

With the valley rafter in place, I turned to the valley jacks. First I made another pattern, this time of the ridge cut of the common rafter. I tacked it on the layout mark on the ridge and stretched the nylon line from it to the valley rafter, being careful to keep it exactly parallel to the common rafters. With the line simulating the top center line of the longest jack, I used the sliding bevel to find the angles of the plumb and side cuts. This time I transferred the angles directly onto the stock and cut it with a handsaw. Each shorter jack was worked in the same way, using the nylon line to find the location and length; the angles remained constant.

Laying out rafters with a framing square is something I do a lot. However, in situations that call for unusual intersections with compound angles, I spend my time dealing directly with the problem. This reduces confusion and allows me to concentrate on the work. □

A Glossary of Roofing Terms

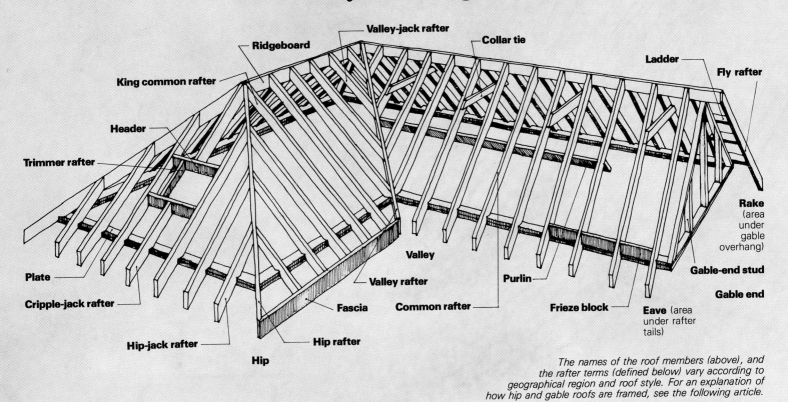

The names of the roof members (above), and the rafter terms (defined below) vary according to geographical region and roof style. For an explanation of how hip and gable roofs are framed, see the following article.

Span—the horizontal distance between the outside edges of the top plates.

Rise—the vertical distance measured from the wall's top plate to the intersection of the pitch line and the center of the ridge.

Run—the horizontal distance between the outside edge of the top plate and the center of the ridge; in most cases, half the span.

Slope—a measurement of the incline of a roof, the ratio of rise to run. It is typically expressed using 12 as the constant run.

Pitch—has become synonymous with slope in modern trade parlance. It is actually the ratio of the rise to the span. A roof with a 24-ft. span and a rise of 8 ft. has a 1-to-3 pitch. Its slope is 8 in 12. Two ways of saying the same thing.

Unit rise—the number of inches of rise per foot of run.

Unit run—this distance is always 12 in.

Common difference—the difference between the length of a jack rafter and its nearest neighboring jack on a regular hip or valley when they are spaced evenly. This is also the same measurement as the length of the first, or shortest, jack.

Rafter pattern—a full-scale rafter template used to mark the other rafters for cutting. It can be tried in place for fit before cutting all the rafters.

Layout tee—a short template cut from the same stock as the rafters and used for scribing repetitive plumb cuts, tail cuts and bird's mouths.

Tail—the part of a rafter that extends beyond the heel cut of a bird's mouth to form the overhang or eave.

Pitch line—an imaginary line, also called the **measuring line**, that runs parallel to the rafter edges at the height of the full depth of the heel cut on the bird's mouth. In common practice, rafters are measured along their bottom edge.

Theoretical length—the length of a rafter without making allowances for the tail or ridge reduction. Also called the **unadjusted length**.

Bird's mouth—also called a **rafter seat**. It is the notch cut in a rafter that lets it sit on the double plate. It is formed by the plumb heel cut and the seat cut, which is a level line.

Plumb cut—any cut that is vertical when the rafter is in position on the roof. Also used as a reference to the top cut on a rafter where it meets the ridgeboard.

Level cut—any cut that is horizontal when the rafter is in position on the roof.

Tail cut—the cut at the outer end of the rafter. If cut at the outside edge of the double plate, it is a flush cut. All the other traditional tail cuts let the rafter overhang the plates—**heel cut** (level), **plumb cut** (vertical), **square cut** (perpendicular to the length of the rafter) or **combination** level and plumb cuts.

Side cut—also called a **cheek cut**, is the compound angle required for the proper fitting of roof members that meet in an intersection of less than 90°, and other than level. This applies to jacks that connect with hips and valleys.

Ridge reduction—rafter lengths are calculated to the center of the ridge of the roof. This doesn't take into account the thickness of the ridgeboard. This allowance reduces the theoretical length of the rafter by one-half the thickness of the ridgeboard. The layout line drawn parallel to the plumb cut that represents this allowance is called the **shortening line**.

Dropping a hip—the amount by which the bird's mouth on a hip rafter must be deepened to allow the top of the rafter to lie in the same plane as the jack and common rafters. This ensures that the roof sheathing will nail flat without having to bevel the top edges of the hip, a process known as **backing**. —*P.S.*

Putting the Lid On

A primer on production cutting and raising a hip and gable roof

by Don Dunkley

One of the most satisfying events in building a house is the completion of the roof. Some builders borrow from European tradition and nail a pine tree to the peak in celebration. At the least, it is usually the excuse for a party. There are good reasons to celebrate. Framing a roof can be perplexing, physically taxing and sometimes dangerous. However, with thoughtful organization of rafter layout, production rafter-cutting techniques and carefully built scaffolding and bracing to help raise the ridge and rafters, your celebrating doesn't have to come out of a sense of relief.

The best way that I know to share my knowledge of roof framing is to describe the steps involved in building a simple hip and gable roof, like the model roof that is shown in plan, below. This article will cover most of the problems that are encountered in a rectangular building—laying out and assembling common rafters, hips and jacks, along with the ridge, purlins and collar ties.

Preparation—The roof is ready to frame once all the walls are built, plumbed up and braced off. The exterior walls must be lined very straight, because any irregularities in the span will show up on the roof frame. Before you start sorting through your framing stock, study your roof plans carefully. They should show an overhead (plan) view on a scale of ⅛ in. or ¼ in. to 1 ft. They will tell you the type of roof (gable, hip or gambrel), the pitch or slope, the length of overhangs (eave and gable end), the layout of the rafters, their spacing (16 in. on center, 24 in. o.c.), and the sizes of the framing members.

Layout—Job-site layout begins with measuring the span of the building. Always measure from the top (double) plate height. There are usually slight variations between the span shown on the plans, the actual span at the bottom-plate level, and the one at the double plate. Since rafter lengths are calculated down to ¼-in. changes in span, use the double-plate measurement. A 100-ft. tape is the tool for this job.

First, as shown in the photo below, the positions of the rafters must be marked on the top of the double plate. This lets you properly locate the rafters when erecting the ridge. The layout is also necessary to distinguish the positions of the rafters from those of the ceiling-joist layout, which should be placed so they can be used as ties to which the rafters can be nailed. Starting with the hipped end of the roof, lay out the positions of the three king common rafters. Strike a line 10 ft. in from each corner down the length of the building, as well as one midway along the width, and write the letter C (for center) on the plate over each of these lines, which will serve as centers for the king commons. Next, lay out hip-jack rafters on 2-ft. centers from the corner of the building toward the king common rafters.

The common rafters are laid out similarly on the plates, starting at the gable end. I usually mark one side of the rafter position with a line across the top plate. If ceiling joists are also on 2-ft. centers, you don't need to lay them out, because they will be installed beside the rafters. If joists are on 16-in. centers, you would start the layout with the tape held 1½ in. past the end of the top plate. This way, a joist will tie into a rafter every 4 ft.

The ceiling joists that sit on the exterior wall will stick up above the rafters, and can be trimmed along the pitch of the roof after the rafters are up, and before the decking is applied. On the hip, the ceiling joists close to the end wall can't be nailed in place unless you notch them or cheat them off the layout, because the hip will interfere. They should be laid flat on their layouts and installed after the hips and jacks are in place.

Layout tees—The layout tee is a handy tool that lets the builder lay out rafters accurately and quickly. It also helps eliminate steps in rafter-length calculations. Layout tees should be made for the bird's mouth and tail of both

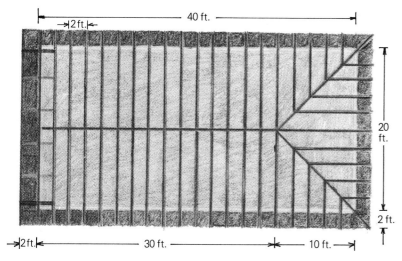

The roof plan of the model, above, shows a gable end using a barge rafter and outriggers for a 2-ft. rake overhang, and a hipped end with a 2-ft. eave. The 2x6 rafters are on 24-in. centers, and the roof pitch is 8-in.-12. The span in this case is 20 ft. Right, a carpenter lays out the joists and rafters by walking the plate, something that should be done only after the walls have been plumbed, lined and well braced.

Illustrations: Frances Boynton

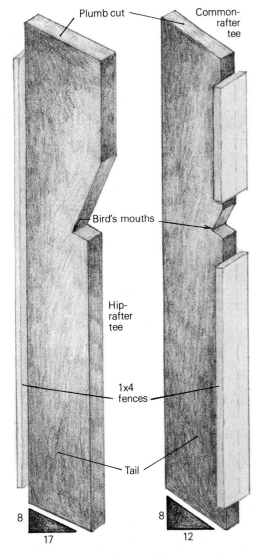

Plumb cut

Common-rafter tee

Bird's mouths

Hip-rafter tee

1x4 fences

Tail

8
17

8
12

common and hip rafters (drawing, above). There should also be a plumb cut at the top of the tee above the bird's mouth to use as a pattern for the top plumb cut.

Tees should be made of the same width stock as the rafters, so in this case the layout tee for the common rafter should be made from a 3-ft. scrap of 2x6. After you scribe the plumb cut at the top of the tee, move the square about 12 in. down from the top plumb mark and scribe out a bird's mouth. Next, mark the tail cut by measuring along the body of the square for the length of the overhang, which is 2 ft. in this case. Then make a mark and scribe the plumb tail cut, for an 8-in-12 pitch in our example. Cut out the pattern and nail two pieces of 1x4 along the bottom of the tee, one staying clear of the bird's mouth and the other not projecting past the top plumb cut, as shown above. When the tee is being used, the 1x4 fence registers against the bottom edge of the rafter stock for marking out the bird's mouth and plumb cuts.

The hip-rafter tee in this example is cut from a 3-ft. 2x8. On one end, scribe an 8 and 17 plumb mark. Move the square down about 12 in. and scribe the seat cut. Since this is a hip, the seat cut must be dropped (cut more deeply) to bring the top edge of the hip into the same plane as the jacks. To determine the amount of drop, lay the square at the top of the 2x8 on an 8 and 17. Where the 17 mark intersects the top of the lumber, measure down the body of the

square half the thickness of the hip (here ¾ in.) and make a mark. Then measure down from the top of the lumber, perpendicular to the edge, to this mark (⁵⁄₁₆ in.). This is the drop needed. Make the hip seat cut ⁵⁄₁₆ in. deeper than the common seat cuts.

The width of the tail of a hip must equal the width of the common rafter, so the wood past the seat cut must be ripped down from a 2x8 to a 2x6. Measure down from the top edge of the rafter 5½ in. (the actual dimension of a 2x6), mark the length of the rafter tail, and rip the excess 2 in. off the rafter's bottom edge. Since this rip creates a step in the bottom edge, both pieces of 1x4 fence should be nailed to the top of the tee. When using this tee, place it on the top edge of the rafter stock.

The next job is working up a cut list for all the rafters. Count the rafters on your plans, and calculate their lengths. The cuts can then be scribed using the rafter tees, and all the pieces can be cut before beginning the actual installation. This approach requires both confidence and intense concentration, but doing all the cutting first speeds up the process by letting you put your head down and frame without having to stop and figure.

The rafter book—I wouldn't want to be without a rafter book when framing roofs. Mine contains 230,400 rafter lengths for 48 pitches. I can look up any building span under the appropriate pitch, and quickly determine the rafter length and angle of cut. This book saves a lot of labor, and eliminates many costly errors.

Calculating common-rafter lengths—In the rafter book under 8 in 12, the common-rafter table shows that our span of 20 ft. requires a rafter length of 12 ft. ¼ in. from heel cut to plumb cut. This measurement doesn't account for the ridge reduction, because ridge thickness is not a constant. With a 2x ridgeboard, the reduction along a level line is ¾ in. But measured along the rafter edge, ¾ in. measures ⅞ in. on an 8-in-12. Rather than laying out a shortening line on the stock, I subtract the ridge reduction measurement from the rafter-book length to get the corrected rafter length down to the bird's mouth—in this case, 11 ft. 11⅜ in.

The overhang length from the heel cut to the tail cut can be taken off the rafter tee or determined from the rafter book by adding the overhang for each run to the span. A 2-ft. overhang will add 4 ft. to the span. In the rafter book, the 24-ft. span at 8 in 12 reads 14 ft. 5⅛ in. Deduct from that figure the full rafter length of 12 ft. ¼ in. This leaves 2 ft. 4⅞ in. in length from the heel cut to the toe of the tail cut. The overall length of the rafter will be 14 ft. 4¼ in.

Calculating lengths of hips and jacks—Use a large pad of paper to organize your calculations for the hips and their jacks, since they involve a bit of figuring. On the job, keep your building plans clean. Don't scribble math all over them. Using the rafter book, an 8-in-12 hip at a span of 20 ft. is 15 ft. 7⅝ in. A ridge reduction is necessary, and this 45° thickness mea-

sures 1³⁄₁₆ in. along the rafter edge at 8-in-12. This reduces the rafter length to 15 ft. 6⁷⁄₁₆ in. from plumb cut to heel cut.

To find the overhang or tail length, add 4 ft. to the span, just as for the common. The rafter book lists 18 ft. 9⅛ in. for a 24-ft. span. This leaves a rafter tail of 3 ft. 2½ in., and an overall length of 18 ft. 8¹⁵⁄₁₆ in.

To calculate the lengths of the hip jacks, look up jack rafters on 2-ft. centers at 8-in-12. Both the square and the rafter book read 2 ft. 4⅞ in. This distance is the common difference, or how much longer one jack will be than the previous one. This is also the length of the first jack before the deductions. If you are using my system of subtracting the ridge reduction (measured along the edge of the rafter) from the rafter-book length, then subtract 1⅜ in. on an 8-in-12 jack to get 2 ft. 3½ in. from plumb cut to heel cut. Only the first jack needs to be figured for the deduction since the rest will automatically follow, as the common difference is added to each one.

Cutting the rafters—With all the calculations complete, the next step is to lay out and cut the rafters. You can use production techniques that save a lot of time without sacrificing accuracy. I use a rafter bench, an oversize, site-built sawhorse that holds ganged rafter stock up off the ground for easy marking and cutting. I try to set up my benches close to the lumber stack, which should be fairly close to the building. You will be worn out before you start if you have to carry a ton of rafters a great distance.

Stack all the rafters of one type on the bench with their crowns down. The crown is a convex edge seen by sighting down the lumber. Crowns should be placed up in construction to help deflect the load placed on the rafters or joists; they are stacked crown down on the rafter bench so you can scribe cut-lines with the

A chalk line snapped across ganged common rafters marks the heel of the plumb-cut line. As indicated by the layout tee (bottom left) the rafters are stacked with their bottom edges up, and their ends even and square. The layout tee will be used on each rafter to scribe the plumb cut, bird's mouth, and tail cut.

Jack rafters stacked on a rafter bench show the common difference of 2 ft. 4⅞ in. on an 8-in-12 pitch using a 24-in. spacing. The diagonal lines indicate the direction of the side cut that will produce pairs of jacks (left and right) for each hip rafter. The bird's mouth and tail will be marked with the common-rafter layout tee.

layout tees. When you stack the rafters on the bench, keep their ends flush so they can be squared up easily with a framing square by drawing a line across their edges. Then use the layout tees to mark the plumb cut at the top, and bird's mouth and tail cuts at the bottom on each of the outside rafters of the stack. Measure the length you have calculated between the plumb cut and the bird's mouth several times, and then connect the marks across the stack with a chalk line (photo facing page, bottom). Lay the first outside rafter down flat on the bench, scribe the plumb, seat and tail cuts with the rafter pattern, and cut them out.

When all the common rafters are cut, they should be dispersed along the exterior walls according to the layout. Before spreading the rafters out, it is a good idea to set a 16d nail at the top plumb cut. This toenail will come in very handy during assembly.

Cut hip rafters using the same procedure, but make double side cuts at the top (for laying out double side cuts, see pp. 26-27). These cuts can be made easily on 2x stock with a circular saw set at 45°. For larger timbers or glue-lams, the angle of the top edge of the stock must be laid out, and the cut made with a handsaw.

With the double side cut complete, measure down from the top of the rafter the distance calculated, 15 ft. 6⁷⁄₁₆ in. Mark this length on the center of the rafter's top edge. Slide the tee to this point on the rafter, and scribe the seat cut and heel cut. Mark the rest of the board, scribing along the tail of the pattern. Rip the tail down to the proper width.

Use the common-rafter tee for the jack layout. Group the jacks on the bench according to length—for the model roof, there will be four sets of four. Load the longest first and work down to the shortest set. Only the tail ends can be squared up. Lay the common-rafter tee on this end and scribe the tail and seat cuts on the outside rafter. Then lay out the rafter on the opposite side of the stack and snap lines.

To lay out the plumb cut at the top of the

jack, measure the length of common difference—2 ft. 3½ in.—up from the seat-cut line for the shortest set, and add 2 ft. 3½ in. progressively to each set of jacks. Square these marks on the top edge of the rafters, and lightly mark two of each set with a 45° line indicating the direction of the angle. Mark the other two with the opposite angle (photo left). The side cut must be laid out this way because the length of a jack is measured from its centerline. Scribe a 45° line in the direction of the light line drawn previously through the center of the plumb-cut line on the top edge of the rafter. Then place the layout tee at the end of the 45° line that intersects the edge of the board farthest down the rafter, and scribe the plumb cut on the face of the rafter. This method creates a slight inaccuracy in the length of the jack on a moderately pitched roof, but it is much faster than marking the precise angle (which can be found in the rafter book or on the square) on the top edge of each rafter.

After cutting all the jack pairs, set them on the roof, paying particular attention to the correct placement of right and left-hand rafters. Drive a 16d nail into the smaller jacks, and hang the head and shank of the nail over the double plate so that the jacks hang along the wall, out of the way but still accessible.

The last roof members to get cut are the ridgeboards and purlins. The 30-ft. ridgeboard on the example here is made from two pieces. Pick straight stock, and cut so the break falls in the middle of a common-rafter layout. The board that includes the hip end of the building should be left long by 6 in., and all cuts should be square.

Assembling the roof—The reward for all the calculating, laying out and cutting is a roof whose members fall right into place once the ridge is up. This is the stage with the largest element of danger, and safety is a primary concern. While nailing joists and laying out the top plate, you'll start to develop "sea legs," gaining confidence in walking around up there. Make sure no loose boards stick out more than a few inches beyond a joist, and keep the top plates between joists free of scraps and nails.

Establishing the ridge height—Before doing anything else, calculate the ridge height to see if you need scaffolding to install it. This is done by multiplying the unit of rise (8 in our example) by the run of the building (10). The bottom of this ridge is 80 in. from the top plate. Ridges 6 ft. or more above the plate need scaffolds. A good scaffold is about 4 ft. lower than the ridge. It must be sturdy, well braced, spanned with sound planks, and running down the center of the building. To make room for the placement of ridge supports and sway bracing, leave a 1-ft. wide space between the scaffold planks.

Raising the gable end and ridge—First, put the tools and materials where you need them. Saws, nails and other tools can be kept handy on a sheet of plywood tacked on the joists. Pull the ridgeboards up on the joists alongside the

scaffolding. You'll need several long 2x4s for braces and legs. Stack them neatly on the joists along with 2x8 bracing for the purlins. To support the gable-end rafters in the initial stage of assembly, nail two uprights to the gable-end wall, perpendicular to the top plate and a foot on each side of the center.

For setting the gable and ridge, you need a crew of four—two carpenters on the scaffold, and one at each end of the span. Starting at the first rafter on the gable end, the carpenters on the outside walls pull up the gable-end rafters, setting the top plumb cut on the scaffold. A small 2x4 block 7½ in. long, the height of the ridgeboard, should be nailed to the plumb cut of one of the rafters. This block temporarily takes the place of the ridgeboard. Make sure the block is flush with the top of the plumb cut. The carpenters on the scaffold pull up the rafters until the seat cuts sit flush on the top plate. The carpenters on the outside walls nail the rafters down, keeping the outside of the gable-end rafter flush with the outside wall. Toenail each rafter to the double plate with two 16d nails on one side and one 16d on the other (back nail). At the plumb-cut end, the rafter with the temporary block must align with the other rafter so that the cuts are nice and tight to the block.

When the gable-end rafters are in position, nail each rafter to the uprights with 16d nails. You'll need to insert a temporary support under the ridge. Measure down from the bottom of the block to the top plate on the gable-end wall to find its length (drawing, below). Nail the leg down to the plate where the 10-ft. center is marked. You'll need another leg under the joint in the ridgeboard, but before you cut it, look for something to set it on. If there isn't a wall directly below the ridge, lay a 2x6 or 2x8 across the joists to carry the leg. After this leg is cut, the block can be removed from the gable end and replaced with the ridgeboard. The carpenter on the other end of the ridgeboard should rest it on the support leg, and scab an 18-in. 2x4 onto the leg and ridge. The scab should stop at least 1 in. below the top of the ridgeboard.

After one end of the ridge is raised, install the common rafter pair that is one layout back from the other end of the first length of the ridgeboard. When you're nailing rafters to the ridge, use three 16d nails to face-nail the first

Supporting the gable end and ridge

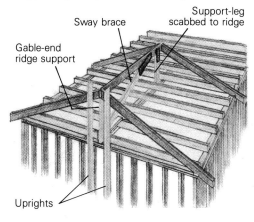

Sway brace

Support-leg scabbed to ridge

Gable-end ridge support

Uprights

The king-common rafter that butts the end of the ridgeboard is try-fitted and used to scribe a line for cutting the ridgeboard in place (left). The rafter in the foreground is a king-common that will be nailed at the end of the ridgeboard, perpendicular to the rafter being used for scribing. The skeleton formed by the three king-commons (center left) supports the ridgeboard so the common rafters can be nailed up with frieze blocks. The vertical 2x8 in the foreground is a temporary gable-end brace.

Bottom left: The underside of a hip rafter shows the jack-rafter pairs in position. The upright brace under the hip is placed over a wall. Hips are cut from stock 2 in. wider than common rafters to accommodate the width of jacks cut at compound angles. The added width gives strength for the long span required of hips.

rafter in the pair; then toenail the second. When these rafters are secured, the gable end should be plumbed, and temporarily secured with a swaybrace, a 2x4 with one end cut on a 45° angle, that reaches from the plate (or a 2x8 nailed to the joists above) to the ridge.

Installing hips—The remaining length of ridgeboard is set next. This is easily done by nailing another support leg at the end of the new ridge piece and setting the two king common rafters that define the hip. The third king common, the one that nails to the end of the ridge, is next (photo top left). The hip end of the ridge should extend about 6 in. beyond the king-common layout to allow for final fitting. Do not nail the third common yet, but slide it up to the ridge and scribe the ridge at the plumb cut when the rafter is flush to the top of the ridge and seat cut is up tight (photo center left). Set the rafter down and cut the ridge off, then nail it in place. The resulting frame should be plumb and strong, and ready for the hips.

Raise the hip, pushing its double side cut into the slot at the ridge, and toenail it at the corner and at the ridge. If the hip is spliced, haul the pieces separately on the roof and nail a 2x4 cleat to the bottom edge of the hip at the scarf joint. Position it on the bottom edge so it doesn't interfere with the jacks. Pull a string from the top center of the hip at the ridge, down to the center of the hip at the seat cut. Nail in a temporary upright under the center of the hip and align it with the string. This should eliminate any sag. If it is spliced, cut a leg to fit under the cleat (photo bottom left). Now you can nail the jacks and their frieze blocks.

Jacks, commons and frieze blocks—Start with the smallest jacks and work up. Nail in pairs, to avoid bowing the hip. Then nail the seat cut.

The remaining common rafters can now be filled in, followed by the frieze blocks that go between the rafters at the double plate. The blocks for a 2-ft. o.c. spacing should be cut 22⁷⁄₁₆ in. and driven tight. Frieze blocks that fit against the hip will have a side cut on one end (photo right). Frieze blocks that are to be nailed perpendicular to the rafters should remain full height. However, if they are to be nailed plumb, they will have to be beveled on the

pitch. This is most easily done on the table saw, but you can do it with a skill saw. In either case, use rafter off-cuts and discarded rafter stock for frieze blocks. For repeated crosscuts, use a radial arm saw, or set up a simple cut-off fixture for your circular saw (see p. 185). The blocks for our example are held to the outside of the top plate, square with the rafter (not plumb) and toenailed flush at the top of the rafter. The next rafter on layout is then pulled up, set in place, nailed at the seat cut and the ridge, and then nailed through the side into the frieze block behind it. Drive two 16d nails for 2x6s; three 16d nails for 2x8s. Whenever a ceiling joist lands next to the rafter, drive three 16d nails through the rafter into the joist.

Purlins—Purlins are required where rafter spans are long. Purlins run the length of the building at the center of the rafter span. They are usually made of the same stock as the ridge, and should be positioned once all the rafters have been nailed in place. If the commons are 18 ft. or over, it's much easier to handle them if the purlins are installed beforehand. To put up the purlin, first string a dry line across the path of the common rafters to check their sag at the center of their span. Start a purlin at one end, and toenail it into the bottom edge of the rafter, while it's being held by two carpenters. It is held square to the edge of the rafter and perpendicular to the rafter slope. Toenail it to the rafters in several places. Then cut legs (kickers) to fit under the purlin (small photo, facing page). The kickers must sit on the top of a wall, and to avoid deflection should not be placed in the middle of a ceiling-joist span.

Finishing up—Gable roofs are also reinforced with collar ties—horizontal members that connect one rafter in a pair to its opposing member. Collar ties should be no lower than the top one-third of the rafter span. Measure down from the ridge along the slope of the rafter and make a mark about one-third of the way down. Now mark the same distance on the opposite rafter. Hold a 2x4 (or wider board) long enough

This hip rafter has been toenailed in place. The frieze blocks required a single side cut for their intersection with the hip. In cutting the bird's mouth for the hip, the amount of drop had to be calculated. This meant taking a deeper cut so that the top edge of the hip is in the same plane as the other rafters.

The purlin in the foreground (above) is supporting the span of common and hip-jack rafters. Braces positioned at interior walls are perpendicular to the slope of the roof.

Standing on the outriggers (right), a carpenter nails the barge rafter. The frame has been notched for the flat 2x4 outriggers, which are face-nailed to the first rafter inside the gable end, and flat-nailed to the gable-end rafter. The rafters are the top chords of Fink trusses.

to span the two rafters at the marks, and scribe it where it projects past the top of the rafters. Using this as a pattern, cut as many collar ties as you need.

The gable ends must be filled in with gable studs placed 16 in. o.c. Each gable stud fits flush from the outside wall to the underside of the gable rafter. You can make the gable stud fit neatly under the rafter by making square cuts with your saw set on the degree that corresponds with the pitch of the roof. For an 8-in-12 pitch, the corresponding angle is $33\frac{3}{4}°$. You can find the degrees in the rafter book under the pitch of the roof. Gable studs are best cut in sets and, like jack rafters, they advance by a common difference.

The example shows a rake of 2 ft., with a barge rafter and outriggers. Unlike the fly rafter and ladder system shown in the glossary (p. 31), a barge rafter usually isn't reduced for the ridge; it butts its mate directly in front of the end of the ridge board. The outriggers support the barge overhang. They are typically 2x4s, 4 ft. o.c. from the ridge down, extending from the barge rafter across the gable-end rafter and beyond one rafter bay. The outriggers are notched into the gable rafter, laid in flat and face-nailed to the common rafter in back, as shown in the photo at right.

To put in outriggers, first lay out the top of the gable rafter 4 ft. o.c., starting from the ridge. The layout should be for flat 2x4s ($3\frac{1}{2}$ in. wide). Then notch the layout marks with several $1\frac{1}{2}$-in. deep saw kerfs and a few quick blows from your hammer. Make these cuts down on the rafter bench. Let the outriggers run long and cut them along a chalked line once they are up to ensure a straight line for nailing the barge rafter. □

Don Dunkley is a carpenter and contractor in Sacramento, Calif.

Roof Shingling

With only a few rules to follow, putting on a wood roof can be relaxing work with pleasant materials

by Bob Syvanen

Shingling is one of my favorite tasks in building houses. Even though roofers may be a little faster, I like to do this work myself. Shingling is the kind of job that requires little calculating and a minimum of physical effort. You can think of other things as you work. You don't have to manipulate unwieldy boards or carry heavy loads. Shingle nails are fairly short, so swinging the hammer is easy on the arm; and if you have the time to invest in the old-fashioned methods of making hips, ridges and valleys, there's just enough cutting and fitting to make the job interesting.

Wood shingles are typically three or four times as expensive as asphalt shingles, but they give a roof a texture and color that you can't get with petroleum products. A wood roof is also much cooler in the summer, and will last nearly twice as long as one covered with conventional asphalt shingles. The only major disadvantage to wood shingles is their flammability; but chemical treatments, along with spark arrestors on fireplace chimneys, can minimize this liability.

In the past, shingles were commonly made of cypress, cedar, pine or redwood. My favorite is cypress, although red cedar is what's most available these days. It too is excellent for roofs because the natural oil in the wood encourages water to run off instead of soaking in, and it helps prevent the shingles from splitting despite wide fluctuations in humidity and temperature year after year.

Wood shingles are a delight to work because they're already cut to length and thickness from the best part of the tree, the heartwood. Shingles are sawn flat on both faces, which distinguishes them from shakes, which are split out along the grain. Wood shingles come in lengths of 16 in., 18 in. or 24 in., and taper along their length. The exposed ends, called butts, are uniformly thick for each length category of shingles. This measurement is always given in a cumulative form—16-in. shingles, for instance, always have butts that are 5/2. This means that five shingle butts will add up to 2 in.

Tools—Tools for shingling are few and simple. I have put on many shingles using a hammer and a sharp utility knife. Most pros use a lathing or shingling hatchet. A shingling hatchet has an adjustable exposure gauge on the blade; however, a mark on your hammer handle works almost as well. Two good features of the hatchet are the textured face on the crown, and the hatchet blade itself. The mill face or waffle head is less likely to glance off a nail onto a waiting finger, especially when the head of the hammer strikes a blob of zinc that hot-dipped shingle nails often have. The sharp blade and heel of the hatchet are useful in squaring shingles, and in trimming hips and rakes. I also use a block plane to trim hip, valley and ridge shingles for final fit.

I prefer to keep my nails in a canvas apron at my waist, but others like leather nailbags hung off a belt. Production roofers use a stripper, a small, open aluminum box that straps to their chest. It has slots that allow the points of the nails to drop into line when it's loaded with a handful of nails and shaken back and forth.

Although it's possible to work from a ladder, proper staging or scaffolding is a big help when starting a roof. Wall brackets (drawing, facing page, left) are my first choice. You can make them from 2x4s, or rent or buy the sturdier steel ones. They should be attached to the wall studs at a comfortable working height below the starter course on the eave. Some brackets will accommodate nearly 30 in. of scaffolding planks, but two 2x10s battened together make

Figuring materials

You can usually buy shingles in three grades. Always use No. 1, the best grade for a roof. No 1. shingles are 100% clear heartwood and edge grain. No. 2s and 3s have more sapwood, knots and flat grain. They're okay for outbuildings where the life expectancy of the structure is shorter, or for starter coursing, shim stock and sidewall shingling.

In order to figure how many shingles you will need, you must first know what exposure you are going to use. Exposure is the measurement of how much of the shingle shows on each course. The longer the shingle, the greater the possible exposure. Maximum exposures are also determined in part by the pitch of the roof. As shown in the chart below, the flatter the pitch, the less shingle that can be left to the weather.

Maximum exposure (in.)			
Roof pitch	Shingle length (in.) 16	18	24
5-in-12 and up	5	5½	7½
4-in-12	4½	5	6¾
3-in-12	3¾	4¼	5¾

Using these maximum exposures, all pitches of 5-in-12 and up will give triple coverage. With lesser pitches, successive courses of shingles will overlap each other four times, so you get quadruple coverage. Wood shingles shouldn't be used on pitches lower than 3-in-12.

When the maximum exposure is used on any length shingle, four bundles of shingles will cover 100 sq. ft. of roof, or one square. To figure your roofing material, multiply the length of your roof by its width and divide by 100. This will give you the number of squares. For starter courses, add one extra bundle of shingles for each 60 lineal feet of eave. A starter course is the first course of shingles, which is doubled to provide a layer of protection at the joints between shingles. In some cases, the starter course is even tripled. For valleys, figure one extra bundle for every 25 ft., and the same for hips. If there is a hip and a valley, figure some of the waste from the valley to be used on the hip. If you are going to use manufactured hip and ridge shingles, a bundle will cover about 17 lineal feet.

When you calculate shingle quantity, figure nails and flashing, and order them at the same time. For 16-in. shingles on a 5-in-12 or steeper pitch, use 3d galvanized shingle nails and figure 2 lb. per square. For 24-in. shingles, get 4d nails, with 5d nails or bigger for re-roofing jobs when you're nailing through other shingles. Hip and ridge caps need nails two sizes larger than the shingle nails used in the field (on the roof slope), because you will be nailing through extra thicknesses.

If you are near a good-sized city, roofing-materials suppliers are fairly common. Their prices are often more reasonable than lumberyard prices, and their inventory is only for roofers, so you're more likely to get the flashings and nails that you need. Ask for the price per square on the shingles that you want, and be prepared to give a figure of how many squares you'll need. Another advantage of buying from a roofing supplier is that many of them deliver on lift-bed trucks and load the roof with the shingles, saving your back and a lot of time walking up and down a ladder. —*Bob Syvanen*

a nice working platform. Make sure the battens (or cleats) extend back beyond the planks to the wall to prevent them from shifting on the brackets inward under the eaves. Scaffolding planks and steel ladder brackets hung from the rungs of two straight extension ladders placed against the siding will also work nicely. Once up on a roof over a 4-in-12 pitch, I use roofing brackets, but I'll get to these later.

Preparation—One decision you make before shingling is what sheathing to use. Where there is no wind-blown snow, or where the weather is humid and wet much of the time, an open slat roof with spaced sheathing is a good choice. Shingles are laid on top of it without roofing felt, so air can circulate freely on the underside of the shingles. The spacing of this sheathing is important. For a 5-in. exposure, the 1x4 sheathing should also be spaced 5 in. o.c. (drawing, below right). The shingle tips should lap over the sheathing at least 1½ in., with two boards butted together at the eaves and ridge for proper nailing of starters and ridge caps.

In snow country, the solid-sheathed roof is best. I have stripped many roofs with solid sheathing and found that they have held up very well. I use CDX plywood and cover it completely with 15-lb. felt. If you are using felt, lay only as much as you need for a day's work. Morning moisture will wrinkle the paper and make shingling difficult. Along the eaves I use 36-in. wide 30-lb. felt. If ice-damming is a particular problem in your area, you can trowel on

Using a shingler's hatchet, Syvanen squares up a red cedar shingle, right. The 1x3 roof stick he is cutting on is a gauge for laying straight courses with the correct exposure. The shingles on the top course at the left of this photo have been butted to this gauge and nailed, each one with two 3d shingle nails.

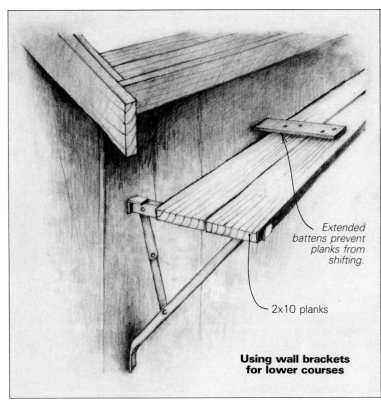

Extended battens prevent planks from shifting.

2x10 planks

Using wall brackets for lower courses

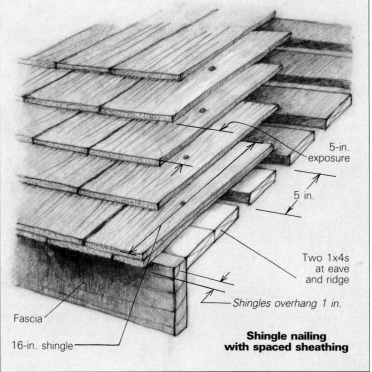

5-in. exposure

5 in.

Two 1x4s at eave and ridge

Shingles overhang 1 in.

Fascia

16-in. shingle

Shingle nailing with spaced sheathing

Photos of author: Fritz Haubner

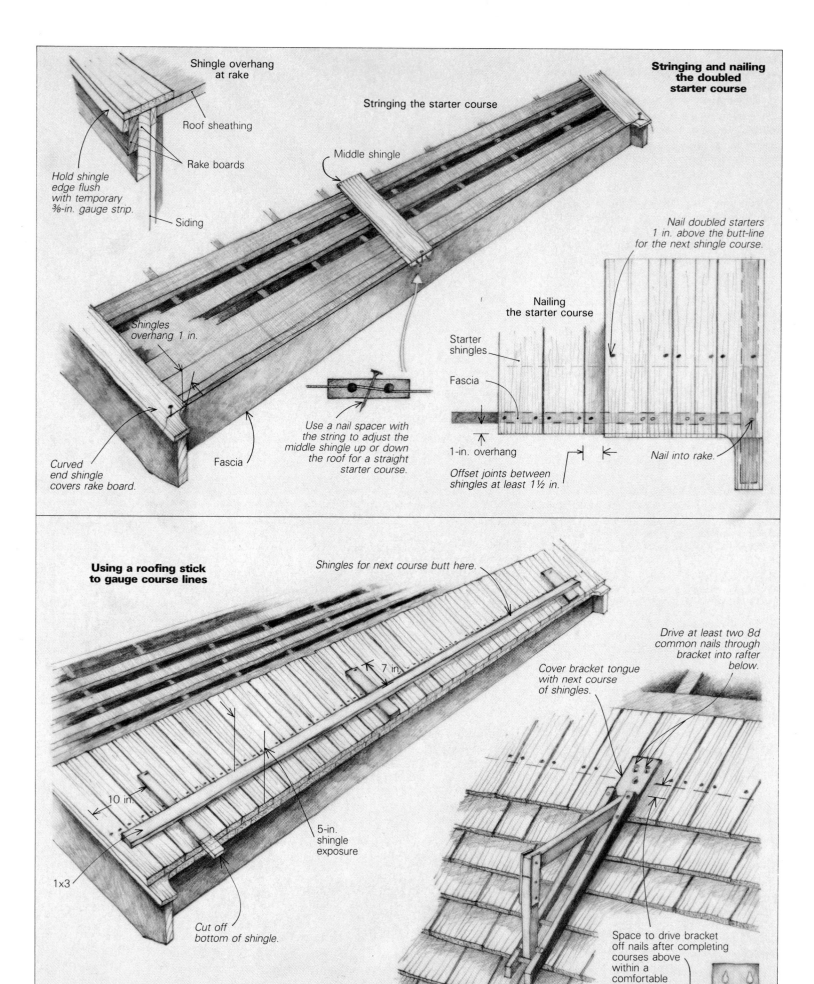

Shingle overhang at rake

Roof sheathing

Rake boards

Hold shingle edge flush with temporary ⅜-in. gauge strip.

Siding

Stringing the starter course

Middle shingle

Stringing and nailing the doubled starter course

Nail doubled starters 1 in. above the butt-line for the next shingle course.

Shingles overhang 1 in.

Curved end shingle covers rake board.

Fascia

Use a nail spacer with the string to adjust the middle shingle up or down the roof for a straight starter course.

Nailing the starter course

Starter shingles

Fascia

1-in. overhang

Offset joints between shingles at least 1½ in.

Nail into rake.

Using a roofing stick to gauge course lines

Shingles for next course butt here.

7 in.

Drive at least two 8d common nails through bracket into rafter below.

Cover bracket tongue with next course of shingles.

10 in.

5-in. shingle exposure

1x3

Cut off bottom of shingle.

Space to drive bracket off nails after completing courses above within a comfortable reach

Installing roofing brackets

Illustrations: Frances Ashforth

When using roof brackets, or jacks, squatting is usually the most comfortable working position. Once you have shingled up the roof to a point where you are working at full arm extension, nail a new line of brackets to the rafters higher up the roof. These brackets are removed when the roof is completed by tapping them at the bottom. The 8d common nails used to secure each bracket remain under the shingles.

a layer of roof mastic over the 30-lb. felt at the eaves, and then lay on another run of felt over this for a self-sealing membrane.

The shingles will need to overhang both the rake and eaves. For the rake overhang, cut some temporary gauge strips ⅜ in. by 1 in. or so, and tack them on the rake board with 4d nails (drawing, facing page, top left). You can then hold your shingles flush with the outside of this gauge board for a ⅜-in. overhang.

Starter course—The first course along the eave is doubled. I like to extend the end shingle of this starter course over the rake board or gutter if there is one. Water runoff will wear away the top of the exposed piece of rake board if it is not covered. This extended shingle can be straight or curved. I like it curved. I use a coping saw and cut several at one time.

To line the starter course, nail the curved end shingles in place. These shingles will be flush with the rake gauge strips and will overhang the fascia on the eave 1 in. Tack a shingle in the middle of the run with the same 1-in. overhang. Because the fascia probably won't be truly straight, the middle shingle will have to be adjusted up or down to straighten the starter course. Do this by stretching a string from one end shingle to the other on nails spotted at the line of the 1-in. overhang.

To keep from butting the shingles directly to the string and introducing cumulative error by pushing against it, fix the string away from the line of the starter course by the diameter of a nail. Then use a loose nail to gauge each starter shingle to the string as you nail it. Begin by adjusting the middle shingle to the string in this way. The drawings at the top of the facing page show how to get the string right.

Once the middle shingle is nailed down, fill in the rest of the starters, nailing them all into the fascia. Nail the end shingles into the rake board (drawing, facing page, top right). Angle the nails away a little to make sure that they don't poke through the face of this trim.

On top of the starters, nail the double starter shingles, making sure that the gaps between them are spaced at least 1½ in. from the gaps in the row below. The shingles in this course can overhang the starters by about ⅛ in., or sit flush. They should be nailed about 1 in. above the butt line for the next course; for a 5-in. exposure, nail about 6 in. up on the shingles.

Roof shingles get a lot of water dumped on them, and they will buckle if placed too close. I just eyeball the distance between edges, but any spacing from ⅛ in. to ¼ in. is fine. The joints between shingles should be offset from the joints in the course directly below by at least 1½ in., and offset from the gaps in the course below that one by at least 1 in.

Roofing sticks and brackets—To lay the second and all successive courses, you'll need a method of gauging the exposure and keeping the courses straight. One way is to use the gauge on your hatchet or a mark on your hammer handle. I prefer to make and use roofing sticks to align an entire course before I have to reposition them. Use as many of these sticks as it takes to span the roof. To make one, take a long, straight length of 1x3 and nail a 2-in. wide shingle to it at right angles every 8 ft. or so (drawing, facing page, bottom left). Then saw off the butt of the shingle so that the distance from this cutoff edge to the top edge of the 1x3 equals your single exposure. To use the roofing sticks, line up the shingle butts on the gauge stick with the butts of the first course. Tack the gauge to the roof with a nail in the upper corner of each gauge shingle. By butting the shingle ends down against the top edge of the 1x3, you get uniform exposure from one course to the next. On a calm day, you can lay in quite a few shingles before nailing them down—a nice feature of this system.

When you reach one of the shingle tips that is part of the roofing stick, just fit the shingle without nailing it. Tuck it under the butt of an adjacent shingle for safekeeping and continue down the course. Then when you've completed the full course, tap the roof sticks free and nail down the individual shingles you left out.

On even a moderate pitch, you'll need roof brackets to work safely, comfortably and at a good pace. These are adjustable jacks that are nailed to the roof to support a scaffolding plank (drawing, facing page, bottom right). Most lumberyards carry roof brackets, and they can also be rented. The ones I use are oak, but metal ones are more common. The first set can be put on after you've shingled up five or six courses and can't reach across the eave from the ladder or scaffolding.

Make sure your roof brackets are nailed into rafters, and use two 8d galvanized nails per bracket. Most brackets will take a single 2x10 plank. Twelve-foot planks are ideal when you

use a bracket at each end. Fourteen-footers will do. You can shingle right over the tongue of the bracket where it attaches to the roof, because the nails that hold it will be left under the shingles when it's removed. When you install the brackets, make sure there's enough space between the top of the bracket and the shingle butts to allow you to remove the bracket easily, by driving its bottom toward the ridge, and lifting the tongue off the nails.

Do's and dont's—Face-grain shingles have a right and wrong side up. Make sure to place the pith side of the shingle (the face that was closer to the center of the tree) down to prevent cupping. Some shingles are cupped or curled at their butts. I put the cup down for better runoff and appearance. I reject hard shingles because they split when nailed and curl up when the sun hits them.

Red cedar shingles can be very brittle, so hold back on that final hammer stroke or the shingle may crack. The nail head should sit on top of the shingle, and not be driven flush. Each shingle, no matter how wide, should get just two nails. If you have a hip or valley to shingle, save the bedsheets (shingles 9 in. or wider) when you're roofing in the field.

If you use a wide shingle in a regular course, score it down the middle with a hatchet or utility knife to control the inevitable splitting. Treat it as two separate shingles, offsetting the scored line from other joints, and nailing as you would two shingles. If you suspect that a shingle is checking, bend it into a slight arch with your hands. If it has a clean crack, use it as two shingles. If it shows several cracks, throw it out.

When you are shingling in the field, don't get so tightly focused on the work that you forget to check every once in a while that the coursing is even and straight. Do this by measuring up from the eave, and adjust if necessary. You may have to snap a chalkline occasionally to straighten out a course.

Once you've completed the roof, spend a few minutes looking for splits that you missed that

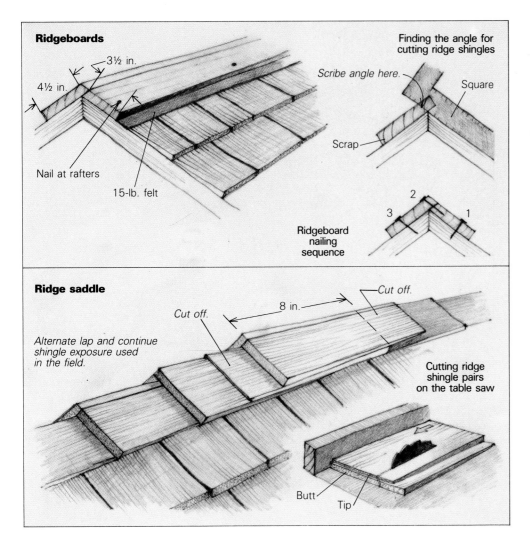

Ridgeboards

3½ in.

4½ in.

Finding the angle for
cutting ridge shingles

Scribe angle here.

Square

Scrap

Nail at rafters

15-lb. felt

Ridgeboard
nailing
sequence

3 2 1

Ridge saddle

Cut off.

8 in.

Cut off.

*Alternate lap and continue
shingle exposure used
in the field.*

Cutting ridge
shingle pairs
on the table saw

Butt

Tip

are not sufficiently offset from joints between
other shingles. Cut some 2-in. by 8-in. strips of
flashing out of zinc, aluminum or copper, and
slip them under the splits.

Ridges—Make sure that the courses are running parallel with the ridge before getting too
close to the top. Since a course of 3 in. or less
doesn't look good at the ridge, measure up
from the course you are working on to the
point where the ridgeboard or ridge shingles
will come, and adjust the exposure slightly so
that the courses work out. If you are using
ridgeboards, staple 15-lb. felt over the ridge for
the full length of the roof. The felt should be an
inch or so wider than the ridgeboard, and will
make painting the final coats on the ridgeboard
easier. Make your ridgeboards from 1x material, and bevel the top edge of each board at an
angle that will let them butt together, and form
a perfect peak. Because the boards butt at the
ridge rather than miter, one side will be narrower than the other (drawing, top left).

To find the bevel angle for your roof on the
table saw, take a pencil, a scrap of wood and a
square up to the ridge. Lay the scrap flat on the
roof on one side of the ridge so that part of the
scrap projects over the peak, as shown in the
drawing. Then lay the square flat on the other
side of the roof so that it projects past the scrap.
Scribe a line onto the wood by holding the pencil against the square. By tilting your table-saw
blade to this angle, you will be able to rip both
the narrow and wide ridgeboards.

To install the ridgeboards, snap a chalkline

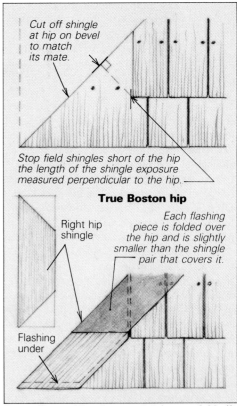

Cut off shingle
at hip on bevel
to match
its mate.

*Stop field shingles short of the hip
the length of the shingle exposure
measured perpendicular to the hip.*

True Boston hip

Right hip
shingle

*Each flashing
piece is folded over
the hip and is slightly
smaller than the shingle
pair that covers it.*

Flashing
under

**A true Boston hip, shown at left in a model built
by Syvanen, uses the hip shingles to complete
each course. The usual way is to superimpose a
line of ridge shingles on the hip. The butts of
Boston hip shingles can run parallel to the
eaves, as shown on the right side of the model,
or perpendicular to the hip (left side).**

the length of the roof where the lower edge of
the narrow ridgeboard will be nailed. Using 8d
common galvanized nails, attach the narrow
ridgeboard to the roof along the chalkline. Be
sure to nail into the rafters. Now nail the wide
ridgeboard to the narrow one. This will give a
nice straight line at the peak. Lastly, nail the
lower edge of the wide ridgeboard into the
rafters. Push down on the ridge pieces to force
the narrow board against the roof. You might
have to stand on the ridgeboards to do this.
After the final coat of paint is on, trim off the
15-lb. felt with a sharp utility knife.

You can also use shingles for the ridges—
either the factory-made kind that come already
stapled together as units, or your own, cut on a
table saw. To make your own, set the blade-to-
fence distance for the width of your exposure,
and set the angle of the blade as explained for
cutting ridgeboards. Saw the shingles in alternating, stacked pairs. The bottom shingle in the
first pair can have the butt facing the front of
the saw table, and the top shingle with the tip
doing the same. The next pair of shingles
should be reversed so that the tip is on the bottom, and the butt is on top. This will give you
shingle pairs whose laps alternate from one
side of the ridge to the other.

To install the ridge shingles, staple a strip of
30-lb. felt down the full length of the ridge. In
this case, the paper should be narrower than
the ridge unit. Nail the ridge pieces along a
snapped chalkline using a double course as a
starter. Alternate the pairs and use the same
shingle exposure you have on the rest of the
roof. Again, the nails should be about 1 in.
under the butt of the next ridge shingle. When
you reach the center, make a saddle by cutting
the tips off two pairs of ridge shingles so that
8 in. remain on each (drawing, facing page,
center left). Then nail them down on top of
each other with their butts facing the ends of
the ridge.

Hips—You can buy factory-assembled ridge
units for hips, or make your own. Most folks
just staple down a run of 30-lb. felt over the hip
and nail the units along a snapped chalkline or
temporary guide boards. Since water drains
away from a hip, this method keeps the rain
out. But when I have the time, I like to cut and
fit a true Boston hip the way the old-timers did
(photo and drawing, facing page, bottom). It
weaves the hip shingles and flashing right into
the courses in the field for a weathertight fit
that doesn't look added on, like standard hip
units. It looks tough to do, but it's not really.

The key to working a true Boston hip is to
stop the shingles in the field the same distance
from the hip on each course. This makes the
hip shingles uniform, allowing you to cut them
on the ground. To find the point on a course
line where the butt of the last shingle should
end, measure the length of the shingle exposure you are using in the field (such as 5 in.), on
a line that runs perpendicular to the hip. You
will have to move this 5-in. line, marked on a
tape measure or square, up or down the roof
(keeping perpendicular to the hip) until it fits
between the hip and the course line. Then

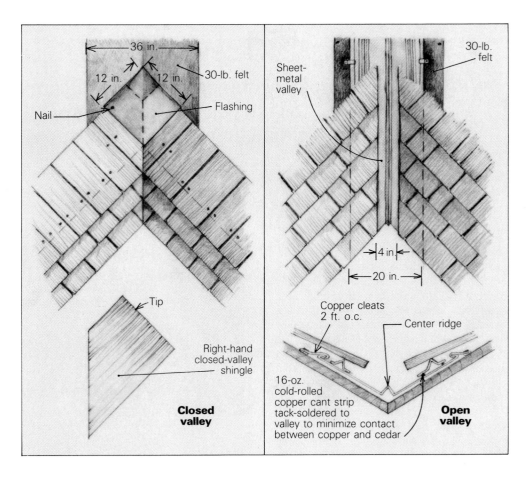

Closed valley

Right-hand closed-valley shingle

Open valley

mark the intersection on the course line and fill
in the shingles in the field to that point.

You will need to make right and left hip
shingles to form pairs. The tip of each hip
shingle needs to be trimmed to fit into the
space left for it by the last shingle in its course
and the one above it. The butts can be left perpendicular to the hip, or cut parallel to the
eave. The long side of the shingle also needs to
be cut. This should be a bevel that alternates
lapping its mate on the other side of the hip.
You can cut these bevels on a table saw or use a
shingling hatchet on the roof.

Flashing pieces, which are used under every
hip shingle, should be cut out with snips on the
ground. If you put a slight crease down the center of each piece, it will straddle the hip easily.
The flashing pieces should be slightly narrower
than each hip unit so they don't show, yet long
enough so that each piece laps the previous
one by a good 3 in.

Valleys—It's critical that valleys get done
properly, since the roof directs water right to
them. Valleys can be open or closed (drawings,
above). The open valley can be shingled faster
and cleaned more easily. In closed valleys,
leaves and pine needles can be a problem, and
some people don't recommend them. But I
prefer their neat look. Shingles butt up tight in
the valley with each course flashed much like
Boston hip flashing.

For a closed valley, start with a 36-in. piece of
30-lb. felt laid in the full length of the valley,
then add a piece of 12-in. by 12-in. flashing cut
diagonally in half. Add the starter course of
shingles. Next comes a 12-in. square piece of
flashing, and the second course of shingles.

Each valley shingle must get a miter cut

along its inner edge where it meets the other
valley shingle. It is often easier to cut these on
the ground, and do your final fitting on the roof
with a hatchet or block plane. Use the bedsheets that you saved. If you are cutting ahead
of yourself on the ground, don't forget to cut
both right and left-handed shingles. The flashing pieces can also be cut ahead. Remember
when you nail both shingles and flashing to use
only one corner nail as far away from the
center of the valley as you can on each side.

For an open valley, also lay in a 36-in. wide
sheet of 30-lb. felt; it's a good bed for the metal
that follows. On pitches under 12-in-12, use
20-in. wide sheet metal, which should be nailed
at the extreme edges with fasteners that are
compatible with the flashing (see page 58).
Though tin, lead, zinc and galvanized steel are
all right for valley metal, I like to use copper. There have been a few cases of copper
flashings and cedar shingles reacting chemically to produce a premature corrosion of the copper. I have never seen this, nor has anyone I
know. Just to be safe, you can use a cant strip
to minimize the contact between the shingles
and the copper (drawing, above right). Whatever flashing you use, it will be improved by a
crimp in the center.

I often begin in the valley by nailing the first
few shingles on five or six courses high before I
carry the courses across the roof. This stacking
allows me to set the valley shingles to a chalkline without being restricted by having to fit
them to the shingles in the field. Just as with
closed valleys, it is important to nail as far from
the valley center as possible.   □

_____

Bob Syvanen is consulting editor to Fine Homebuilding, *and a builder on Cape Cod.*

Sidewall Shingling
Simple techniques and a steady nailing rhythm can give you a natural, no-maintenance siding that will last 40 years or more

by Tim Snyder

In spite of the vast selection of residential siding materials available today, cedar shingles are still an attractive choice for builders and renovators. Left unfinished, they will outlast boards and battens, textured plywood and clapboards. Shingling a house with cedar doesn't demand anything more elaborate than a few hand tools and some basic carpentry skills. It is a labor-intensive job, and can be done without a helper, two reasons why you can save a lot of money if you do the work yourself.

Quality, size and exposure—Shingles are smooth on both sides because they've been sawn from the log, unlike shakes, which are split. Shingles come in random widths and are usually ⅜ in. thick at the butt end, tapering down to a sharp edge at the tip. Red cedar shingles will weather to a brown color; white cedar will turn silver-grey.

To order your shingles, you'll need to decide on shingle length and exposure. Exposure is the distance between the bottoms of successive courses. Figured with the total square footage of your walls, it determines the number of bundles you'll need for the job, as explained at right. Shingles come in 16-in., 18-in. and 24-in. lengths, packed in bundles that may vary in size, depending on shingle type. Exposure must be less than half the shingle length (7½ in., 8½ in. and 11½ in., respectively, are acceptable maximum exposures). Roof exposures are usually less. The exposure for the re-siding job shown in the photos is 6 in. with 18-in. No. 1 grade Perfections, a designation of the Red Cedar Shingle and Handsplit Shake Bureau (515 116th Ave. N.E., Suite 275, Bellevue, Wash. 98004). Using a narrow exposure requires more shingles, but it means the siding will last longer and provide better insulation.

Use only No. 1 or No. 2 shingles to side or re-side a house. With a lower grade, you'll have to contend with wavy grain, sapwood and knots. You can't easily split or trim these shingles to size and using them involves more waste, so you really don't save anything in the long run. No. 3 and No. 4 shingles do have their uses: barns, outbuildings, interior walls and undercoursing. I wouldn't use them on the outside of a house. Premium-grade shingles are easy to work, uniform in appearance and they're cut only from cedar heartwood—far more weather resistant than sapwood.

Before you buy, shop around for a good deal on No. 1 or No. 2 shingles. Prices fluctuate, as

How many shingles to buy?
First calculate the total square footage of the walls you plan to shingle, say 1,300 sq. ft. Shinglers speak in terms of squares, with one square equalling 100 sq. ft. However, shingles are packed in bundles of various sizes. Find out from your building supplier what size you can get. A size designation of 18/18 means two 18-shingle stacks sandwiched together, as shown above. Now refer to the chart below. Read down the appropriate bundle-size column to the row corresponding to the exposure you plan to use, for example, 6 in. The number on the chart (27.8) tells you how many square feet of wall each bundle will cover. Dividing the square footage of your walls by this figure gives you the number of bundles you need (1,300 ÷ 27.8 = 46.76); add 15% for waste, and your order is 54 bundles.

		Bundle size				
		10/10	12/12	14/14	18/18	20/20	28/28
Exposure (in.)	5	12.8	15.4	18.0	23.1	25.7	35.9
	6	15.4	18.5	21.6	27.8	30.8	43.2
	7	18.0	21.6	25.2	32.4	36.0	50.4
	8	20.6	24.7	28.8	37.0	41.1	57.6
	9	23.1	27.8	32.4	41.6	46.3	64.8
	10	25.7	30.8	36.0	46.2	51.4	71.9
	11	28.3	33.9	39.6	50.9	56.5	79.1

does quality. Check with a local builder for a good brand, or purchase a few bundles and see how they nail up. Contractors usually order about 15% more than their square-footage requirements. You should do the same, although with a little care you can waste as little as 5% to 8%. Save the split-off sections when you trim wide shingles. You can use these for undercoursing at the base of the wall, or to fit in narrow spaces later on in the job. Don't discard shingles whose butt ends aren't square; throw them in a separate pile and when you get enough, you can true up the stack on a table saw or radial-arm saw and use them for undercoursing. Most lumberyards will let you return unused bundles.

Surface preparation—On a roof, shingles are usually installed over purlins for good air circulation behind the cedar. Circulation behind vertical shingles isn't as important, so sidewall shingling is generally done over solid sheathing. On some old houses, like the one shown here, the underlayment is a mix of clapboards, roughsawn boards and plywood.

No matter what you're shingling over, it's good practice to protect all corners and windows or door openings from water infiltration. To do this staple 6-in. to 8-in. wide strips of builder's felt around corners and openings. If possible, force strips of felt under door and window casings. In re-siding, don't cover the entire wall surface with felt before shingling; this can create an exterior vapor barrier and trap moisture inside the wall.

Still, you have to consider wind penetration since even the tightest shingling job is subject to infiltration. To cope with this problem you can cover most sections of the wall with red resin paper, which is similar to builder's felt but not asphalt impregnated. It lets moisture escape from the wood but stops the wind.

There should be a groove along the underside of each window sill for the shingles to tuck into; otherwise they'll pop away from the wall. Make the groove ⅝ in. wide and ½ in. deep.

The next step is painting the trim. Window and door casings and cornerboards, if you're using them, should be sealed with a primer coat. If you've got the time, paint the finish coat too; this job is tedious and time-consuming once the shingles are on.

Drip edges over all windows and doors need to be flashed. To protect your casings, leave enough flashing to be bent over their edges

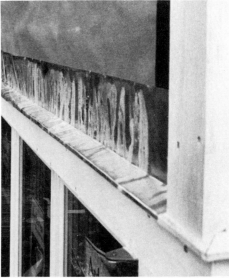

Above, flashing drip edges above windows. The flashing should be bent over the edge and extend 5 in. to 6 in. up the wall. Here it is also tucked under the cornerboard in the foreground.

At left, preparing sidewalls. Strips of builder's felt protect all corners and casings from water infiltration. Red resin paper stapled over the sheathing will stop wind infiltration without trapping moisture in the wood.

Right, custom-fitting. Whenever the course doesn't line up with the casing, shingles have to be trimmed and notched to fit over the window.

The fastest way to mark course guidelines between short stretches of wall is by snapping a chalkline, as shown below.

Corner boards on outside and inside corners

Woven inside corner

Mitered outside corner

Corners. Of the three corner treatments for shingle siding, corner boards stand up best to the weather. They also make shingling work go faster, as at left, since no woven or mitered joints are required.

line is eight courses up, compensate ⅛ in. on each course. It's almost impossible for the eye to detect such small adjustments.

Nailing up—This is the easiest part of the job. Use only zinc-coated (hot-dipped galvanized) or aluminum nails, since common nails will rust long before the shingles wear out. For 16-in. or 18-in. shingles, 3d nails are fine; use 5d or 6d on 24-in. shingles and for all re-siding jobs.

There are several nailing rules to remember as you put up your shingles. Use only two nails, no matter how wide the shingle. Place each nail ¾ in. to 1 in. in from the shingle edge, and about 1 in. up from the successive course line. Don't drive the nail head below the surface of the wood; this causes splitting and creates a depression where moisture will collect. You don't need to leave a gap between adjacent shingles, although some folks do. The shingles you buy today are green and will shrink. Gaps or no, don't allow these spaces or joints to line up vertically for more than two courses. There should be at least a 1½-in. vertical overlap to prevent water infiltration.

The bottom course of shingles has to be doubled. This is also true for the first course above a door or window as well as to the course at the base of a wall. Then you can snap a line or tack up a guideboard and start on the next course. Some shinglers find it easiest to build up their corners for several courses before working horizontally across the same section of wall. If you're not confident enough to rely on a shingler's hatchet alone for splitting shingles to width, have a sharp block plane handy for trimming and truing up your hatchet work. A sharp utility knife also makes a good trimming tool for white cedar shingles.

Corners—You can deal with corners in one of three ways: they can be woven, mitered or butted to corner boards as shown in the drawing, above left. Mitering is by far the most time-consuming corner to make, and is probably more susceptible to weather damage than the woven corner. Whether mitered or woven, shingled corners won't stand up to hard wind and driving rain as stoutly as a solid corner board.

Cutting the corner shingles to the alternate overlap for the woven corner can be done accurately enough with a handsaw. I like to use a sharp keyhole saw for notching and crossgrain trimming. It's easy to carry around—especially when you're up on scaffolding—a quick-draw item. Mitered corners have to be cut on a powersaw. All that measuring and trimming or running back and forth to your table saw might be worth it while you're on the ground, but you won't feel that way about climbing up and down for second-story work.

I prefer to use corner boards. Once you've nailed up the vertical boards and prime-coated them, shingling work can go really fast. If you decide to use cornerboards, buy either No. 2 pine in 5/4 thickness (1 in. actual) by whatever width you want, or 2x lumber. Outside corners are usually no less than 3 in. on a side, and inside corner boards should be at least ⅞ in. square in section. ☐

slightly. Install the flashing with as few nails as possible and try to keep them up and away from the drip edge.

Chalklines and storypoles—One more point before you start nailing up shingles: try to calculate your exposure so the butt edge of the nearest course of shingles lines up with the bottom edge of your window sills. This not only makes the job look nice, but also saves you the trouble of cutting shingles to fit around a casing. If the window sills on all four sides of the house aren't on the same level, some cutting to fit is inevitable; you'll have to choose which casings to align. Let's take an example or two. Say the bottom of the window sill is 30 in. up on the wall. With 6 in. of exposure, your fifth course will line up right on the casing edge. If you want a 7-in. exposure, four courses will bring you up 2 in. short of the casing height. Your best bet in this situation is to make up the difference by adding ½-in. extra exposure to each of these lower courses.

To map the course layout, professional shinglers often use a storypole, a straightedge

held vertically with the course lines marked on it. Using this guide to transfer course lines from corner to corner, you can then tack horizontal guideboards across wall sections to align each course as you nail it up.

The storypole and guideboard arrangement works best when you have broad expanses of wall unbroken by corners or casings. When you shingle small sections (between windows, for example), it's troublesome to cut and nail up guideboards. Work will go faster if you just snap chalklines to line up your courses.

If you're re-siding an old house, there's a good chance that some settling has occurred and that the house isn't level. Rather than shingle on these skewed lines, snap a level chalkline around the house, several courses up but below window sills so your line is uninterrupted. Nail down the first course following the existing contour of the house. For second and all successive courses up to your reference line, compensate in small increments so that by the time you reach the line, you'll have a level course all the way around the house. For example, if one corner is 1 in. lower than its neighbor and your

Installing a Factory-Built Skylight

Careful selection and minor structural modifications make for a good job

by Jim Picton

Many home owners and craftsmen who are willing to cut into floors and walls will steer clear of retrofitting skylights because of their reputation for developing leaks. This notoriety is largely a result of the failure of older or improperly flashed units, which rely entirely on a chemical seal like asphalt roof cement to keep water out. One such variety, still available today for about $25, is the plastic bubble—a sheet of acrylic with a bulge in the middle—which is plopped down in a bed of cement, with shingles laid over the edges. Successive layers of cement are then applied to the edge surfaces as often as necessary.

We can thank the Arab oil embargo for improving the standards of skylight construction. Interest in alternative energy sources has brought solar heating to the fore, and with it the need for reliable roof windows. While there are still times and places when you will want to build your own units, competition in the marketplace has produced a number of well-designed skylights that are as weatherproof and problem-free as conventional vertical windows. You should still shop carefully, though. Prices vary widely, and are generally an indication of quality.

There are several things to look for when you're shopping for a unit. First, decide whether you want a fixed or operable skylight. Second, consider the flashing package. Step flashing will last the life of the roof. Strip flashing won't. Third, see if the skylight has a thermal break. On some, metal extends from the outside surface to the inside. This can cause a lot of condensation trouble in a cold climate. Fourth, check to see if you need tempered glass. You can save $50 or so by not using it, but building codes specify it where standard glass might easily be broken. Finally, examine the screen setup if you're buying an operable unit. Some are easy to use. Others are a bit quirky: For example, my Velux screen opens and closes with Velcro strips.

Most skylights, whether site-built or manufactured, have curbs 4 in. to 6 in. high, which raise the glazing above the level of the shingles and divert water around the unit. The glazing is spared the cascade that develops as water from the rest of the roof flows to the gutters. In addition, curbs keep granules from the shingles and other debris off the glass. This prevents scratching, and prolongs the life of the glazing seal.

A fixed-glass unit will run between $200 and about $600, depending on flashing and other

Picton's operable skylight was installed between reinforced rafters in a room with a ceiling that followed the roof pitch.

materials. You can get operable skylights for as little as $200, but most cost a great deal more. Crank, spring-loaded and center-pinned units are available, some with strip flashing and some with step flashing. With my contractor's discount, my center-pinned Velux and its flashing package cost me $500.

Structural considerations—If you have a flat ceiling with an attic above, you'll have to build a shaft between the roof and the ceiling. When roof and ceiling are separated only by rafters, as in the installation shown above, the job is more straightforward. In either case, before cutting a hole in your roof, think about the effect of cutting through one or more rafters. As a start, consider how roof openings are framed in new buildings. Double framing is conventional wherever rafters or headers are used to support additional roof-framing members. Rafters and headers that frame the opening are doubled. Cripple rafters run from the headers to the top and bottom of the roof.

Installing double headers is hardly more trouble than installing single ones, but doubling the rafters can be a problem. To be most effective, the double rafters should extend all the way to the points of support—usually to the ridge above and to the double top plate of the wall below.

You can usually do most of your cutting and framing from inside without disturbing the exterior roof surface. This lets you work without worrying about the weather. In some installations though, the whole point is to disturb the interior finish as little as possible, and you may decide to remove some of the sheathing from the outside of the roof, and install double rafters from above. This involves additional re-roofing, but may save you the grief of having to

live in the dust and rubble caused by tearing up the ceiling.

The added load taken by the rafters on each side of an opening can also be offset by reducing their span, or the distance between points of support. If there is a small attic crawl space above the opening, a purlin can be snugged into the area between the collar ties and the rafters. Although collar ties themselves reinforce the rafters, a purlin will cut down the span of the rafters and transfer the load to the walls.

Another alternative is to double up only a portion of the side rafters. A piece of lumber the same thickness and width as the rafter, extending a few feet above and below the skylight opening, and nailed solidly to the rafter, can have a stiffening effect.

When you're selecting a skylight, consider the width of the unit in relation to the rafter spacing. Choosing the next smaller size may mean cutting through one less rafter. Get the advice of a professional, a structural engineer or the local building inspector, if the problems you foresee do not have direct solutions.

Roughing out inside—Once the location for the skylight is selected, mark the rough opening for the unit on the ceiling. It should be about ¼ in. larger all around than the outside dimensions of the skylight unit you've settled on. Measure up 3 in. from the top line you've marked, and down 3 in. from the bottom one. You'll need this extra space for the new double headers. Next, find the rafters at either side of the opening and mark cut-lines that follow their centers. When the opening is completed, the ceiling will be patched at these lines, and the rafters will provide a nailing surface for drywall or finish trim.

Cut the ceiling using a utility knife, or a skillsaw set to cut only the thickness of the ceiling material. You don't want to cut through hidden wiring. When the ceiling panel is removed, pull out any exposed insulation and re-route the wiring if necessary.

To prevent the roof from sagging slightly when the rafters are cut through, support the rafters above and below the opening with temporary braces made by knocking together two 2x4s in the shape of a T, and wedging them between the rafters and ceiling joists. Then cut the rafters to be removed square, 3 in. back from the rough opening line.

Install the double headers one at a time. Cut the first piece to fit between the side rafters,

Picton removes a rafter, above, having already reinforced those on either side by doubling them up. The T-brace to his left keeps the roof from sagging before headers are installed.

Both boards of the double header are cut to fit between the side rafters, left, then installed one at a time. They are toenailed between the side rafters, then face-nailed. In this installation, the center nailing surface for the top header is the end of a collar tie.

Once the roof has been opened up, the skylight unit can be lifted into place, below. After it is checked for square, it can be fastened to the roof. A helper comes in handy for this part of the job.

and toenail it with 16d nails. Check to make sure you've got a square opening that allows about ¼ in. of clearance on all sides of the sky-light unit. Then face-nail through the header into the end grain of the cripple rafters. Toenail the second header to the side rafters, and face-nail it to the first header. After this framing is complete, stuff any remaining openings with insulation, and then make the necessary repairs. If the ceiling is drywall, this is a good time to get a first coat of tape on the patch.

If you are contemplating re-shingling your roof, now is a good time to do it. If not, you will need to remove some of the shingles at the top and sides of the opening in order to flash the new skylight. This isn't a big problem, and it has the advantage that most of the shingles at the sides and bottom of the skylight will be cut and left in position, eliminating the need to mark and trim each shingle individually.

The roof opening should be cut from the outside. Locate the opening by driving nails up through the roof at the four corners of the opening you've just framed up inside. String chalklines and snap the perimeter on the shingles. Then pound the nails back through the roof, and remove them. Now you've got a chalked outline of the rough opening. The shingles should be cut back to about ¾ in. from the edges of the unit, so you need to mark a cut-line ½ in. outside the line you've snapped. Double-check the measurements of both the skylight and the rough opening.

Asphalt shingles are easy to cut with a circular saw and an old carbide-tipped blade. The carbide tips can be dull or chipped, as long as the teeth are widely spaced. Asphalt material quickly gums up ordinary sawblades and makes them useless. Set your blade depth to avoid cutting into the sheathing, then cut the shingles along the outside lines. Remove and discard them. Now mark the rough opening on the sheathing, using as guides the holes left by the corner nails. Cut the roof sheathing out along these lines with a better, sharper blade.

Installing the unit—At this point, the skylight can be set in place. If you're not re-shingling, you'll have to lift some shingles from around the edges of the opening or remove them by popping nails with a flat bar, so that you can mount brackets or install flashing, depending on the design of the skylight. When the unit is centered in the opening, check it for square and be sure it operates correctly. Then attach it securely according to the manufacturer's instructions or the design specifications.

If you are installing a new roof, the new shingles should be applied up to the course whose lower edge is within 10 in. of the opening. The bottom flashing for the skylight can then be installed.

Most skylight manufacturers offer a flashing package with their units. It's usually more expensive than flashing you can make yourself, but if you have a problem with the skylight, the manufacturer could void your warranty if you haven't used his flashing, even if you've done a good job with your own. If flashing has been provided by the skylight manufacturer, install it

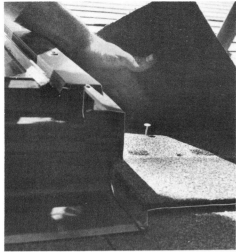

Flashing is provided with prefabricated skylights, or can be purchased. Above, the first
piece of step flashing is actually an extension of
the bottom flashing. It's trimmed so it won't extend below the side shingles, and fastened in
place with a nail in its upper outside corner—
the proper technique for all step flashing.

With the side shingles trimmed and the top two
courses along the skylight's head removed, the
top flashing is set in place, right. The top
flashing supplied with this unit has a bent
return along its upper edge to keep backed-up
water from getting under the roofing.

Full shingles are used as the starter course at
the top flashing, below right. Like the shingles
that will overlay them, they are worked up
under those already on the roof, and are nailed
high enough so the flashing won't be punctured.

according to the instructions provided. If the instructions merely tell you to nail the flashing to
the roof, do so with two nails only, at the upper
outside corners of each end.

Some manufacturers supply decent base and
head flashing, but rely on strip flashing and
mastic along the units' sides. I have little faith in
strip flashings, and if that was all the kit included, I would probably make and install my
own step flashing, as explained on pp. 52-53.

The inside surfaces of the framed skylight
opening can be trimmed with wood or covered
with drywall. Depending on the design of the
skylight, drywall butting into the bottom of the
curb may require a small wood molding to finish it off, or you may decide to flat-tape this gap
or rabbet the curb to accept the thicknesses of
the finish material.

Your pleasure in looking at the installed skylight for the first time is balanced by your anxi-
ety during the first good rain. No leaks are likely in a proper installation, but you might see
wet corners or dripping glass under certain
conditions. Condensation is a problem not limited to skylights, and humidity and cold can
turn any large expanses of glass into a marvel-
ous condenser. A number of methods exist for
controlling humidity in a house, and skylights
require no special treatment. If you notice
dampness around the skylight, check its source.
If it's condensation, forget it. □

_Jim Picton is a carpenter and contractor in
Washington Depot, Conn._

Site-Built, Fixed-Glass Skylights

An energy-efficient, watertight design that you can build with standard materials

by Stephen Lasar

Fixed-glass, site-built skylights can provide many cost and thermal benefits for new construction, as well as for additions to existing structures. And having a reliable design that can be built on site gives the architect or builder the flexibility to meet functional and aesthetic requirements. The skylight design that we use most often is based on standard techniques and materials, and any skilled carpenter can install them. So far, our skylights have withstood a wide range of weather conditions without failing. And the prepainted aluminum flashing and battens we use give these skylights a clean, unobtrusive appearance.

The curb—In holding the glass above roof level, the curb keeps runoff and debris off the skylight. The curb is made from straight-grain 3x10 Douglas fir that we rip in half to yield two 3-in. by 4½-in. curb members. We then rabbet them along one edge to accept the glass. Center curbs, if any, hold two glass panels, and are rabbeted along two edges. The depth of the rabbet is 1¼ in., which is ¼ in. less than the combined thickness of the insulated-glass panel with its two glazing strips. We take up this extra ¼ in. by compression when we screw down the battens to form a weathertight seal (drawing, facing page). The rabbet is ¾ in. wide. We size the curb enclosures so that ½ in. of this width holds the glass and glazing tape. This leaves ¼ in. of open float space between the edges of the panel and the wood.

We assemble the curb box on the roof, using butt joints at corners and a housing (dado) to let each center curb into its top and bottom curb. Toenailing the curb to rafters and headers is usually enough to hold it securely, but if the roof pitch is steeper than 12 in 12, then we use metal clips, too. For structural reasons, we use 3x10 or 3x12 rafters on most spans beneath central curbs (photo below left).

Site-built skylights call for flashing that is also cut and formed on site. The system shown on p. 65 works well with this type of curb. The aluminum should be .019 in. or thicker, and at least 8 in. wide. I always specify prepainted flashing stock because it blends in with a new roof. Shiny, unpainted aluminum draws needless attention to the skylight.

Tape and glass—The inside tape we use is Pre-shim 440 ¼-in. by ½-in. spacer-rod tape made by Tremco, Inc. (10701 Shaker Blvd., Cleveland, Ohio 44104). It's a butyl-base material and is compatible with the silicone second seal of the insulated-glass units we use (for more on caulk-sealant compatibility, see *FHB* #8, p.16). The spacer rod, or shim, is continuous, and limits the compressibility of the tape, so that the glass won't settle away from its seal over time. It also helps to prevent tape squeeze-out at pressure points. Sticky on both sides, the tape comes in rolls, with one side faced with paper. It is applied to the bottom of the curb's rabbet, set slightly back from the vertical edge. It's important to lay the tape carefully, so that it has room to expand without touching the inside of the rabbet or protruding over its outer edge. Butt each tape so there are no airspaces between sections, and lay the tape with the paper facing up (photo below). Don't strip off the paper yet, because you may have to shift the glass panel after you lay it down.

The glass is installed from the outside, so be sure you rig a secure scaffold. The insulated panels used here are standard, double-pane tempered sliding-door units, 92 in. long, 34 in. wide and 1 in. thick. Don't try to lift and position the glass by yourself; get at least one helper. You won't need suction cups for getting the glass onto the roof, but you will need them to lower the panels down into the curb and to shift them into final position.

Before setting the glazing in place, put a neoprene setting block (¼ in. thick, 4 in. long and 1 in. wide) one-quarter of the way in from each bottom corner of the rabbeted curb. These spacers hold the panel away from the bottom edge, giving the glass room for expansion. Now you can set the bottom edge of the panel against the setting blocks and lower it carefully into the curb. Center the panel and make cer-

With the rabbeted curbs nailed to rafters and flashed to the roof, the next step, above, is to lay the glazing tape. The paper should be left on the tape until the glass panels are centered. Then the top layer of glazing tape is applied, right. A batten joint will hide the flashing and tape that cover the curb between panels. Tape is laid against the edge of the glass to allow ¼ in. of float space on all sides.

tain that you've got your ¼ in. of float space around all four edges. To do this, you'll have to pull up the panel at the top, adjust it, and set it down again. Once the position is right, pull up the panel one last time, strip the paper facing off the glazing tape and set the glass panel down for good.

The next step is to apply the outer glazing tape. We use Tremco Polyshim tape, a butyl-base, compression-type tape with good adhesion and elastic qualities. As with the tape beneath the glass, this glazing strip shouldn't butt right against the wood. Leave about ¼ in. of space between wood and tape for expansion. Once you've taped the glass, you should also flash and tape all corners of the curb as shown in the small photo on the facing page. Wherever aluminum battens will intersect, cover the curb with a strip of flashing and a length of tape. Remove the shim from the tape so that when the battens are screwed down, the glazing tape will be squeezed into the joint to form a weathertight seal.

For battens over the curb's perimeter we use 3-in. by 3-in. by ⅛-in. pre-painted aluminum angles. We cut the battens for center curbs from flat aluminum stock 3 in. wide and ¼ in. thick. The side angles should be cut with ears at both ends to create an interlocking joint with top and bottom battens, as shown in the drawing at right. All top and side pieces also need to

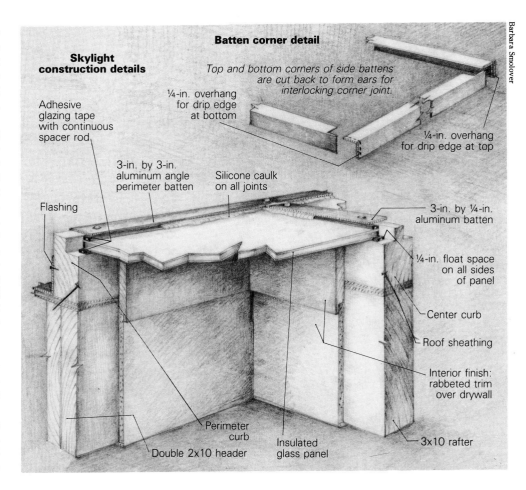

Skylight construction details

Batten corner detail

Top and bottom corners of side battens are cut back to form ears for interlocking corner joint.

Adhesive glazing tape with continuous spacer rod

¼-in. overhang for drip edge at bottom

¼-in. overhang for drip edge at top

3-in. by 3-in. aluminum angle perimeter batten

Silicone caulk on all joints

3-in. by ¼-in. aluminum batten

Flashing

¼-in. float space on all sides of panel

Center curb

Roof sheathing

Interior finish: rabbeted trim over drywall

Perimeter curb

Insulated glass panel

3x10 rafter

Double 2x10 header

Below, a flat aluminum batten is laid down over a central curb. It has been predrilled for screws, and will cover the top layer of glazing tape at the edge of the glass panel. Details of the skylight construction are shown in the drawing above.

Top, stainless-steel screws with neoprene and stainless-steel washers pull the battens tight, compressing the glazing tape for a weathertight seal. Above, the last step in building the skylight is to caulk all batten and glass-to-batten joints. For best results, warm the caulk before you start to apply it and keep the bead continuous.

be cut slightly long to create drip edges. The perimeter battens cover the curb flashing. Install them first, over a generous bead of silicone caulk. Where battens meet, leave about $\frac{1}{16}$ in. between metal edges, for caulking and for the glazing tape to squeeze into. We use $1\frac{1}{2}$-in. long stainless-steel No. 10 Phillips panhead screws with stainless-steel and neoprene washers to pull the battens down tight over the glazing. As for screw spacing, 16 in. o.c. is good for angled battens; 12 in. o.c. for flats.

Replacing a defective panel of insulated glass is an irritating and expensive job. Not considering breakage, failure of the seal or spacer between panes is the most common problem with insulated-glass panels. It's a good idea to specify a double seal that has a good rating from the Insulating Glass Certification Council. This group publishes "Certified Product Directory," a 28-page booklet that lists the names, addresses and phone numbers of over 50 American manufacturers of insulated glass, along with details on their products' corner construction, spacer materials, dessicants and

sealants. Though intended mainly for manufacturers, it may be of use to architects, contractors and builders. It costs $1 from the IGCC (Industrial Park, Route 11, Cortland, N.Y. 13045).

Caulking is the last step. The joint between the glass and the aluminum batten and the joints between battens should be sealed with an even, continuous bead of exterior-grade silicone, carefully lapped at corners. For best application, get the caulk warm before you start using it. Fill any voids or uneven spaces, and smooth the caulk with your finger where necessary.

There are several options for finish trim on the interior of the skylight. We use either veneered plywood nailed directly to rafters and headers, or a trim strip, rabbeted to cover plaster or drywall. In either case, it's important not to butt the plywood or trim against the glass surface. Allow $\frac{1}{4}$ in. to $\frac{3}{8}$ in. all the way around the skylight so the glass and wood have room to move. ☐

_____

Solar architect Stephen Lasar practices in New Milford, Conn.

Flashing a curb

by Jim Picton

Most manufactured skylights can be bought with flashing kits, but if you're installing a site-built unit like the one described in the previous article, cutting and forming your own flashing is the only way to go. Even if you're working with a factory-built unit, you may choose not to buy the optional flashing kit, but to make your own instead. This is an especially attractive alternative when the manufacturer supplies strip flashing for the unit's sides instead of the superior step flashing. Buy a roll of .019-in. thick aluminum 16 in. or 20 in. wide, and follow the directions below.

Base flashing—First unroll a length of aluminum, and cut off a piece about 8 in. longer than the width of the skylight. Slice it to a width of 8 in., and save the scrap for step flashing. Bend the aluminum lengthwise along a straightedge to form a right angle, with one side about 3 in. across and the other, 5 in. Then set the 3-in. side against the bottom curb of the skylight, and the 5-in. side flat on the roof. Hold the angled aluminum so that an equal amount of flashing extends beyond each side of the skylight. Now mark a line on the vertical side of the flashing flush with the corner of the curb on each side of the skylight. This is a fold line. Next, draw a line to make a diagonal cut across the vertical side of the flashing, ending exactly at the point where the fold line meets the bent corner of the flashing, as shown in step 1 of the drawing on the facing page.

When you've cut both sides, fold the vertical half of the flashing around the skylight, and flatten the rest onto the roof. If you're not re-shingling, slide the angled tab of flashing under the first course of shingles on each side of the skylight. This is your base flashing. Folded and cut, it's ready to tack to the roof with a single nail in each upper corner.

Step flashing—Most manufactured flashings include the first two pieces of step flashing as an integral part of the bottom flashing. Each piece is bonded to the bottom flashing with a soldered or locked seam joint. You can approximate this detail using an 8-in. square of aluminum sheet from your roll, bending it at a right angle with 3 in. vertical and 5 in. flat, and making a V-cutout along its fold so that it can be bent around the base (step 2).

The rest of the step flashing, except for the two top corner pieces, consists of square or rectangular sheets with a single 90° bend. The precut sheets sold in hardware stores as step flashing are typically 5 in. by 7 in., a size that offers minimal protection. Buy larger pieces if you can, or make your own from 8-in. squares.

If you're putting on a new roof, lay the next course of full shingles when the first piece of step flashing is in place. Then mark the shingle that will fit next to the skylight and cut it, leaving about $\frac{3}{4}$ in. of space between curb and shingle edge. Position the shingle and nail it down at the top edge farthest from the skylight. Now slide the second piece of step flashing under the shingle adjacent to the skylight curb, so that the bottom edge of the flashing is about $\frac{1}{2}$ in. higher than the bottom edge of the shingle. The flashing should lie directly under the bottom half of the upper shingle, and over the entire concealed upper half of the lower shingle (step 3). If you're step-flashing a roof that's already been shingled, be sure to lace each piece of flashing over and under the shingles before nailing its upper corner to the roof (see p. 59).

Head flashing—The last pieces of step flashing will be cut and bent around the top corners of the curb (step 4), just like the step flashing at the lower corners. Once you've done this, the head flashing can go on. Bend, cut and fit it just as you did the base flashing, but use a full-width piece (at least 16 in.) from your roll. Before nailing it to the roof, nail one thickness of shingles to the bare roof above the curb. This will shim the roof up to the level of the shingles on each side, and prevent the head flashing from sagging in the middle and collecting water and debris. Remember that the upper part of the sheet must go under the first course of shingles above the curb, while the edges rest on top of the corner step flashing (step 5).

An exception to this procedure is when you are using a manufactured skylight such as Roto Stella, which requires installing the top and bottom flashings at the time the unit is installed. In this case, you have to slide the last piece of step flashing up under the preformed corner of the top flashing, and forget about bending the step flashing around the top.

Manufactured flashings usually won't extend too far up under the shingles, so they often have a bent return on the upper edge of the top flashing. This feature prevents water that has backed up under the shingles from traveling beyond the top of the flashing and back under the roof (see *FHB* # 1, p. 5 for more on this idea).

Site-constructed flashings that are run 16 in. or 17 in. under the shingles do not need a bent return, but you can make one as an added precaution. Using a straightedge and your fingers, bend the top of the flashing up, about ½ in. from the edge. Fold it all the way over, then place a board over it and hammer it flat. Finally, use a screwdriver to pry the bent edge back up a little, so a space is visible between the top edge of the bent return and the rest of the flashing below.

The most difficult part of fitting homemade flashing is working around the four corners of the skylight curb. The points of these corners are the one area in a flashing job where, unless you are using soldered, welded or otherwise pre-formed components, no overlap is possible. The way to solve this problem is to do what carpenters and roofers always do when faced with the impossible: caulk it.

Counterflashing—Sometimes called angled or perimeter battens (drawing, p. 51), counterflashing is a piece of metal (usually a heavier gauge than roof flashing) bent lengthwise at right angles. Half of this flashing covers the top of the curb; the other half protects the sides of the curb and covers the top edges of the roof flashing. On some prefabricated skylights, counterflashing is permanently affixed to the curb, and the step flashing has to be slipped under and worked into position as the roof is shingled. Other units have a removable counterflashing that is attached quite simply with screws once the roof flashing is complete. In either case, it completes the exterior seal. □

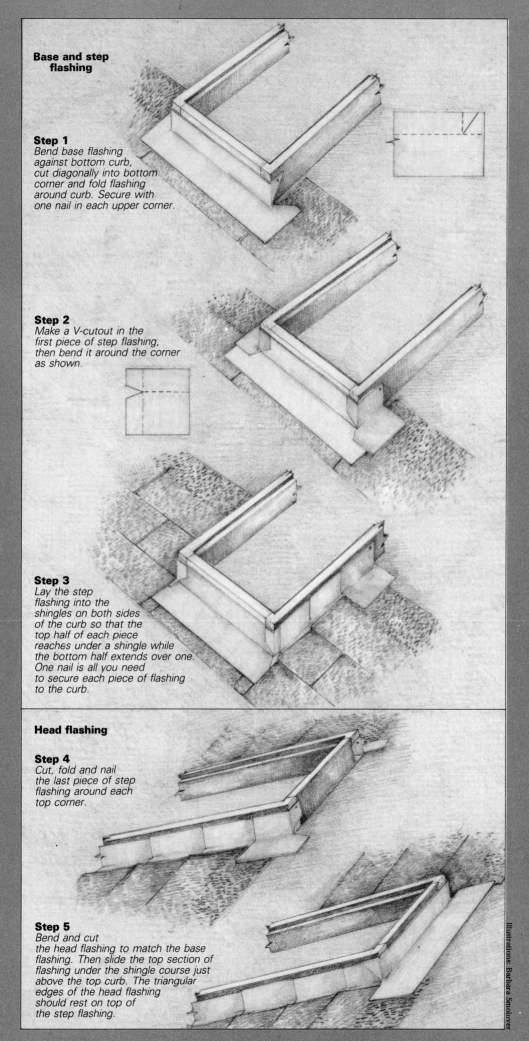

Base and step flashing

Step 1
Bend base flashing against bottom curb, cut diagonally into bottom corner and fold flashing around curb. Secure with one nail in each upper corner.

Step 2
Make a V-cutout in the first piece of step flashing, then bend it around the corner as shown.

Step 3
Lay the step flashing into the shingles on both sides of the curb so that the top half of each piece reaches under a shingle while the bottom half extends over one. One nail is all you need to secure each piece of flashing to the curb.

Head flashing

Step 4
Cut, fold and nail the last piece of step flashing around each top corner.

Step 5
Bend and cut the head flashing to match the base flashing. Then slide the top section of flashing under the shingle course just above the top curb. The triangular edges of the head flashing should rest on top of the step flashing.

Illustrations: Barbara Smolover

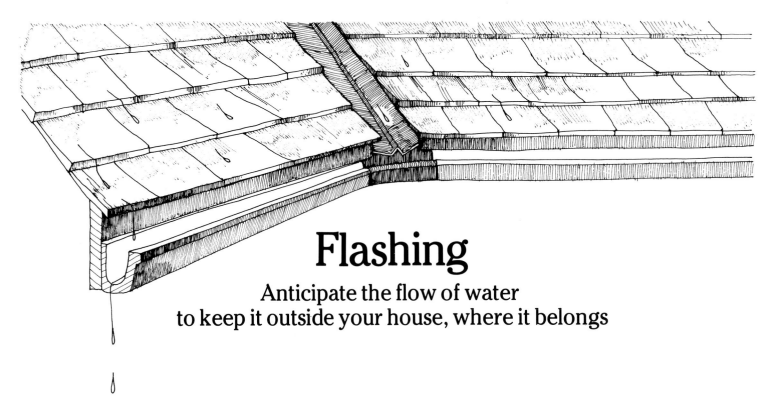

Flashing

Anticipate the flow of water
to keep it outside your house, where it belongs

by Bob Syvanen

Flashing is that ounce of prevention worth a pound of cure. I can't begin to count the repair jobs I've done because of faulty flashing. All the mystery was taken out of flashing for me when an old-timer advised that I take a ride on a drop of water as it runs down the roof or sidewall. In so doing, you quickly realize where that drop wants to go, and you understand how to change its direction. Anything that stops the flow will do it. A gutter is a good example. A strong wind will also cause a change in direction, and leaks often result. Other problems oc-

cur at junctures of dissimilar materials, where, for example, a chimney meets a roof, or an aluminum window frame joins a sidewall. Any vertical crack is also an open invitation to water—door and window jambs, casings and corner boards.

Flashing used in these and other locations will keep water out of your house. The material used for flashing can be copper, lead, aluminum, galvanized steel, plastic or paper. What material is used where depends on cost, how severe the condition is, and how long it should

last. Copper is the best and the most expensive flashing material. It is strong, long-lasting and can be easily shaped. Lead is right up there in quality and cost; it can be shaped quite easily, but you have to be careful because it punctures and tears. Galvanized (zinc-plated) steel comes next, followed by aluminum, plastic and paper. Unformed metal flashing comes in 10-ft. lengths, in sheets or rolls of various widths and gauges.

Bob Syvanen is consulting editor to Fine Homebuilding *magazine.*

Make your own

Although sheet-metal shops are equipped with power shears and brakes that produce crisp lines in a hurry, it is often cheaper and quicker to make and fit your own flashings. This will cut down on travel and turn-around time. The one exception is galvanized steel flashings. Because they are so stiff, it is usually better to purchase them preformed.

Most flashings require a 90° fold **(1)**. The easiest way to get one is to bend the flashing material over a sharp wooden corner. Use a 2x6 with its eased edge ripped off. Then set up a simple jig **(2)**, with nails set along a guide line to position the metal quickly for bending. For narrow bending, a pair of wide-jawed, lock-grip pliers is useful **(3)**. Thin-gauge metal flashing can be cut easily with a sharp utility knife. For heavier metals use tin snips **(4)** or aviation snips **(5)** for even easier cutting.

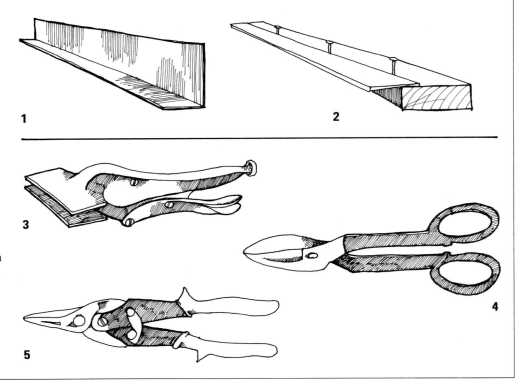

Doors and windows

I like to flash window sills with paper, but I use metal on door sills **(6)**, because doors have moisture problems with snow, rain, wet leaves and the like. With window sills, tuck the flashing up into the groove that receives the siding **(7)**. For door and window jambs, flashing paper or 15-lb. felt, 6 in. or more in width, is stapled around the frame **(8)**. The siding felt is tucked underneath this flashing. Make sure that these paper splines run over the sill flashing, which overlays the siding paper. For shingle siding, the sill flashing and the bottom of the jamb spline should overlap the tops of a shingle course, and then be covered by the two remaining courses of shingles underneath the window sill **(9)**. This way the felt won't show through.

Window and door headers must also be flashed, and for this, I think thin metal is best. Nail it in place, keeping the nails high, and continue with the siding. Let the siding felt lap over this head flashing **(10)**. After the siding is on, the ½-in. overhang can be bent over the head casing, using a 2x4 block about 1 ft. long **(11)**. Finish it off by holding the block against the flashing and beating it with a hammer to make a nice flat surface **(12)**.

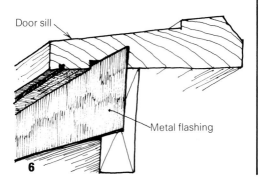

Door sill — Metal flashing

6

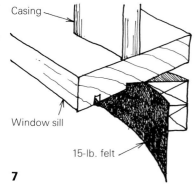

Casing — Window sill — 15-lb. felt

7

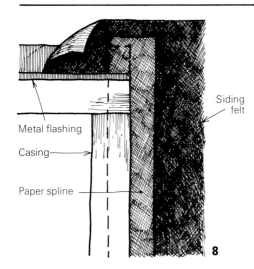

Metal flashing — Casing — Paper spline — Siding felt

8

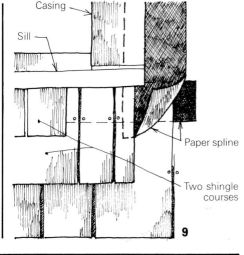

Casing — Sill — Paper spline — Two shingle courses

9

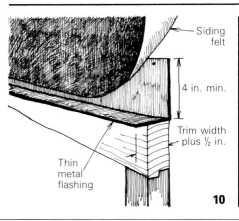

Siding felt — 4 in. min. — Trim width plus ½ in. — Thin metal flashing

10

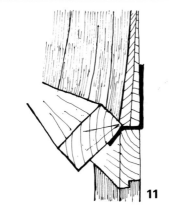

11

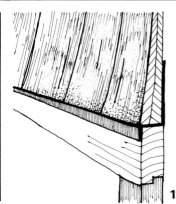

12

Gutters

Another kind of flashing that is made most easily on the job protects wood gutters. Gutter ends that butt against a rake board must be flashed with an oversized piece of lead **(13)**. The flashing can be coaxed into place by gentle tapping with the end of a rubber-handled hammer. The final trimming can be done after the flashing is formed. A wide bed of caulking on the lead edge will seal it, and copper tacks hold it in place. A gutter splice **(14)** is leaded the same way, but there are no compound curves.

13

Lead flashing — Bead of caulking under this edge — Wood gutter — Rake board

14

Caulk under — Lead — 3 in. — 3 in. — Joint

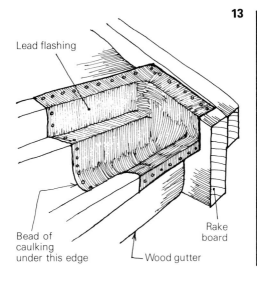

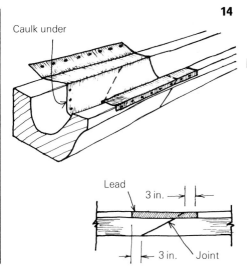

Wall and wall-to-roof junctures

Brick-veneer walls require flashing at their bases to keep water away from the wood framing (15). Where a sheet of plywood butts another, the flashing for a horizontal seam is similar to that over a window (16). It is called Z-flashing, and is typically preformed aluminum or galvanized metal in 10-ft. lengths.

Flashing an attached shed roof is hard to do right, and a prevailing wind makes it even tougher (17). Flashing at the peak of a freestanding shed is quite simple (18). One of the toughest conditions to flash is where a pitched roof butts into a wall (19),

but with careful work, water can be kept out. Where a shed roof tucks into an inside corner, the answer is a copper-flashed, tapered cant strip (20).

Step-flash where a roof meets a sidewall (21). Each roof shingle course requires a separate flashing shingle underneath, so I take a roll of galvanized steel, aluminum or copper and cut enough 8-in. squares to do the job. Fold each one 90° over a wood block, and they are ready to nail up. The shingle exposure is typically 5 in., so an 8-in. step-flashing shingle will give you a 3-in. overlap. To keep the shingles in place, nail their upper corners only.

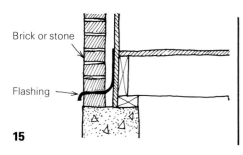

Brick or stone

Flashing

15

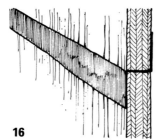

16

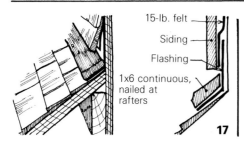

15-lb. felt

Siding

Flashing

1x6 continuous, nailed at rafters

17

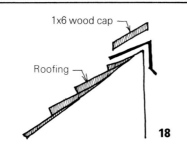

1x6 wood cap

Roofing

18

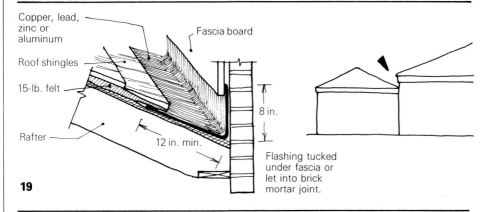

Copper, lead, zinc or aluminum

Roof shingles

15-lb. felt

Rafter

Fascia board

12 in. min.

8 in.

Flashing tucked under fascia or let into brick mortar joint.

19

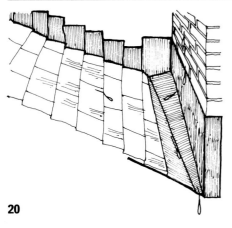

20

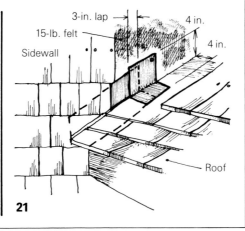

3-in. lap

15-lb. felt

Sidewall

4 in.

4 in.

Roof

21

Chimneys

I think chimneys cause almost as much damage as floods. It's a miracle when they don't leak, sitting up there, buffeted by wind and rain, with at least one big hole on top. The hole on top can be protected with a stone cap, but this doesn't always look good. The very best flashing is thru-pan flashing, and copper is the best material (22). The pan redirects water that gets in at the chimney top. The roof shingles are base-flashed and step-flashed. The brick is cap-flashed over the tops of the base and step flashing (23).

Another common system is to incorporate base flashing and cap flashing with each piece of the step flashing.

Another thru-pan system uses one large lead sheet with a hole cut for the flue, and carefully formed and folded to lie flat over the masonry and on the roof (24).

For chimneys less than 2 ft. wide, a simple flashing bent to the roof pitch will shed rain and snow (25). If the chimney is wider, a cricket is the best solution. It can be a one-piece sheet-metal cricket if it's not too wide (26). A wood-framed cricket can be built for wide chimneys, and handled just like a dormer roof (27).

Beams

For exterior decks, pressure-treated lumber is best, but if untreated lumber is used, flashing the deck beams will help prevent water damage (28). Intersecting beams should be cap-flashed (29). A beam projecting from a sidewall should be flashed and capped (30), particularly if it is a two-piece beam. Water just loves to sit in that crack between the beams. This solution takes care of the top, but shrinkage at the sides makes caulking the best way to seal this joint. A better solution (31), although not cheap, requires soldering of all joints.

Members fastened to the sidewall are flashed the same way as windows (32). A post through a flat roof is no problem if it is flashed (33). The tops of untreated posts, if left flat, will suck up water like a sponge. Sloping the top will help. A better way is to cap-flash the top. This used to be common practice.

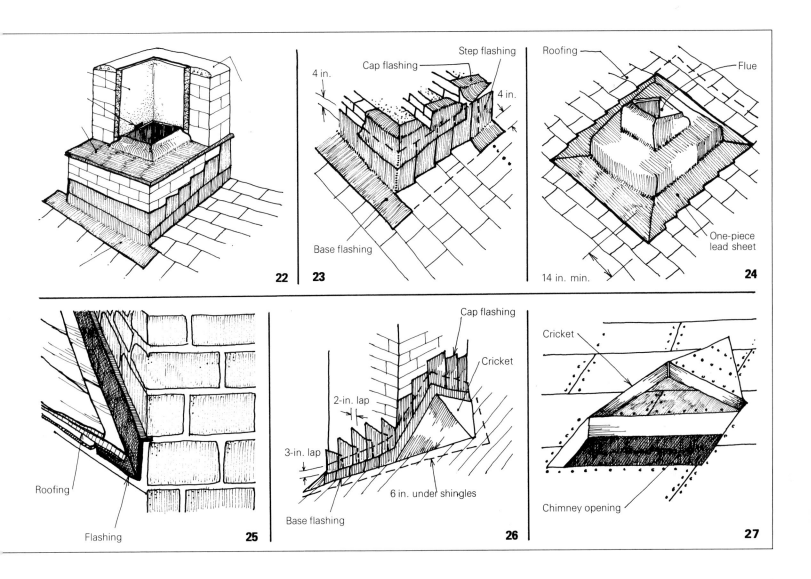

22 | **23**

4 in.

Cap flashing

Step flashing

Base flashing

4 in.

Roofing

Flue

One-piece
lead sheet

14 in. min.

24

Roofing

Flashing

25

Cap flashing

2-in. lap

3-in. lap

Cricket

6 in. under shingles

Base flashing

26

Cricket

Chimney opening

27

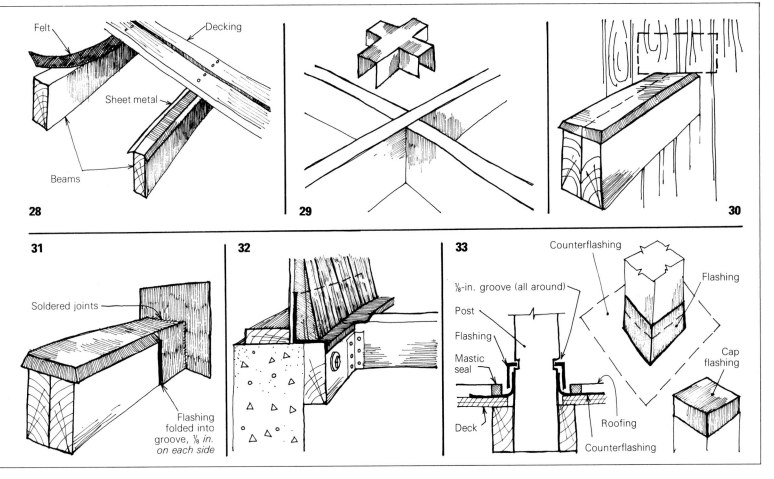

Felt

Decking

Sheet metal

Beams

28

29

30

31

Soldered joints

Flashing
folded into
groove, ⅛ in.
on each side

32

33

⅛-in. groove (all around)

Post

Flashing

Mastic
seal

Deck

Counterflashing

Flashing

Roofing

Counterflashing

Cap
flashing

Roofs

Since the neoprene sleeve came on the scene, roof stacks have become easy to weatherproof. You just select the right-sized flange, slide it on the stack, and shingle around it **(34)**. If the roof is wood shingled, check for splits that might leak. If there are any cracks that line up, slip a 2-in. by 8-in. flashing strip, sometimes called a tin shingle, under the split **(35)**. I consider this part of the job whenever a roof is shingled.

Another tough area to flash is a valley. There are basically two kinds: open valleys and closed valleys **(36)**. I like the look of the closed valley, but it is more work. The closed valley starts with a base of 30-lb. felt its full length. The shingles, whether wood or asphalt, are laid to a chalkline snapped up the middle of the felt. The flashings are 12-in. squares of zinc, aluminum or copper, folded diagonally to fit the valley **(37)**. The first piece is a rough half piece. Each shingle course gets one piece held in place with a nail in one outer corner. The shingles are then angle-cut for a nice tight fit.

The open valley is just what it says: open. The valley material can be 90-lb. roofing felt, or copper or galvanized steel. Ninety-lb. roofing felt makes a fine valley if it's installed in two layers **(38)**. Always watch for nails falling into the valley as you work, and keep it clean. The valley should be 5 in. or 6 in. wide **(39)**. It looks better if the valley is the same width on top and bottom, but some recommend widening the bottom to take care of the increase of water. Take your pick, but be sure to cut off the upper inside corner of the asphalt shingles. If the tip is left on, it will catch water, which will ride down the upper edge of the shingle course until it finds a way out, often 15 or 20 ft. away.

Metal valleys are extremely durable, but more expensive. They can be flat in the center, or crimped **(40)**. Some have dams for severe runoff conditions **(41)**. The valley can be nailed at the edges to keep it in place, or you can use clips every 24 in. o.c. **(42)**.

A caution about mixing metals. Galvanic action can be a problem, so don't combine metals from the opposite ends of the following list: aluminum, zinc, tin, lead, brass, copper, bronze. The safest way to proceed is to use clips or nails of the same metal as the flashing.

There is no need for guesswork in flashing. A little thinking goes a long way toward a good job. Just ride that drop of water through the area to be flashed, and you'll know what to do. □

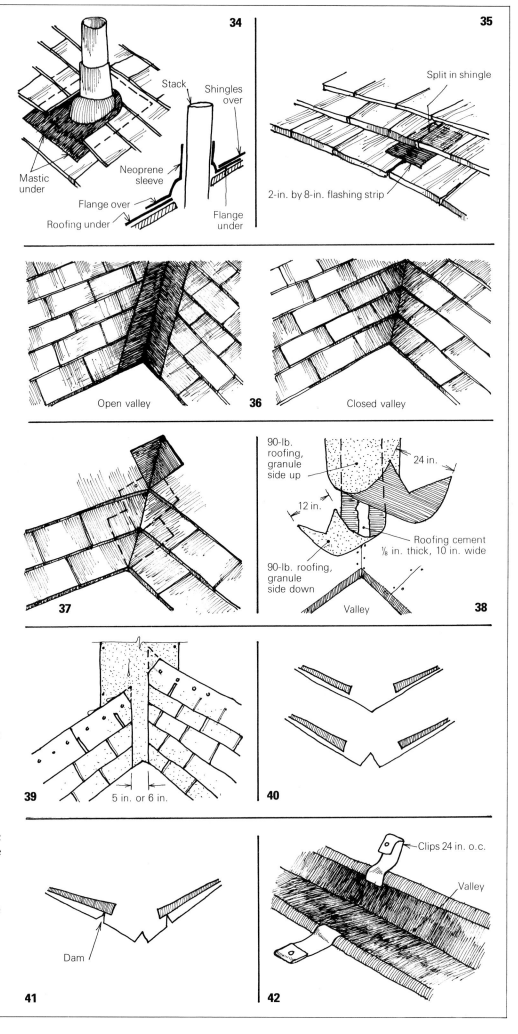

34

Stack
Shingles over
Mastic under
Neoprene sleeve
Flange over
Roofing under
Flange under

35

Split in shingle
2-in. by 8-in. flashing strip

Open valley   36

Closed valley

37

90-lb. roofing, granule side up
24 in.
12 in.
Roofing cement ⅛ in. thick, 10 in. wide
90-lb. roofing, granule side down
Valley   38

39   5 in. or 6 in.

40

Dam   41

Clips 24 in. o.c.
Valley
42

Valley Flashing

Successful valley flashing is one of the fine points of making a wood-shingle roof weathertight. I favor the old-fashioned method of nailing shingles into 1x3 horizontal stripping; fixing shingles to a solid sheathing such as plywood (even with a layer of felt paper between) doesn't allow air to circulate around shingles. Because most of the shingles I use

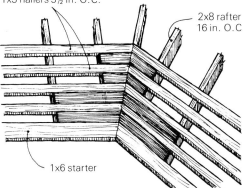

1x3 nailers 5½ in. O.C.

2x8 rafter 16 in. O.C.

1x6 starter

have 5½ in. to the weather, my 1x3s are also spaced 5½ in., except in the valleys, where I butt strips together to create a solid base for the flashing that follows.

I use 20-in. wide copper roll flashing, which I cut to 4-ft. lengths on a shear. I bend the copper on a brake as shown. The middle folds first and then the outside crimps. The 1-in. fold in the center acts as a kind of levee to keep water driven in a storm from sloshing up under shingles on the opposite side; the ½-in. crimps are back-up protection in case some

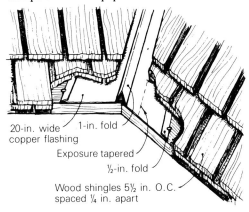

20-in. wide copper flashing

1-in. fold

Exposure tapered

½-in. fold

Wood shingles 5½ in. O.C. spaced ¼ in. apart

water does make it up underneath—the water can't jump the crimps so it runs back down the valley toward the gutter.

To secure the flashing, I overlap 4-ft. sheets about 6 in. for weathertightness. I fasten the sheets by snugging the heads of large-headed roofing nails tight enough against the outside of the crimp to pinch the fold. I put a nail at every point that the copper passes over a rafter. Be sure when putting the lowest course of shingles in the valley not to punch through it with a shingle nail. Careful work forms a very functional and lasting valley.

When placing the first course of shingles around the valley, I vary the amount of exposed copper from 3½ in. at the upper end to 5 in. at the lower end, a 1½-in. taper from top to bottom on each side of the center fold. This wider exposure at the bottom allows ice and leaves to slip out much more easily.

—William C. Barthelmess, Woodbury, Conn.

Wiring in old walls

Here are some tricks to help you use a fish tape when you are trying to run new wiring through old walls. A fish tape is a thin, spring-steel wire with a hook on one end. It is mounted on a spool, and most frequently it is used to pull wiring through sections of conduit. It can also snake through hidden wall and ceiling cavities.

It can be hard to hit a target with a fish tape. So I catch it with a string. When you are working on a vertical run, such as an attic to a wall-mounted outlet, lower the string from above. It should be tied into a series of loops, one pushed through the other, and the string weighted with a piece of solder. The fish tape is then introduced into the lower access hole, and manipulated until its hook catches one of the loops. The same trick can also work horizontally—just push the weighted string along the flat run with a stick until it falls into the vertical cavity.

Another way to get a string from one place to another in an empty conduit is to tie it around a small wad of cellophane and use compressed air to blow it down the run. Then tie it to a heavier string and you're ready to haul on the tape. For a really convoluted run, where a normal fish tape won't follow the necessary path, a speedometer cable can do the job. When you are doing this kind of work, make sure that you know the location of any live boxes and unprotected wiring, like knob and tube work. And to make things easier in the event of future wiring changes, leave a piece of nylon string pulled through the runs.

—Norman Rabek, Burnsville, N.C.

Double-beveled rafters

Carpenters in central California use this technique for cutting double bevels on hip or valley rafters fashioned from conventional 2x framing lumber. Set the tongue of the framing square on one end of the rafter as shown and

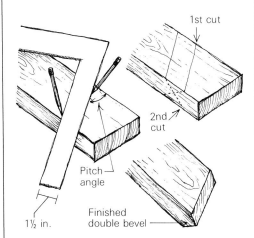

1st cut

2nd cut

Pitch angle

1½ in.

Finished double bevel

draw two parallel pitch lines (they will be 1½ in. apart). Then adjust the foot of the circular saw to cut a 45° bevel, and make the first cut in the direction indicated in the drawing. The second cut starts from the opposite side of the board. The resulting double bevel allows the carpenter to tuck the rafter between two perpendicular common rafters.

—Andrew Kujawa, Santa Cruz, Calif.

Chalkless nailing gauges

These two site-built gauges for marking plywood and board siding for nailing come in handy where nail lines and spacing have to be uniform, or where visible chalk lines cannot be left on the siding.

The nailing gauge for plywood (drawing below) is like a very long T-square. The blade of the square has nail points on its centerline at the specified nailing schedule. We found that the last inch of an 8d galvanized nail, driven blunt end first, worked best when the point was left to protrude about ⅛ in.

Allowing for the offset between the nail points and the edge of the gauge, lay out stud centers on the plywood top and bottom, hook the marker over the top edge of the plywood (plywood should be face up) and along the layout marks, and strike the device with a hammer along its length. The perforations that will be left represent the nail holes to be used in attaching the siding.

The gauge for board siding is made from 2x framing lumber (as wide as the boards to be marked) with plywood fences nailed to one edge and to one end. The nail points in the 2x are spaced on stud centers along its length, and as desired across its width.

The siding is marked by positioning a pre-

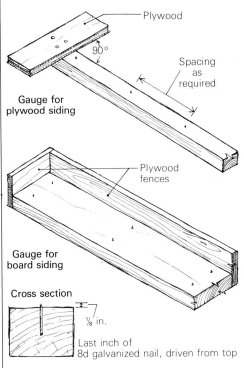

Plywood

90°

Spacing as required

Gauge for plywood siding

Plywood fences

Gauge for board siding

Cross section

⅛ in.

Last inch of 8d galvanized nail, driven from top

cut board against both fences face down applying pressure; once again, the perforations indicate nail locations. Boards can be positioned and marked at various points along the gauge to allow for joints in the siding, or openings such as windows or doors. Nail temporary stops on the gauge to locate these positions. These easily made jigs enable you to produce accurate, evenly spaced nailing lines at production speed.

—Malcolm McDaniel, Berkeley, Calif., and Paul Spring, Oakland, Calif.

Masonry

Laying Brick Arches

A masonry inglenook becomes the warm center of an architect's new house

by Elizabeth Holland

When Ed Allen and his wife Mary started thinking about designing their own house seven years ago, they knew they wanted something big and barnlike, yet New England simple and cozy. Ed, an architect, had been fascinated by domes, vaults and arches during his studies, but it wasn't until the couple spent six months living in Liverpool during the winter of 1975-1976 that an inglenook became part of their house plans.

"We spent a lot of cold evenings huddled around a fire, trying to keep warm in an unheatable English house," remembers Allen. "We kept sketching our ideas and dreaming about how we wanted to have a house where we wouldn't be so cold."

The sketches showed the influences of the English and Welsh houses they had seen, particularly of their visit to the Welsh Folk Museum at St. Fagan's. This is where the Allens encountered the inglenook.

It is thought that the word *ingle* comes from the Gaelic *aingeal*, meaning fire or light, and was originally applied to open fires burning on primitive hearths. In medieval times, it came to mean a fireplace. The inglenook was a corner or a small room near the chimney where the family would gather before the heat of the flames. This became the central idea in Ed Allen's plans, and the inglenook ultimately became the dominant design element of the house he was to build.

Allen spent a long time working out the exact dimensions of the inglenook, and was working on a 4-in. module. Then the house was designed around that, from the inside out. "I was always trying to keep it as small as I could," Ed says, "but it kept growing as I made sure all the spaces around it were the proper size."

The Allens' inglenook is all brick, part of a 50-ton masonry mass that encompasses flues and fireplaces, and divides the house into two equal parts on each floor. The inglenook was designed to accommodate four or five people comfortably. The interior dimensions are 7 ft. 4 in. wide and 7 ft. deep from front to back, not including the fireplace.

According to Allen, it was all worked out logically—the dimensions of what he wanted things to be, plus the dimensions of the necessary brick. For example, the archway is 16 in. thick because the walls contain an 8-in.-square flue plus 4 in. of brick on each side.

To make the layout easier, Allen designed the brickwork to be modular, with 7⅝-in. long bricks and ⅜-in. mortar joints, for an overall length of 8 in. Later he discovered that only about two-thirds of the assorted Full Range Belgian bricks he ordered were the right size. The rest were up to ¼ in. too long. This required some cutting when he got to his closers, the last bricks in each course (for cutting brick, see *FHB #*3, p. 43).

Mortar color and tooling can significantly affect the look of brickwork. Allen chose a standard dark masonry cement. On the horizontal mortar joints, he used a flat-joint finishing tool to make a weathered joint, flush with the brick at the bottom and cut back at the top. This joint casts a shadow on the mortar joint, accentuating the pattern of the brick. The vertical joints, however, are gently concaved.

Choosing a bond—Bonding, the overlapping patterning of the bricks, knits the various wythes (thicknesses of brick) together. For a single wythe, a simple running bond can be used. But structural brickwork is usually at least 8 in., or two wythes, thick. "An 8-in. thick wall can use any of a variety of bonds, and some of them are quite beautiful," says Allen. "I was planning to use an English Garden Wall bond because if you're doing an 8-in. wall where one side is going to be concealed, you can save a lot of time by laying up the concealed wythe in 4-in. concrete blocks."

An English Garden Wall bond consists of three courses of stretchers (bricks with a long edge showing) and a fourth course of headers (bricks with their short ends facing out and

---

Bricklaying tips for amateurs

After years of studying and teaching brickwork, and laying bricks, Ed Allen is convinced that interior brickwork—including arches—is well within the grasp of a reasonably careful amateur.

"Once you get into masonry, it's just like putting up a wood frame for a house—absolutely routine, very secure, very simple," he explains. "There's just no end to what you can do with it."

Bricklaying is a relatively straightforward concept that gets fairly complex in practice. Here are some tips for beginners:

- You can't learn brickwork on your own. Learn from someone who knows how to do it. You can pick up the rudiments in a day or two. "Probably 90% of the success in laying brick is getting the mortar the right consistency and using the trowel properly. Learning to mix mortar to the proper consistency and to use a trowel are things that simply can't be gotten from a book."
- Plan in advance and know your dimensions. Determine the heights at which the arches will spring, and the heights of the arches. Consider the placement of flues when determining thickness.
- An arch supports a vertical load by transforming it into a diagonal load. Make sure your design has enough mass on either side of the arch to absorb the thrust and keep the arch from spreading.
- The labor involved in laying brick is almost directly proportional to the number of corners. Called leads, the corners are laid up first, four to six courses at a time. Care in laying up the leads pays off in level courses of the right height. The bricks that fill in the flat stretches between corners, aligned with strings pulled taut between the leads, are laid relatively quickly. It's best to eliminate as many corners as possible. You've always got to decide, of course, whether it's worth the extra labor to get it the way you want it.
- Practice by building the foundations for whatever you are going to build. This gives you a chance to develop techniques before laying up courses that show. If you're still having trouble, get some help.
- It's easiest to work at waist level, or roughly between your knees and shoulders. Arrange the scaffolding accordingly. As you go higher, your work slows down because it becomes more cumbersome to transport heavy and bulky materials.
- Brickwork is not as precise as carpentry. The irregularity of it is part of the charm, and a real plus for amateurs. Nevertheless, you should strive for precision, so things don't get too far out of whack. —*E.H.*

their lengths extending in across the two wythes). On the concealed wall, a course of concrete block takes the place of three courses of brick plus their mortar joints. The fourth course is composed of the headers, lying across both the brick and block wythes.

A header course is laid so the bricks straddle the vertical mortar joints in the stretcher course below. Since Full Range Belgian bricks are narrow, spacing the bricks properly would have required extra-wide joints. This led Allen to change from a full header course to what's called a Flemish header course (drawing, right), where headers and stretchers alternate.

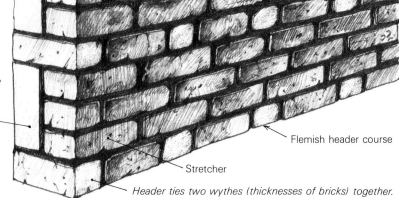

Concrete block is 4 in. thick and as high as three courses of brick with mortar joints. Using block where it won't show saves time and money.

Brick bonding

Flemish header course

Stretcher

Header ties two wythes (thicknesses of bricks) together.

Building the arch.

Allen and a friend did all the masonry work themselves, in two stages. First, in May, they laid up the block foundation for the masonry core. The brickwork began in July, and from then until mid-autumn they laid 7,000 face bricks, an undetermined number of concrete blocks and concrete bricks, and several hundred sections of flue tile. The graceful curve of the inglenook's archway is inviting, a welcome contrast to the sturdy straight lines of the brick walls. Within the inglenook and overhead, the arch repeats, each time in a slightly different form.

"Arches are really fun—they're a wonderful structural form," Allen reflects. "I think everyone has an immediate, positive emotional response to them. By the time we finished the arches we decided they were quite easy to build and not terribly time-consuming, a conclusion quite contrary to our initial expectations."

Step one: the centering form—The brick walls are laid up to the level called the springing of the arch, the point where its curvature begins. Now the centering is built—the wooden form over which the bricks in the arch will be laid.

There are many ways to build a centering. Mine consists of two identical curved trusses with a single rectangular piece of ¼-in. Masonite nailed securely to the top of them (drawing **A**, below). You should check the centering's fit by holding it between the brick sidewalls.

Step two: patterning the arch—Lay the centering on its side on the floor. Stand the bricks you're going to use on end, all around the curve (**B**). You should have an odd number of them. Using an even number results in a mortar joint positioned at the crown of the arch. This looks bad and makes the structure weak. Because of the curvature of the arch, the mortar joints will be wedge-shaped. They should be about ³⁄₁₆ in. wide at their narrowest point, next to the centering. Shuffle the bricks around until the spacing looks good. You could have wedge-shaped bricks specially made, but these are usually expensive, and must be ordered well in advance of when you'll need them.

With a pencil, mark the thicknesses of the mortar joints on the centering itself. Take away the bricks and use a square to run the joint lines across the curved surface of the form. When you're finished, the centering will be marked to show you where all the bricks belong, and how thick the mortar joints should be. This step is crucial—without it you will end up with uneven joints, and you'll have to trim bricks to fit odd spaces.

Step three: placing the centering—Lift the centering between the existing brick walls. The bottom of the centering should be just a couple of inches below the spring of the arch (**C**). Support it by four lengths of 2x4, cut just a little longer than the distance from the floor to the spring of the arch so they can be angled under the four corners of the centering and wedged in place. It's best to align the centering so its front edge is even with the brick walls; then bricks in the arch can be laid to the front edge of the form. If your centering is wider than the arch, pencil a line on it to indicate where the front ends of the arch bricks should be laid. Level up by tapping the bottoms of the wedged 2x4s. Once the centering is level and in the right position, drive shims into the small gaps between the centering and the walls at the four corners, to hold the form firmly in place.

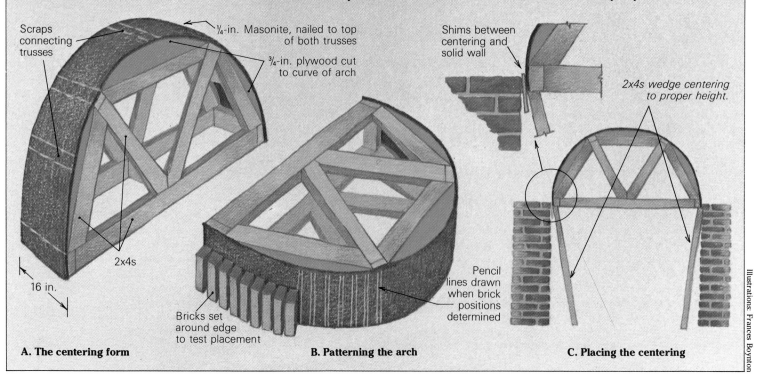

Scraps connecting trusses

¼-in. Masonite, nailed to top of both trusses

¾-in. plywood cut to curve of arch

2x4s

16 in.

Bricks set around edge to test placement

Shims between centering and solid wall

2x4s wedge centering to proper height.

Pencil lines drawn when brick positions determined

A. The centering form

B. Patterning the arch

C. Placing the centering

Illustrations: Frances Boynton

Putting Down a Brick Floor

The mason's craft is easier when the bricks are laid on a horizontal surface

by Bob Syvanen

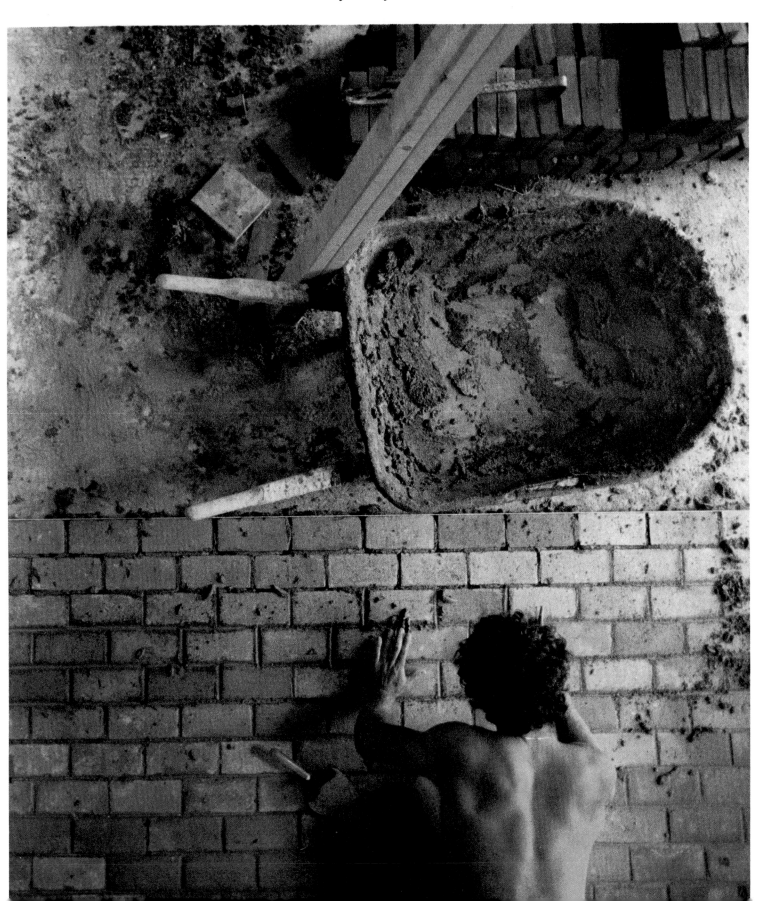

Photos: Edward Allen

D. The ends of the bricks in the arch must be in the same plane as the wall. Check for alignment with a trammel board tacked to the center of the horizontal truss member.

E. Bricks are set between the pencil lines that were drawn on the centering during a test-fitting on the ground. This guarantees a proper fit and mortar joints of uniform size.

F. Bricks have been trimmed with a mason's hammer where wall meets arch. It looks best if the width of the curving joint between wall and arch remains constant.

Step four: laying up the arch—Lay bricks from both lower edges until they meet at the top. As they go up, check the bricks with a level to make sure their ends are in the same plane as the wall, or with a board tacked to the center of the truss (**D**). If you follow your pencil marks, you'll end up with the right number of bricks, and uniformly wide joint spaces (**E**).

Step five: finishing the wall—With the centering still in place, lay up the flat walls around the sprung portion of the arch. If the centering is removed too soon, the arch might collapse, because the mortar is still fresh and because a semicircle is not the strongest arch form. It could bulge out at the sides and drop in at the center. Once the walls are laid up around the arch, however, it can support a lot of weight.

In the Allens' house the vaulted ceiling over the inglenook carries a concrete slab floor above it.

Always work in from the leads at the outer end of the wall, keep the vertical joints lined up properly and use taut string lines to keep the courses perfectly level.

On each course you will have to cut the last brick to fit, where it intersects with the arch. A diamond saw is the most efficient tool for this, very precise and sure. But a diamond saw is expensive and to Allen's eye the cut is too sharp-edged and cold. An abrasive masonry blade ($5 to $6 in a hardware store) in a circular saw is a good alternative. "It's a messy operation and it doesn't cut the brick as well as a diamond saw," he says, "but you can score the brick deeply and then crack it with a hammer to get a pretty clean break."

For his own arches, Allen used a mason's hammer to cut the bricks. This takes some skill, he cautions, and results in a somewhat ragged break and a lot of wasted bricks because you can't always get the bricks to break exactly as you want. The flatter angles that are required near the top of the arch are particularly difficult to make with a hammer.

A ragged cut on a few bricks, though, is less important than making sure that the curved mortar joint between the arch and the entire brick wall around it is a constant thickness (**F**). A curved joint that varies in thickness is unattractive, and once it's there, you can't do much about it.

Once the courses are laid up around the arch, tap out the 2x4s carefully, drop the centering gradually, then remove it and admire your arch.

Tooling the joints—In typical brickwork, after the brick is laid in the wet mortar, the mason cuts off the excess mortar at the face of the brick. In one to three hours the mortar will be thumb-print hard, the proper consistency for tooling with a V-shaped or rounded metal rod called a jointer or striking tool. This produces a clean and attractive joint. (For exterior brickwork, tooling is doubly important because it helps compact the mortar at the face, making it much more weather-resistant.)

But with an arch, the bottom mortar joints can't be tooled when the mortar is thumb-print hard, because the centering is in the way. By the time it is removed, the mortar is too hard for tooling.

Yet old arches and vaults have well-tooled joints. Allen was puzzled. Books on the subject

were no help. Several masons suggested rubbing the joints with full-strength muriatic acid. The acid quickly dissolved the excess mortar that had been stuck between the brick and the centering, but the joint remained undefined and fuzzy.

Allen eventually learned that before the era of portland cement, arches and vaults had been laid up with lime mortar, which sets very slowly. He says that even if the centering were kept in place for several weeks, when it was removed the lime mortar would still be soft enough to be tooled. Although lime mortar is not as strong as portland cement, the arches themselves are structurally stable enough to be able to compensate for the weaker mortar. Allen recommends either using lime mortar alone for the arches, or using it for only the bot-

tom part of the joint that meets the centering, and then using mortar made with portland cement for the work above this.

Living with the inglenook—The Allens' inglenook can hold up to six people within its candlelit confines. The snug spot sits off a spacious country kitchen, and is used more often on social occasions than on weekday evenings. "It can get a little crowded," Allen says, " but in that kind of space it doesn't feel crowded. I think people are accustomed to drawing close around a fire." □

Elizabeth Holland lives in West Shokan, N.Y. She writes about the design and construction of energy-efficient buildings, and is contributing writer for this magazine.

B rick floors are experiencing a revival. The material used only for patios or walkways 10 years ago is moving inside the house. There are good reasons for this. The color, texture and pattern that brick brings to a room can't be duplicated with vinyl or wood. And brick is a logical choice for a passive-solar building, because it increases the thermal mass while offering a more finished look than a concrete slab.

Laying a brick floor sounds risky if you're not skilled with a trowel and a brick hammer. But after watching master mason John Hilley lay the floor in a house I'm building in Brewster, Mass., I'm convinced that it's not too difficult. As with any job, if you know the tricks, the battle is half over.

A brick floor laid with mortar is different from a wall or a patio. First, you can ignore plumb and just concentrate on level. Second, you don't have to contend with the shifting, subsiding backyard quagmire that is the substrate for most patios. A brick floor should be laid on a concrete slab, which is flat and solid; or on a wood floor that has been beefed up to take the extra weight of the bricks and stiffened so that its flexibility won't crack the mortar joints.

Brick—If you think that brick is brick, and that your only decision will be how many to buy, you'll change your mind when you hit your first masonry-supply yard. Bricks come in many styles, colors and prices. They run anywhere from 30¢ each to 50¢ each or more. If your floor extends out into the weather, use paver bricks. The surfaces of pavers are sealed, so the bricks won't spall when it freezes. If your floor is inside, anything that strikes your fancy will do.

Used bricks make a very nice floor but they're getting scarce, and as a result, expensive. You never know exactly how much waste you'll get when you scale the old mortar off used brick, but plan on buying at least 3,000 used bricks to get 2,000 usable ones. For new

4x8 bricks, you should figure 4½ bricks per square foot of floor and then add at least 5% for waste.

Since a standard brick weighs about 4½ lb. and you're going to need a lot of them, you'll want them delivered to a convenient place. Most yards bring the bricks on pallets to your job site, and if the delivery truck is equipped with a hydraulic arm, you're even better off. The arm lifts the wood pallets off the truck and sets them down anywhere you say. A skilled driver can just about put the bricks in your back pocket.

Use brick tongs for hauling the bricks from the pallets. Brick tongs are simple tubular-steel contraptions that will carry between 6 and 10 bricks at a time by holding them in compression. At about $16, using tongs beats weaving around the site with a pile of bricks stacked up against your forearm.

Layout—Although there are many patterns or bonds that bricks can be laid in, the *running bond* is still the easiest and one of the most attractive. The joints between the bricks in each course are offset from the joints in adjacent courses by a half brick. This means that each course begins with either a half brick or a whole brick. After sweeping the slab absolutely clean, determine which direction the brick courses will run. You can then begin the rough layout of the floor with a tape measure.

To avoid having to cut a course of narrow bricks at the end of the room, you may need to adjust the width of the joints. Laying out to a full course at the end of the room is time well spent. Do your figuring on paper by adding an ideal joint width, for example, ¼ in., to the width of your brick and dividing the total length of the room by this sum. Then confirm your calculation by *dry coursing*—laying a full row of bricks along the length of the room without mortar to test-fit the layout. Pick carefully for representative bricks, since they can differ considerably in length and width. Remember that thin joints look better than fat

ones when you're making adjustments between courses; and that you'll get another inch at each wall to play with because the wall finish and baseboards that will be installed later will cover that much more of the floor.

Once your dry coursing has been adjusted so that the joints are even, nail 1x or 2x layout boards to the wall studs along each side of the room so that their tops are even with the top of the finished brick floor (drawing, below). Mark the leading edge of each course on the board, and drive a nail into the top of the board at each of these marks for a string line. After laying a course, move the string forward one nail on each side of the room for the next course. This line represents both the finished height of the floor and the leading edge of the course, leaving very little for you to eyeball.

A layout board can also be cantilevered off the top of bricks that have already been laid. This setup works well when an exterior wall takes a jog, making the room narrower, and ending a run of layout boards. Course lines are marked on the end of the board that projects out to the unlaid part of the floor, and the course string is attached so it rides on the bottom edge of the board. This maintains the same finished floor height as layout boards held flush with the top of the bricks.

You should dry-lay a test course along the width of the room, too. You will be starting with a half brick or a whole brick on one end, and adjusting the joint width between the ends of the bricks to determine the length of the brick on the other end. It won't always work out to half and whole bricks, but the less cutting you have to do, the easier the job will be. Don't end a course with a very short brick.

For anything wider than a closet or a hallway, use *control bonds* to make sure that the bricks are being spaced uniformly. These are bricks whose ends are laid to a string as a reference every 6 ft. or so (about 10 bricks) within each course. This in effect breaks a long course up into several small courses.

The control bonds and end bricks are the

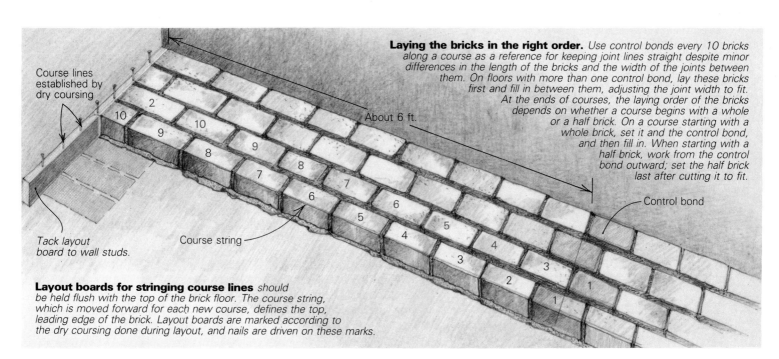

Course lines established by dry coursing

Tack layout board to wall studs.

Course string

About 6 ft.

Laying the bricks in the right order. *Use control bonds every 10 bricks along a course as a reference for keeping joint lines straight despite minor differences in the length of the bricks and the width of the joints between them. On floors with more than one control bond, lay these bricks first and fill in between them, adjusting the joint width to fit. At the ends of courses, the laying order of the bricks depends on whether a course begins with a whole or a half brick. On a course starting with a whole brick, set it and the control bond, and then fill in. When starting with a half brick, work from the control bond outward; set the half brick last after cutting it to fit.*

Control bond

Layout boards for stringing course lines *should be held flush with the top of the brick floor. The course string, which is moved forward for each new course, defines the top, leading edge of the brick. Layout boards are marked according to the dry coursing done during layout, and nails are driven on these marks.*

Photos: Bob Syvanen; Illustration: Frances Ashforth

first bricks to be laid in each course. The rest are filled in to fit. This way, the joints of every other course at the control bond will form a straight line, and the cut bricks at the end of courses, as well as the width of the mortar joints at the ends of the bricks, can be kept fairly consistent from course to course. Try to place control bonds in highly visible spots such as stairways and entries, and string them just as you did the course lines.

Stock the floor once you've completed your layout. This way you can stretch strings and get to know the peculiarities of the room and the slab you'll be working on before you begin littering it with bricks. Using brick tongs, distribute the bricks so that they will be within easy reach when you begin to work. Keep in mind that several layers of bricks on a pallet, or even the whole thing, can be a very different color from the rest of the load. Mix these colors and tones as you stock them on the floor so that your floor doesn't end up with big patches of only dark or light bricks.

Tools—A brick floor is laid with standard mason's tools (photo below). In addition to a 4-ft. level, a tape measure and nylon string, you'll need a brick hammer to break the brick to length at the end of a course. It has a square, flat head and a long, flat chisel peen on the other end, and is made of tempered steel. Brick hammers come in various weights. They have either steel, fiberglass or wooden handles, and cost about $15.

Brick trowels also cost about $15, and are made with wood or plastic grips. There are two basic shapes: the London and the Philadelphia. Most brick masons prefer the London pattern, an elongated diamond shape with its heel farther forward on the trowel than the Philadelphia. London patterns come with either a narrow or a wide heel. The narrow heel is fine for brick since less mortar needs to be carried by the trowel than for stone or concrete blocks.

You'll need a jointer for smoothing and shaping the mortar between the bricks. This tool looks like an elongated steel S, and is gripped in the middle. Each end of the tool has a different profile.

Tools of the trade. **Brick tongs, top, make stocking the floor with bricks much easier. The trowel, jointer and brick hammer below it are the basic tools used to lay the floor.**

Mortar—The mortar between the bricks in your floor makes it permanent, and provides a visual relief from the brick itself. In this case, it is a mixture of masonry cement, sand and water. Use a shovel and buckets to proportion the ingredients for the mortar. Mix your mortar with a hoe in a mortar box or a wheelbarrow. An easier way to mix is with a mechanical cement mixer. Do your mixing outside where you are free to hose out your mixer at the end of the day, but keep the fresh mortar out of the hot sun.

The amount you'll need depends on two things: how much bedding is required for the slab you are working on, and the thickness of the joints between bricks. If your slab is flat and level, the bed of mortar under the bricks will be fairly uniform. A good bed is ⅜ in. thick, and no less than ¼ in. However, a serious hog, or hump, in the floor can double the amount of mortar, because you will need to bring the bricks for the rest of the floor up to this level with a much thicker bedding.

The width of the joints between bricks is the other factor that affects how much masonry cement and sand to order. Joints look bigger than they really are because bricks are molded and don't have hard crisp edges. A ¼-in. joint after finishing will look ⅜ in. wide, which is a nice size. Figure on one bag of masonry cement for every 100 bricks if your slab is uniform and your joints are ¼ in. wide. You will need 1½ cu. ft. to 2 cu. ft. of sand per bag of masonry cement.

The consistency of your mortar will have a lot to do with your success in laying bricks. Mortar should be firm enough to support a brick as bedding, yet soft enough to compress easily at the joints. A soupy mix will lay down, or self-level, in the wheelbarrow. A mix that's too stiff will support itself even when it's stirred into the shape of a breaking wave. A workable mix is the consistency of whipped cream, and the secret to mixing it that way is adding water to the dry ingredients in small amounts, and lots of practice.

A good mason mixes only the amount of mortar that can be made with one bag of masonry cement at one time. With this amount, the consistency is easy to control. A batch will last about three or four hours before setting up. Temper the mortar every 15 minutes by working it briskly with a hoe or shovel and adding a little water if necessary. The books say not to add water, but all masons do. Whether the mud is mixed with a mechanical mixer or in a tray with a mortar hoe, it's best loaded into a wheelbarrow because it is convenient to move around the work area, and it's easy to scoop out of.

Laying the bricks—You are ready to begin laying bricks when the mortar is mixed and the control-bond strings and the first-course string are stretched. For courses that begin with a full brick, lay the control bonds first, the full end brick second, and then fill in between, as shown in the drawing on the previous page. Start from the control bond and work outward. If the course starts with a half

brick, lay the control bond first, and then work from this brick out toward the ends, cutting the bricks on either end to fit. A little gain or loss that accrues as a result of the inconsistent length of the bricks can be offset by adjusting the size of the last brick.

You cut the last brick in a course with the mason's hammer. Hold the brick in your hand, and hit it sharply once or twice directly over your palm. This usually does it, but some bricks need a shot on both sides. Sometimes they break where you want, and other times you end up with a handful of brick shards. You'll get better with practice, but until you do, order enough extra brick so that each blow isn't critical. If the brick fractures at an angle, set it down on a hard surface and use small, chipping strokes with the hammer to straighten out the line of cut.

Another tool for cutting brick that requires a less practiced stroke is the brick set. This wide chisel is placed on the brick where you want it to fracture, and struck with a hammer (for more on breaking brick, see *FHB #3*, p. 43). A brick set will cost you about $6. Plan to waste a few bricks using this tool as well.

You can tell a journeyman from an inexperienced mason just by watching his trowel hand as he picks up a load of mortar and places it. There is a familiarity with the material that is unmistakable. First, using the back of his trowel in the surface of the mortar, he will stroke away from his body. This creates a mound of mortar in the wheelbarrow. With the face of the trowel he scoops a load of mortar in an upward motion, then drops his trowel arm abruptly, ridding the trowel of excess mortar. What remains won't slide off the trowel even if it's turned upside down. The technique is stroke away, scoop up, and settle.

The first trowel of mortar should be placed on the floor for bedding. Thrown is more accurate, though this too takes a bit of practice. Then choose a brick. If your bricks are water struck on one face, this face should be laid up because its slightly glossy surface is less porous and will wear better. With brick styles that are water struck on all sides, or not at all, just choose the best face.

With the brick in one hand and the trowel in the other, pick up a thin line of mortar using the same stroke as before, and wipe it on the edge of the brick. As you get more experienced, it will almost look like you're throwing the mortar on. Unless you overload the brick, the mortar shouldn't slip off while you are handling it. Trial and error will tell you how much mortar to use. Finish buttering the brick by loading up the end that will butt the previously laid brick.

To set the brick in place correctly, all of its joints need to be in compression. This is one of the secrets of good brickwork, and the way to accomplish this is the *shove joint*. The brick should be held out from its ultimate resting place and square to it. As you begin to bed it down, the brick should sit slightly above the string line on the mound of mortar underneath it (photos facing page). Using your thumb, push the side of the brick toward the

The shove joint. Above, mason John Hilley beds a brick in a running bond. It is buttered on one side and one end with mortar, and will sit well above the course string on the bedding mortar until it is pressed and tapped down. This string indicates the top of the floor, and the leading edge of each course. It is moved ahead one nail on the layout boards at each side of the room after a course is completed. The brick is leveled to the string by a gentle tapping with the butt of the trowel (above right), while pressure is applied with the left hand. This compression is the key to a tight, permanent brick floor. The lines of mortar between bricks are jointed, right, when they reach the consistency of putty, using excess mortar on the surface of the brick. The mortar is fed into the joints and compressed with a jointer. This is done twice on each course to produce smooth and compact mortar joints.

previously laid bricks and use the butt of the trowel to tap the brick down until its face is just below the string. If the brick is too low anywhere, lift it out and load the bed with more mortar. Then press the brick into place again. Also use the trowel handle to tap the end of the brick until the width of the joint between it and the previously laid brick is correct. During all of this you should be pushing the side of the brick with your thumb. This pressure keeps the brick from settling unevenly, and keeps the joints in compression. Don't worry if the mortar doesn't rise all the way to the surface in every spot along the joints. The holes will get filled in later.

Scrape away any excess mortar with the edge of your trowel. This mortar can be allowed to dry a little and be used for jointing. Try not to smear the mortar on the surface of the bricks, but if you do, don't worry. It can be washed off later after it sets up.

Jointing—The brick joints are ready to be filled and jointed when the mortar between the bricks has the consistency of putty. Test it by pressing the mortar with your finger. You can use the scrapings on the surface of the brick to fill the joints if they hold together when you squeeze them in your hand. They shouldn't be soft like fresh mortar, or they won't compress when jointed. Dried or crumbly mortar shouldn't be used either.

The joints should be filled and compressed in two stages. For the first fill, feed the mortar into the joint and press it with the jointer

(photo above right). Work your way along the course, filling and pressing. Then go back to the beginning of the course for a second fill and a final jointing. Jointing not only increases the strength and durability of the mortar joints, but it also gives the floor a smoother, more uniform appearance. Do the final jointing with smooth, level strokes. Use your whole arm, not just your wrist.

Cleanup—Let the mortar set for a week before cleaning. You can wait longer, but the brick will get harder to clean. First, scrape off heavy spots of mortar with the chisel peen of a mason's hammer. To remove any mortar smeared on the surface of the bricks, you will need to use a 20% solution of muriatic acid. Mix the acid in a bucket by adding one part acid to four parts water. Always add the acid to the water, and make sure that you are wearing eye protection and heavy-duty, acid-resistant rubber gloves. You will also need running water from a garden hose close by. Open as many doors and windows as you can for ventilation. Muriatic acid is nasty stuff—treat it with respect and caution.

The acid will soften the mortar spots, enabling you to scrub them off the surface of the bricks. For this you will need a short, stiff-bristled brush for hand-scrubbing, and a 9-in. stiff-bristled brush on a long handle. Before you begin, flood the whole floor with water until the bricks are saturated. This keeps the acid action on the surface where it can be controlled, and then washed off as soon as it

has penetrated the mortar smears. The bricks will look shiny when they are saturated, dull when they are beginning to dry out. Keep the bricks saturated during the entire process so the acid won't burn the mortar. Get a helper to operate the water hose.

As soon as you mix the acid solution, test its action in some hidden spot—under the stairs or behind the chimney. The joints will have a soapy appearance as long as muriatic acid is present. It's best to do this job systematically and work small areas. Dip the brush into the acid solution and then begin scrubbing the saturated bricks. As soon as you have covered a few square feet, hose it down until all the foaming stops. Then hose it a bit more, saturate the next area, and move on with the bucket and brush. After the entire floor has been scrubbed and rinsed, hose it down one more time. If any muriatic acid remains, it will continue to break down the mortar.

Brick floors should be sealed, but not until all the moisture has left the bricks and the mortar joints. Wait at least a month. If you get impatient and seal a brick floor with moisture in it, the sealer won't bond. The best product I've used for sealing brick is Hydrozo Water Repellent #7 (Hydrozo Coatings Co, Box 80879, Lincoln, Neb. 68501). Five gallons will do the average floor and will cost a little over $80. Put on two coats, letting each one dry for 24 hours. I follow this with a coating made of equal parts of turpentine and Valspar polyurethane varnish (The Valspar Corp., 1101 South Third, Minneapolis, Minn. 55440).   □

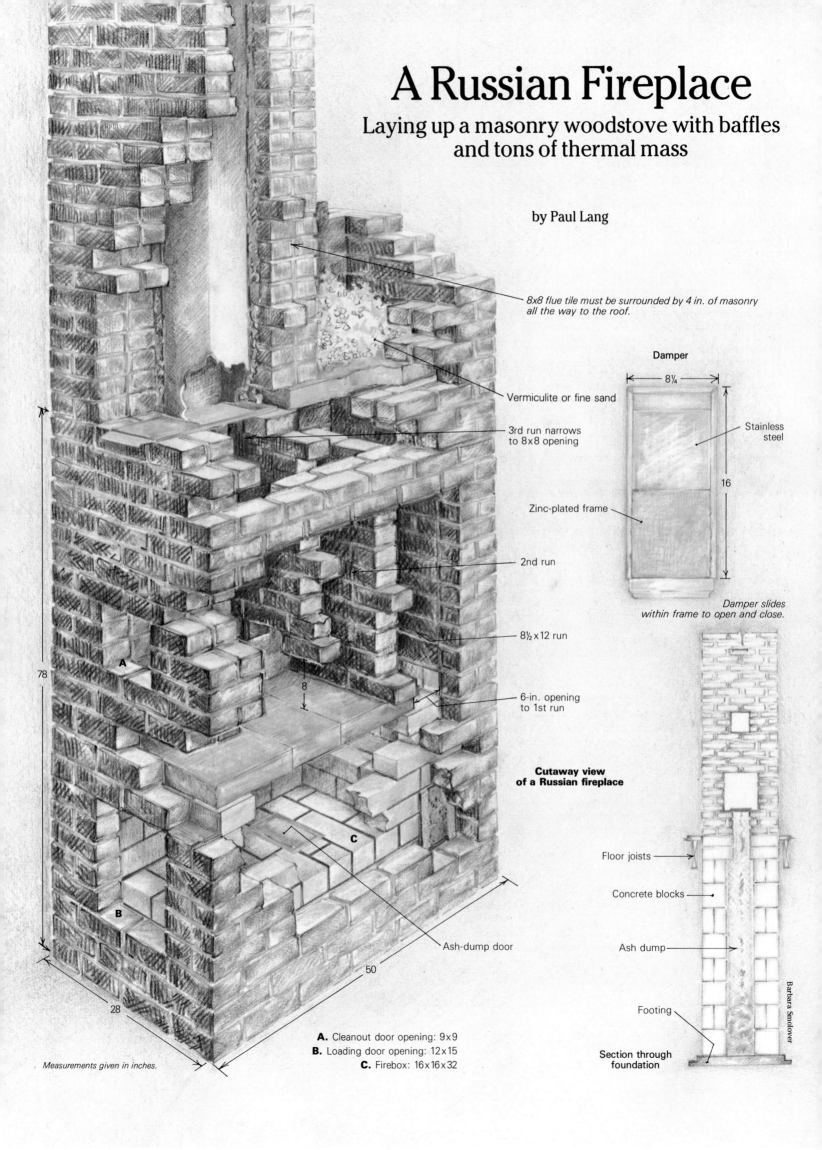

A Russian Fireplace

Laying up a masonry woodstove with baffles and tons of thermal mass

by Paul Lang

8x8 flue tile must be surrounded by 4 in. of masonry all the way to the roof.

Vermiculite or fine sand

3rd run narrows to 8x8 opening

2nd run

8½ x 12 run

6-in. opening to 1st run

Cutaway view of a Russian fireplace

Ash-dump door

Damper

8¼

16

Stainless steel

Zinc-plated frame

Damper slides within frame to open and close.

Floor joists

Concrete blocks

Ash dump

Footing

Section through foundation

Barbara Smolover

A

B

C

8

78

50

28

Measurements given in inches.

A. Cleanout door opening: 9x9
B. Loading door opening: 12x15
C. Firebox: 16x16x32

The revival of wood burning has kindled new interest in getting as much energy as possible from every log. Though there are lots of cast-iron and sheet-metal stoves on the market, the masonry stove may well be the most efficient means of heating with wood.

I first became interested in masonry stoves after reading an article in the February 1978 issue of *Yankee* magazine. It was through this article that I came to meet and work with Basilio Lepuschenko of Richmond, Maine, who had designed a masonry heater that is popularly called a Russian fireplace. Basilio and Albie Barden of the Maine Wood Heat Company have been responsible for generating most of the recent interest in masonry stoves. For a history of masonry stoves, see the box on p. 73. The masonry heater in the drawing on the facing page is a modified version of the stove designed and copyrighted by Lepuschenko.

Masonry stoves have little in common with the traditional open fireplace. With their enclosed fireboxes and baffle systems, they are more like modern airtight metal woodstoves. But there are functional differences. Most obvious is that the masonry heater can store heat. Even the smallest unit contains about 64 cu. ft. of mass, and once the external walls heat up to their maximum 150° to 200°F, the brick will slowly radiate warmth for nearly 48 hours.

A stove of this design has to have a minimum of three runs and two baffles. The smoke must go up, pass down over the hot firebox, and go up again. Larger units have four baffles, and smoke makes an additional pass down and up before exiting up the stack.

Another difference is that a masonry stove is not designed to burn continuously. It is fired up only once or twice a day. A quick, hot fire with relatively small, split wood is ideal. After the wood has been reduced to ashes, a sliding damper located at the end of the last baffle is closed, trapping all of the heat inside, allowing it to warm the room throughout the day or night. Temperatures of up to 1,200°F ensure the secondary combustion of gases in the baffle system and eliminate the danger of creosote buildup. (The only time I've ever seen creosote build up in a masonry heater was in one with baffles much longer than usual. Baffles longer than about 48 in. let the temperature drop enough for creosote to begin to form.)

Firing up this way, wood consumption is drastically reduced. Even the most ardent wood-burners should be willing to forsake their dream of a 24-hour burn for a 60% reduction in the amount of wood required to heat their homes.

Masonry stoves are relatively simple to build, but designing a heater into the floor plan can present some difficulties. Proper clearance from combustible material like framing and siding is a must. The minimum is 36 in. on all sides.

In order to take full advantage of the stove's ability to radiate heat, it should be centrally located. The stove shown here is 50 in. long and 28 in. wide. It is complete at about 6½ ft. above the floor, and from there, a single 8x8 flue tile extends to join up with those from other fireplaces.

I like to see floor plans designed around the masonry stove rather than awkward attempts to squeeze one in somewhere like an additional closet. Masonry heaters work so well that I dislike compromising their efficiency for mere architectural considerations, but my feelings in this regard border on masonry madness and should be viewed as such. Suffice it to say that space and clearance can be a problem. Ten tons of hot brick can't go just anywhere.

There is no single correct way to build a masonry stove. Flue configurations are limited only by the builder's imagination, and many stoves differ in this regard. The basic design of Lepuschenko's Russian fireplace has proven itself efficient. Since my introduction to Basilio I have built four stoves to his general design, though no two were exactly alike.

Every masonry stove has its own design problems and requires some modification to fit within a given house or masonry system. Brickwork isn't nearly as easy as it looks. If you haven't got the patience to plan precisely and proceed carefully, or if you have a low frustration threshold, you may want to leave the construction of your masonry heater to a mason.

One of the main differences between this stove and those of Lepuschenko's design is the thickness of the walls. Because of strict building codes in Connecticut, where I work, the walls of this stove are 8-in. double-brick thickness throughout. The ones I learned to build with Lepuschenko in Maine were a single 4-in. brick thick in the baffles above the firebox. The double thickness increases the thermal mass substantially but also lengthens the time the unit takes to heat up. I personally think a single 4-in. brick wall is the way to go, and that additional thickness is unnecessary for either safety or thermal storage.

Another variation in this particular unit is the use of high-temperature refractory cement in the laying of the firebox. I lay all the firebrick used in the construction with a premixed, air-setting, high-temperature cement, which provides a strong bond with little deterioration after extended firing. I also use either precut arched firebrick or 12-in. by 24-in. by 3-in. refractory slabs to form the ceiling of the firebox. The arched firebox ceiling provides greater strength than the refractory slabs, and I like to use it in five-run masonry stoves.

The dimensions of a masonry stove are based upon the length of the individual bricks, the number of baffles and the location of the stove. There are some key factors to consider, however, when designing a masonry heater.

Firebox dimensions are important. The firebox of this stove is 12 in. wide by 16 in. high by 32 in. deep. Height and width should remain the same even on larger stoves, although the depth can be increased to 40 in. Substantially larger fireboxes and the resulting larger fires could cause increased thermal stress and minor hairline cracks in the mortar joints. In masonry heaters, bigger is not always better.

A three-run Russian fireplace built on Lepuschenko's model will heat an area of 600 sq. ft. or more. Significantly larger areas would require a five-run heater. Both work well.

Simplicity and availability of materials are important features of Lepuschenko's design. The

firebrick slabs and precut arched firebricks are available only from a refractory-materials supply firm, as is the high-temperature cement. All other materials are sold by masonry suppliers. No special tools are required, although I often use a diamond-blade brick saw to make the necessary cutting faster and more precise. In most cases, a mason's hammer or a brick set will do the trick (for how to break brick, see *FHB* #3, p. 43). The standard cast-iron cleanout and loading doors used on the stove can be bought at most masonry supply yards. The sliding damper is stainless steel in a zinc-plated frame. This is one item that is not readily available. I have a friend with access to sheet-metal fabricating equipment, who can make one up for me in about two hours.

Construction—First make sure the floors have been framed so there's enough room for the masonry to pass through. A plumb line, a ruler and a little thought are all you need. Once the layout has been established from basement to roof and all the necessary carpentry has been done, it's time to start.

Foundation work for a masonry stove is conventional—using the same materials (concrete blocks) and procedures you'd follow for a traditional fireplace. For this job, we poured a footing 12 in. deep in the basement, 5 in. larger all around than the external dimensions of the heater. Then we built up to the first floor level with 8-in. block laid to the same dimensions as the heater. The whole center of the foundation becomes an ash dump, so we left an opening for a cleanout door. If the stove is being built on a slab, there's a lot less work, but you still need that 12-in. thick footing.

Once the block foundation is at floor level, you need to cap it. Often you can just span the gap with 4-in. solid block, leaving an opening for the ash dump, and building up courses to floor level.

The courses of brick begin above the floor joists. I use plumb lines to keep the brickwork true. I think this makes the construction both quicker and better looking. This is also why I lay corners first and use a horizontal line to lay the brick between them.

The firebox sits about 12 in. above the floor. Its firebrick base is laid and leveled in standard mortar (I mix mine with one part portland cement to one part lime to four parts clean sand, adding water as needed). The opening for the ash dump should be close enough to the loading door to allow access, and it must fit flush with the top of the firebox floor. To accomplish this, I notch the firebrick ¼ in. deep, using a contractor's saw with a carborundum blade.

Next, I lay six to eight courses of the external brick walls, leaving an opening for a 12-in. by 14-in. loading door at the front. It's important that this brick be laid with full joints, and that any excess mortar be removed, and the remainder jointed smooth on the inside as well as out. This will ensure a tight fit between the firebox and the external brick.

The firebricks of the firebox walls should be

Paul Lang of Newtown, Conn., has been a mason for eight years.

laid out dry to determine its exact depth. They are then laid in refractory cement, which can be thinned and the bricks dipped, or left as is and applied with a trowel. I trowel it on because it is easier for me.

The firebrick must be set tight against the exterior walls, because any airspace will act as an insulator. There should be no cement or mortar between interior and exterior bricks. In this stove, I used a layer of ⅛-in. mineral wool as an expansion joint between the firebrick and the external brick. This was an experiment. I've read that master Finnish stove-builders employ joints of this type to allow for some movement of the firebrick without hindering the conduction of heat.

At the rear of the firebox, I build a shelf two or three bricks high and about one brick's length deep (photo left). On this job, I used cement block to form the shelf, then covered it with firebrick. This is the beginning of the flue passage. This first vertical run is the hottest spot in the stove. It must always be built to double brick thickness, even if the rest of the stove is only one brick thick. I raise the external brick walls another few courses—enough so I can install the 3x3x¼ angle iron 20 in. long that spans the top of the loading opening. Then I continue the firebox walls to a height of about 16 in. before laying its ceiling.

The ceiling of the firebox can be formed with cantilevered firebrick, refractory arches or firebrick slabs. I chose slabs for this stove because they are quick to install, they fit nicely within its dimensions and they are strong enough for a heater this small. I use refractory cement for this, too. The first slab is set tightly in place against the front wall just above the loading opening, and all three slabs butt together to form a continuous firebox ceiling that stops 6 in. to

Above: Lang sets firebrick at what will be the hottest part of the masonry heater: the beginning of the first run. The mineral wool between the exterior common brick and the firebrick along the side of the stove is experimental—an attempt to eliminate stress cracking. The ash-dump door is set in the floor of the firebox, while the loading opening is spanned with angle iron. A cast iron door will be hung after the brickwork on the stove is completed.

Left: The entire heater is of double-wall construction. Lines dropped from above ensure plumb at each corner. The two cantilevered bricks at the center are temporarily supported by scrap. A third brick will span the gap to form the base of the second baffle, allowing the smoke to pass beneath. The opening just begun in the front of the stove, top, is the cleanout.

Facing page: The first two vertical runs are sealed off with refractory slabs, surrounded by mortar and common brick. Several courses of brick will cover the slabs, then the outer walls will be laid up to the ceiling. Sand or vermiculite will be poured into the cavity as a firestop.

8 in. from the back wall to leave room for the first vertical run.

Next I lay six to eight more courses of external brick, incorporating a cleanout opening in the front wall just above the firebox ceiling. As with the loading opening, the cast-iron door can be installed later. This opening is helpful in cleaning up fallen mortar during construction, though its primary function is to allow access for inspection and cleaning of the first two runs during use. Properly built and used, masonry heaters should require no cleaning at all. Only the repeated burning of green wood could cause creosote to build up inside.

I lay out the base of the baffles dry to position them so each of the three vertical runs will be the same size. Runs should be 8 in. by 12 in., though small variations are all right. The first baffle is built right on top of the firebox ceiling. The second requires an 8-in. opening at its bottom. The easiest way to form this opening is with cantilevered bricks (photo facing page, bottom). Both baffles must be tied into the side walls of the heater. I make this bond by laying a baffle brick into the interior side wall every four or five courses. In a heater with walls only a single brick thick, the 4-in. butt end of these bricks will show on the outside. External and internal walls go up several courses at a time, with the baffles continuing upward within them.

This masonry stove will draw well with an ordinary 8-in. by 8-in. flue tile. The final run is narrowed to this dimension with cantilevered brick, and the damper is set in place. The damper and frame are only $\frac{3}{8}$ in. thick, and fit into the mortar joint of the external brickwork.

A flue tile is set atop the damper, and the brick is laid around the flue to ensure a tight fit. In this stove, the clay flue tile had to angle sharply to clear the framing of the second floor. Ordinarily,

the flue would continue straight up, and be surrounded by 4 in. of masonry.

The unit is ready to be capped off about 4 ft. from the top of the firebox. The first baffle ends 8 in. from the top, and the second continues all the way up. At this point, the first two runs of the heater are sealed off. I use firebrick slabs, as shown in the photo below, but cantilevered brick would do just as well. The slabs are quicker to install but more expensive.

The firebrick slabs should be covered with a couple of courses of brick so that there is at least 8 in. of masonry over the vertical runs. At this point the brick can stop, but I usually like to continue the external walls up to the ceiling because the stove looks more complete that way. The resulting interior space can then be filled with fine sand or vermiculite to act as an additional firestop.

After the stove has been completed, stoutly resist the temptation to fire it up immediately. A great deal of moisture is trapped inside the masonry and must be allowed to evaporate completely. The mortar must cure slowly. The ideal curing period is three or four months, though few owner-builders can be so patient. Firing early can cause the moisture to be driven out quickly, causing possible stress cracking. Fires should be small at first, allowing a break-in period of about three days.

Whether you choose to build a masonry heater or oversee the construction of one, try to keep one thing in mind. The internal brick of the heater should be laid with the same care as the external walls. Though the flue passages will never be seen after the heater is completed, they are really the heart of the stove. The reward for careful craftsmanship can be an object of functional beauty that will heat your home for many years to come. □

A History of Masonry Stoves
by Albie Barden

In Europe the shift from a hunting-gathering existence to an agricultural society produced both permanent shelters and bake ovens. The primitive European oven was also a wood-fired masonry heater. A fire was built directly in the clay and stone oven. Once the fire burned out, coals and ashes were removed, and bread was put in to bake.

In Central Europe and the Alps, masonry heaters and bake ovens have been built for several hundred years. These have a firebrick core and a ceramic exterior, often surrounded by warm benches reserved for the elders of the family. Sometimes the heaters are fired from a kitchen and project into a dining-living room. Sometimes they are fired from a hall or some other service area.

In Scandinavia, the tall rectangular and cylindrical styles of the heaters that evolved serve no baking function. These massive, tile-covered radiation heaters were placed in rooms that required heating. In Russia, Lithuania, Poland, Czechoslovakia and Finland, heaters more commonly used brick or whitewashed stone than the fancier tile used for heaters in Germany, Austria and Switzerland.

The brick heaters currently becoming popular in the United States and Canada have their origins in the simple brick heater, or *grubka*, found in nearly every home of the great Russian land mass. Their efficiency and rustic charm attracted American attention at least once before, nearly 200 years ago when Ben Franklin and John Adams saw them in Europe.

In the late 1700s, models of the European brick stove were brought to Salem, Mass., by John Dodge, an enterprising sea captain who quickly patented and began marketing the idea, until his business was brought to an abrupt end by a heart attack. Apprentices carried on his work for a brief time, and elsewhere in New England other heaters of brick, stone or soapstone were built.

In the 20th century, a modest tradition of Russian-style masonry heaters was sustained by a small White Russian community in Richmond, Maine. At the same time Sam Jaakkola, a Finn in western Maine, built several Finnish-style heaters, the first masonry heaters I ever saw. In contrast to the Russian-style heaters, which are end-loading and work well as room dividers, the Finnish-style heater is front-loading and doubles as an open fireplace, with large doors you can open to see the flames.

The best research on brick-and-mortar heaters is being done in Finland, where a leading architect-designer, a major foundry, brick and firebrick manufacturers and a cement firm are all working together to design and test projects using vocational schools as construction sites and students as their labor.

In the United States, the New Mexico Energy Institute in Albuquerque and the Southeast Community College at Beatrice, Nebraska, both have home-grown research programs on masonry-heater construction.

Albie Barden's Maine Wood Heat Company sponsors workshops on masonry heaters, sells plans and materials,and publishes the Masonry Stove Guild Newsletter. *(For information, send a self-addressed, stamped envelope to RFD 1, Box 640, Norridgewock, Maine 04957.)*

The north wall of the building seen from the inside as it neared completion. Concrete from the upper courses dripping down the wall will be removed after the entire wall is constructed. The interior has a uniform and plumb appearance.

Form-Based Stone Masonry

A method for constructing cast-in-place stone walls

by Richard MacMath

The skills needed to build a stone wall are simple enough for self-taught masons to get beautiful results after a short learning period. Of course, construction must be done with care, and a few rules kept in mind.

The cast-in-place method we used to build the walls shown here is probably the easiest for the novice. We've had "first timers" who were able to learn the technique in one day. Many were producing beautiful results in a few days.

This method is best suited to construction with cobblestones—round-shaped stones in random sizes. Often these stones are found in fence rows between fields and in sandy or gravel soils. For such stones, using formwork is simpler than laying a freestanding wall, especially if one wants a

Richard MacMath is a partner in Sun Structures, Ann Arbor, Mich. The wall is part of Upland Hills Ecological Awareness Center in Oxford, Mich.

durable, true, residential wall. Using formwork as a guide also makes the work go faster. The stacking and leapfrogging of forms ensures plumb and level results.

The job's most difficult aspect, and one impossible to avoid, is getting the stones to the construction site. Once you have found an adequate supply, hauling the stone becomes a task for as many helpers as you can assemble. Although this seems like a troublesome job, remember that transporting free stone—even over a few miles—is inexpensive when compared to the cost of other building materials.

Even if you have stone and labor in abundance, it's important to be selective in the use of stone masonry. It is best suited for retaining walls and below-grade foundation walls with outside insulation covered with earth. Foundations, basements and earth-sheltered homes are appropriate applications. In such uses horizontal

and vertical steel reinforcing bar (rebar) is required to strengthen the wall against the extra pressure exerted by the earth mass. Stone is also a good choice for interiors, especially for thermal mass walls. A passive solar design rule-of-thumb calls for 1.0 to 1.5 cu. ft. of masonry exposed to direct sun for every 1.0 sq. ft. of south-facing glass.

As a note of caution, above-grade stone walls that require insulation are a problem. They must be insulated in one of two ways: either by constructing a cavity wall or by adding an insulating layer to one side of the finished solid wall. Cavity-wall construction is difficult, time consuming and not recommended for the novice. The other option—adding insulation to the interior or exterior face of a solid masonry wall—requires studs or furring strips, insulation and a finish surface (usually gypsum board) that will completely cover one side of the stone wall. In both cases it would be easier to construct a well-insulated stud bearing wall first and add the stone veneer later.

Rules of stone masonry construction—The first rule of sound wall construction is to set the stones so that gravity rather than mortar holds them in place. Each stone must be set in a firm, relatively flat bed formed by the stone below and its covering layer of mortar. I have seen large stones fall out of the wall when formwork was removed because the novice stonemason relied too much on the concrete and formwork to hold them in place. This is the most common mistake in form-based stone-masonry. Place the stones as if you were laying a dry wall; use the concrete only as joint fill. If a level, flat bed cannot be prepared for a stone, then place it so gravity forces it inward. In this way, adequate beds can be made for angled and rounded stones.

The second rule is to place each stone so that its weight is distributed over at least two other stones below. This is the principle of crossing joints for maximum strength. Crossing joints ensures that the wall will work as a unit rather than as individual parts. This rule doesn't apply when setting small stones on top of large ones, but it's best to avoid a continuous joint from top to bottom in a wall. Otherwise you invite cracks, and the finished appearance lacks the random overlapping pattern that makes stone masonry so attractive. In long walls, however, a continuous joint is constructed intentionally, every 30 ft. or so. This is called a control joint, and it allows for expansion and contraction of the wall under changing thermal conditions.

The third rule concerns both structure and aesthetics. The mason must not only look at each stone carefully, but also at how each stone fits into place and contributes to the structural integrity of the wall as a whole. For example, a particular stone may have one face that is flat

and colorful. The beginner's impulse is to expose this face on the finish side of the wall. Unfortunately, its best use may be as a flat surface, providing a firm bed for succeeding stones. In this case, aesthetics must yield to the structural requirements of the wall.

Remember that these rules are easy to overlook when employing this cast-in-place method because the formwork holds the wall in place while the concrete is setting, hiding the finished surface from view.

Designing the wall—If the stone wall is going to be a bearing wall or a retaining wall, then the structural loads must be calculated. These loads will determine wall thickness, footing size and the amount of steel reinforcement required. Fortunately, we have engineering training so we perform the necessary calculations ourselves. Also, construction details must be figured out and drawn to scale to illustrate connections to other structural elements. For example, anchoring the roof to the top of the wall requires proper planning. Our design called for 12-in. long anchor bolts set into the top of the wall on 4-ft. centers. These secured a top plate that served as a nailing surface for 2x12 roof rafters. Since we were constructing a sod roof, we had to continue

the waterproof membrane down the outside of the wall and over a layer of rigid insulation.

In our building the north enclosing wall is a below-grade bearing wall that supports both roof and earth loads. Vertical loading from floors and roof is easily transferred to the footings because a masonry wall is very strong in compression. However, the major structural load below grade is the lateral force of the earth pushing on one side of the wall. These lateral loads force the wall to react in tension on the interior side. Pure masonry walls have almost no tensile strength, so steel reinforcement must be added. As a rule-of-thumb, earth backfilled up to 6 ft. high against a 12-in. thick wall requires no vertical reinforcing. We designed our north wall to support the lateral load of almost 10 ft. of earth. The vertical loads from the roof and the weight of the wall itself provide some resistance to the force of the earth, but we still had to add substantial reinforcement. Note that the rebar is placed on the tension side of the wall—in this case the inside—and additional vertical reinforcing was added to the bottom half of the wall where the earth loads are greatest. Horizontal rebar is laid down between stone courses at 2-ft. intervals during wall construction. The amount of reinforcement is determined by structural calculations.

Footings are generally twice as wide as the wall is thick and equal in depth to the thickness. Here, in a retaining wall situation, the wall is placed close to the interior side of the footing so that the weight of the earth on the outside helps to keep the wall from overturning.

There is one place in our building where the wall changes from a retaining wall enclosing the building to a freestanding exterior wall. Because these two sections of wall experience different loading and temperature conditions, we built a continuous control joint in between them. At the control joint the two walls are tied together with horizontal rebars wrapped with felt paper. The

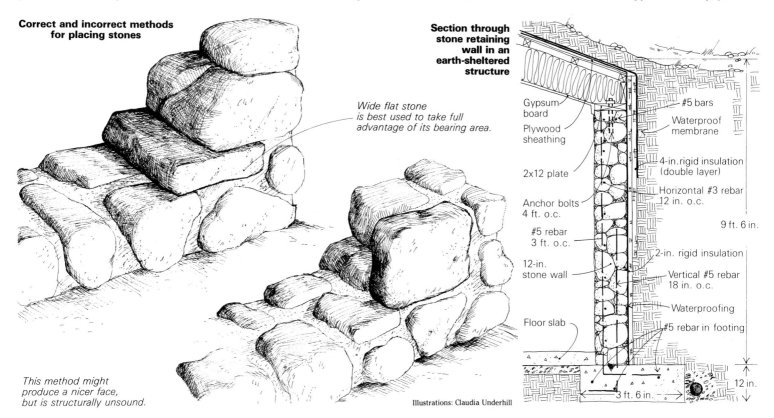

Correct and incorrect methods for placing stones

Wide flat stone is best used to take full advantage of its bearing area.

This method might produce a nicer face, but is structurally unsound.

Section through stone retaining wall in an earth-sheltered structure

Gypsum board

Plywood sheathing

2x12 plate

Anchor bolts 4 ft. o.c.

#5 rebar 3 ft. o.c.

12-in. stone wall

Floor slab

#5 bars

Waterproof membrane

4-in. rigid insulation (double layer)

Horizontal #3 rebar 12 in. o.c.

9 ft. 6 in.

2-in. rigid insulation

Vertical #5 rebar 18 in. o.c.

Waterproofing

#5 rebar in footing

3 ft. 6 in.

12 in.

Illustrations: Claudia Underhill

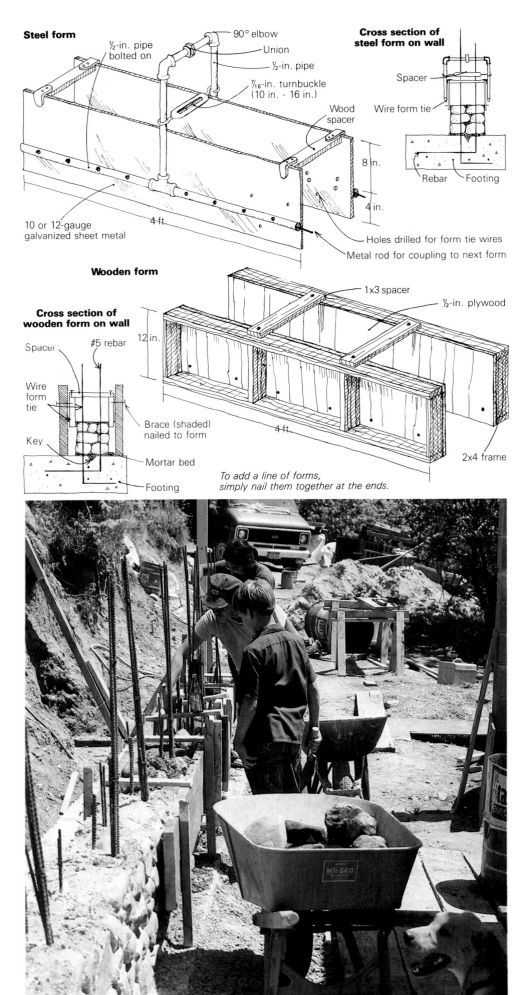

Steel form

½-in. pipe bolted on

90° elbow

Union

½-in. pipe

7/16-in. turnbuckle (10 in. - 16 in.)

Wood spacer

10 or 12-gauge galvanized sheet metal

4 ft.

Holes drilled for form tie wires

Metal rod for coupling to next form

Cross section of steel form on wall

Spacer

Wire form tie

8 in.

4 in.

Rebar

Footing

Wooden form

Cross section of wooden form on wall

Spacer

#5 rebar

Wire form tie

Key

Mortar bed

Footing

1x3 spacer

½-in. plywood

12 in.

4 ft.

2x4 frame

Brace (shaded) nailed to form

To add a line of forms, simply nail them together at the ends.

Wood forms work just as well as metal ones. Concrete, stones and water must always be on hand. Working with at least three or four people at a time keeps the work progressing quickly. Here two people are setting stones in a bed of concrete, layer by layer. The wood forms shown are heavier than metal ones, but last longer and are easier to clean.

rebar holds the retaining wall in place laterally while allowing for expansion and contraction in a direction parallel to the wall.

Formwork—We used both metal and wood forms, both serving the same purpose. When set on the footing or clamped to the top course of the wall, the forms held the concrete and stone in place while drying. Metal forms were lighter and easier to assemble, but sometimes bulged at the bottom as they were being filled. Wood forms, though heavier and clumsier, better resisted the weight of concrete and stone. Both types were coated inside with used motor oil to keep them from sticking to the concrete.

Design for the metal forms came from Ken Kern's book, *The Owner Builder's Guide to Stone Masonry* (Owner Builder Publications, Box 550, Oakhurst, Calif. 93644). Kern uses galvanized pipe, pipe fittings and sheet metal, together with a metal turnbuckle and wooden spacers. The forms are 12 in. high and 4 ft. long and can be interlocked at each end with other forms. We assembled 4 forms, each weighing 20 lb., but they gave us some problems. First of all, we had difficulty keeping them clean. Concrete was hard to remove from the threads on the turnbuckle and from the pipe fittings. Also, tightening the turnbuckle sometimes bent the pipe.

Generally we preferred the wooden forms even though they were heavier by about 10 lb. They were made in two parts and connected when clamped to the wall. One advantage of using wood was that the forms could be nailed to stakes and other forms for alignment and stability. We always used the wood forms when building the curved north wall. Bending pipe to the proper radius seemed difficult to us. Spacers and wire were used with these forms as with the metal ones.

Required materials and tools—Besides the stones and forms, you'll need the following materials and tools to build a form-based stone wall:

Sand, gravel, portland cement.	Mortar boards
Water	Hose, containers of water (for soaking stones)
Reinforcing steel (rebar)	
Waterproof coating (asphalt or cement base)	Wood stakes (for layout work)
Polyethylene sheets	String
Used motor oil	Gloves (lots of pairs)
Cement mixer or deep wheelbarrow and mason's hoe (has holes in it)	Wire
	Hacksaw
	Level (preferably 4-ft. length)
Wheelbarrows (for hauling stone)	Chalk line
	Paintbrush
Shovels	Wire brush
Buckets or pails	Plumb bob
Mason's trowels: large size (for filling forms with concrete), small size (called pointing trowels, for finish work)	Nails
	Hammers
	Sledgehammer
	Mason's hammer and chisels (if you want to shape stone)

We used a standard 3-2-1 concrete mix: 3 parts gravel, 2 parts sand and 1 part portland cement. To estimate the amount, assume that the concrete mix makes up approximately one third of the wall's volume. Whenever we were asked how much water to add to the dry concrete mix, my friend Wayne would reply, "to taste." There are so many variables—weather, wet or dry sand, gravel size—that the right amount of water will often vary from day to day. We preferred to use a mix that was dry enough to stand on a trowel, but wet enough to fill all the gaps between the stones. A dry mix can be held in by the form, but must be scooped out of the wheelbarrow with a trowel or shovel.

Reinforcing steel (rebar) is available in diameters from $\frac{1}{4}$ in. (#2) to $1\frac{1}{4}$ in. (#10). After calculating the amount required, you have a choice of using a small number of large diameter rods or large number of small-diameter rods. Small-diameter rebar is easier to bend and cut to the required length. For example, we often substituted three #3 bars ($\frac{3}{8}$ in. diameter) for one #5 bar ($\frac{5}{8}$ in. diameter). The cross-sectional area is the same, but the #3 bars are easier to bend into the footing and cut with a hacksaw. Since rebar is sold by the pound, there is no additional cost in using bars of smaller diameter.

There are many different products for waterproofing below-grade walls (see pages 13 through 15). The standard products are asphalt-base and cement-base coatings that are applied with a brush or trowel. We brushed on an asphalt base coating and added a sheet of 6-mil polyethylene over that. Polyethylene sheets can also be used to protect the masonry during cold weather and to keep various floor and wall surfaces clean during construction.

If you're using a cement mixer (ours was driven off a wind-powered electric system), then have a few large buckets on hand for filling it quickly and easily. Using a wheelbarrow and mixing by hand requires measuring by the shovelful—a slow process. Add water by the bucketful or with a hose.

Be sure to have some large containers on hand for soaking the stones. We used old 55-gal. drums. The wall cures more uniformly when the stones are holding water as they are set into the wet concrete. Concrete and stones then dry together, and a better bond results.

Always have plenty of work gloves on hand. After the fabric of the glove is worn away, (sometimes in one day), handling the stones removes skin from your fingertips. It hurts just thinking about it. Prepare yourself for seeing a lot of useless gloves lying around with holes in the fingers.

Spend the money on a 4-ft., 3-tube level; the time saved and accuracy provided more than justify the expense. Wall surfaces will never be perfectly flat, and when you check for level and plumb, the level must span four or five stones. This isn't possible with a shorter level.

The construction sequence—Building the wall requires proper design and construction of the footings. Since we had continuous vertical reinforcement from the footings up into the wall, we placed the rebars before pouring the footings. As the concrete began to set, we chiseled a key way—a groove 1 in. wide by $1\frac{1}{2}$ in. deep—down the center of the footing. This provided a better bond between the footing and the first course of stone.

As the wall begins to go up, make sure there are enough stones stacked within reach of the

With the first course of stone and concrete in place, top, the forms have been removed. Note the width and depth of the concrete footing and the size and spacing of the reinforcing steel. The string line is a horizontal reference 4 ft. above the top of the footing supported by stakes 8 ft. high. A plumb bob is dropped from this string to plumb the wall. For a proper structural connection between the two poured layers, the top of each layer is left rough, with some stones and reinforcing steel projecting up. Each stone must be placed so that a relatively flat bed is created for the stones above. After the second course is laid, above, the metal forms are still in place on the projecting retaining wall to the left rear. The concrete has already been scraped away from these stones with a trowel. Each stone is set so that its weight is distributed over at least two stones below. Unless a control joint is required, joints are never placed directly above each other because this leads to cracks.

55-gal. drums, which should be convenient to the section of wall being worked on. One or two people should mix concrete while others fill the drums with water and stones. At the end of the day, we refill the drums for the next day.

We often worked on two different parts of the wall at once, a team completing one course while another started the next. We find the optimum team to be two people filling the forms, with one strong assistant to lug the stones, mix and haul concrete, keep the drums filled, and point the finish side of the wall once forms are removed and shifted to another section.

After the footing has cured, use a chalk line to mark the inside and outside surfaces of the wall. The first row of forms is set on these lines. We used 8-ft. 2x4 stakes with string running the entire length of the wall as a plumb guide. Adjust the stakes so that a plumb bob held from the string touches the outside form. Using 8-ft. stakes allows you to move the string up the wall as construction progresses.

The curved part of our wall was more difficult to lay out. When we were drawing the building floor plan, we calculated the radius of the inside

of the wall. After locating the center point of this curve on the ground, we marked the footing with string and chalk. We relied on the level for keeping this part of the wall plumb.

Whether you are starting at the footing or on top of a previously completed course, you should begin by hosing down the surface and removing any loose material. This ensures a strong bond. When the surface is clean and wet, set the forms in place. For the first course, this simply means setting them down on the footing with the wooden spacers in place to make sure the form is tightened to the proper dimension. For subsequent courses, the forms have to be clamped in place.

Metal forms can be erected quickly. Hold the form in place, and tighten down the turnbuckle that is just above the metal sides. Wooden spacers help prevent over-tightening. After many setups, our forms began to bend at the vertical pipe, causing the bottom to bulge out slightly. To prevent this we drilled holes in the sheet metal and improvised wire form ties that were tightened by twisting the ends around nails on the outside of the forms. Thicker sheet-metal

The exterior of a masonry wall must be kept as smooth and clean as possible for waterproofing later. In the foreground, mortar is being applied for a smooth finish. In the background, wood forms are in place on the curved part of the wall. Temporary wood forms made of 2x12s can be used when a lot of people are willing to help and additional formwork is needed.

forms, 10 or 12 gauge instead of 14 gauge, would also prevent bulging.

Wood forms take a little more time to set up. The two sides are connected by furring strips nailed into the top and by wires threaded through the form at the bottom.

After the forms are in place, use a trowel or shovel (trowels are easier to handle) to lay approximately 1 in. of concrete on the footing or previous course. Begin a layer of stones by packing them against both sides of the form and then filling in between with small stones and rubble. Make sure each stone has a face set into the bed of concrete. Some stones may be large enough to touch both sides of the form so that center-filling isn't necessary. Pack the stones as tightly as possible, allowing minimal space for concrete. This is the best way to save money, since cement is your most expensive construction ingredient. Once a stone is laid in place, do not move it. Movement weakens the bond. Always have stones of various sizes and shapes on hand, so that placing stone around the vertical rebar is no problem.

Once the first course of stone is complete, add another layer of concrete. We often shoveled concrete into the form, and then agitated it with a small trowel, to fill gaps between the stone. Don't worry about concrete running out of gaps at the bottom of the form. This may happen at first if the mix is wet, but as you pack stones and more concrete firmly against the form this waste will be minimized. Pack concrete firmly around all of the rebar, both vertical and horizontal, to bond the masonry and steel.

Fill the entire form in this fashion until the stone projects a few inches above it. This provides a better joint with the next course and makes it easier to clamp the form to the protruding rocks when it is moved. When proceeding smoothly, three people were able to fill a form 12 in. high and 4 ft. long in about half an hour.

If you build a below-grade wall as we did, each side of the form should be packed differently. For the earth side of the wall, try to achieve a smooth finish for a waterproof coating by packing the concrete in tightly. On the finish side, consider final appearance as you select and pack stones. Packing the concrete is not as critical on this side since the wall will be pointed after the forms are removed. However, you won't be able to see how the faces of the stone fit together on the finish side of the wall until the forms are removed. As part of the learning process you should start the wall at an end that will not be highly visible, because until you've gained some experience you won't be able to visualize the fit of the stones.

How long the forms must be kept in place varies with the weather and the concrete mix, but they must be removed soon enough to tool the joints to a finished state. We never keep our forms in place longer than two hours, but you will have to test this timing for yourself, since the variables of mix, weather and stone may differ. Detach the metal forms by loosening the turnbuckle and removing the nails from the wire end. Pull the wire through the wall and lift off the form. Removing the wood forms requires only pulling out the wire at the bottom, since we keep the top wood strips in place. Scrape the form clean, give it another coat of oil, and set it again atop the wall. The forms need be hosed down only at the end of the day.

Once the wall surface is exposed, it begins to dry quickly, so the joints should be tooled tight to the stone immediately. On the finish side, remove concrete around each stone with a small pointing trowel. Recessing the joints in this way exposes more of each stone, making them the prominent elements in the wall and giving it a laid masonry appearance. On hot days we keep the wall wet by hosing it down with a fine spray as we work and at the end of each day. The slower the concrete dries, the stronger it cures.

To give an idea of required construction time, it took us six weeks to complete a wall 60 ft. long and 8 ft. high. The wall cannot be thoroughly cleaned until it is completed, because concrete continues to spill down the sides during construction. We waited until the building was enclosed before sandblasting the inside of the wall. This is a quick method, but you must be equipped with gloves, goggles, proper clothing, a compressor and fine sand. Scrubbing with muriatic acid and a stiff brush is another method of cleaning that removes most of the surface dirt and excess mortar. Once the wall is clean, everyone is surprised by the beauty of the color and texture in the stone. □

The Point of Repointing
Renewing tired masonry joints can bring an old brick building back to life

by Richard T. Kreh

Brickwork deteriorates. This is inevitable but not, fortunately, irreversible. Replacing bad bricks and repointing with new mortar can restore a building's original integrity. And repointing, known as tuck pointing in the trade, is a job that can be handled by a careful homeowner. The resulting pride and satisfaction can be as substantial as the money saved.

In any repointing job, the first step is to determine exactly what the problem is. In some cases, the mortar simply has worn out. Old mortar often contains no portland cement, just lime and sand, with animal hair as a binder. These mortars can soften and break down over time, losing their ability to seal joints or adhere to brick. When this happens, they need to be replaced.

In other cases, improper flashing or structural damage to the roof has allowed water behind the brick. This water can be absorbed by the brick and mortar. As the wall dries out, the moisture escaping can cause the bricks to chip and the mortar to pop out. This problem is even more severe in colder climates where the expansion and contraction of freezing and thawing water accelerates the process. Sometimes the ground beneath the wall has settled. This can cause cracks in the wall, which might eventually require rebuilding. Before any actual repointing

Richard T. Kreh has over 33 years of experience as a mason and author of masonry books.

is done, find out why it needs doing, and correct these conditions first. If you don't, all the effort and expense of repointing will be wasted.

Preparation—Spring and fall are usually ideal times to do repointing work, because the best temperature for curing mortar is 70°F. Repointing can't always be done when conditions are perfect, but if you have a choice, arrange to do the work when they are at their best.

Never repoint when temperatures are likely to dip much below 40°F unless you can protect and heat the wall. Mortar that freezes before it cures will be brittle, and it will pop out when the wall warms up. Don't fall into the trap of repointing on a warm winter day and leaving the mortar to freeze overnight. If you hire a contractor to do the job, remember that he will schedule it to suit himself. He has a living to make year-round. Be sure to agree in writing that he will not attempt to work in freezing temperatures.

In hot weather, try to work on the shady side of the building to keep the mortar from drying out too fast. You might need to keep the wall damp (not soaked) with a fine mist from a garden hose or tank sprayer. (Never wet a wall in cold weather, because a freeze would cause the mortar to pop.)

For years, sandblasting has been used to remove dirt and old paint before repointing. Sandblasting is the continuous bombardment of

a masonry surface with abrasive particles sprayed through a nozzle in a high-velocity stream of air. Silica sand, the aggregate often used, cuts away part of the brick's surface. Cutting through this outer surface can cause permanent damage. Most disasters occur when inexperienced operators do their own sandblasting. If your brick face needs to be cleaned and you decide to use this method, hire a reputable contractor who specializes in such work.

Clear, liquid silicones can be applied to brickwork after it has been sandblasted and repointed. When reapplied periodically, they will protect the brick face for years. This treatment is especially recommended for historic buildings. Silicones are available from building-supply dealers. Follow manufacturers' instructions when applying them.

Repointing—You can't repoint without making a mess. There will be dust, dirt and mortar droppings to contend with. If you plan ahead, though, you can make cleanup a lot easier. Construction-grade rolls of 4-mil polyethylene plastic are ideal for protecting windows, doors, and flowerbeds and shrubbery near the house. Work on one area of the house at a time, and clean up as you go along. If you have to rent scaffolding, do all the high work first, so you can save money by returning it as quickly as possible.

The traditional method of removing mortar

Repointing Mortar

You don't need an especially strong mortar for repointing work. Repointing mortars need to bond with the old brick, seal the joints and match the original mortar as closely as possible in both strength and appearance. This is best done by a mortar with a much greater percentage of lime than of portland cement. Mortars with a high percentage of portland cement are often harder than brick; the brick gives under stress rather than the mortar. The result is _spalling_, a cracking and flaking of the brick face. High-lime mortars form a resilient cushion on which the brick can rest. They shrink very little, hold water well during repointing, and are easily worked. They can also heal small hairline cracks when the wall is moistened. During this process of autogenous healing the hydrated lime dissolves in water and is then recarbonated by the atmosphere's carbon dioxide, sealing the crack.

For an excellent all-around mortar with good bonding ability, use Type S hydrated lime, Type 1 portland cement, washed bank sand and potable water. Mix 1 part Type 1 portland cement to 2 parts Type S hydrated lime to 8 parts washed building sand. If the sand is coarse (sharp), add an extra half-part lime to improve workability and bonding. This mortar will test out at about 300 psi after it has cured—strong enough. —_R.K._

The author repoints a brick wall.

A

B

C

D

Removing mortar joints—Mortar joints can be removed using a power grinder (A), but it takes a skilled operator to avoid damaging the brick. To remove the joints by hand, use a plugging chisel (B, top) specially shaped for the job. The mason's all-purpose chisel (B, bottom) has other uses. Cut the joint out squarely (C). Brush out mortar particles with a soft-bristled brush (D). Use the same brush to dampen the joint before repointing.

joints is to use a hammer and chisel. Today, power grinders are often used to speed up the process (photo A). Even with a trained operator using the machine, however, it is all too easy to damage brick edges. The old way is best; repointing is not a job that can be rushed.

To remove mortar by hand, use a joint (or plugging) chisel (at the top of photo B). It has a specially-tapered blade that cleans mortar from the joint without binding or chipping the brick surface. It should be available at any store that stocks masonry supplies. Cut the joint out to a depth of ¾ in. to 1 in., depending on the depth of the deteriorated mortar (photo C). Cut it out square, not in a V shape.

After cutting, measure with a rule to be sure you've reached the correct depth. Brush all dust and mortar particles out of the joint with a whitewash or wallpaper paste brush (photo D). Next, use a wet brush to dampen the joint. Don't soak it, or the new mortar will not adhere. The idea is to prevent the repointing mortar from drying out too quickly.

Tuck-pointing mortars should be prehydrated to improve workability and reduce shrinkage. To do this, mix the cement, sand and water thoroughly in the proper proportions. Then mix again, adding enough water to produce a damp (but unworkable) mix that will retain its form when pressed into a ball in your hand (photo E). Leave the mortar in this dampened condition for an hour or two, then add enough water to bring it to a consistency that is somewhat dryer than

regular masonry mortar, as shown in photo F.

When the mortar is ready, apply it to the joint. Use a thin, flat piece of steel, called a striking tool, to press the mortar firmly into the moistened joint, filling it out (photo G). If the joint has been cut out to a depth greater than 1 in., apply new mortar in two layers. Allow time for the first layer to set before beginning the second.

Don't tool a finish on the joint until the mortar has set enough so that it won't smear. If your thumb leaves a clear impression on the mortar, it's time to tool. Doing so too early will cause the joint to streak or sag. Waiting too long will result in a black mark, which is caused by a reaction between the slicker and the chemicals in the mortar. (This is known as "burning the joints.")

Finish the joint by tooling a flat, concave or grapevine shape on the surface. (A grapevine is formed with a steel tool that leaves an indented line in the mortar.) With a little practice, it won't take you long to get the hang of finishing. After the mortar has set well enough to prevent smearing, brush off the wall to remove all mortar particles (photo H).

Replacing bricks—In most repointing jobs some deteriorated brick will have to be cut out and replaced. Use a plugging chisel to remove the mortar joint, and an all-purpose mason's chisel to cut out the soft brick (photo I). Brush away all dust particles. Use mortar to butter the replacement brick and all sides of the wall cavity in which it will be placed. This prevents any

voids in joints after the new brick is inserted. Push the new brick in place and point the mortar joints (photos J-L).

You should make an effort to match the original brick. Look around the site or in adjacent outbuildings for bricks left from the original construction. If they haven't softened with age, they would be the best choice. You can also try your local building-supply dealer. Many companies make new bricks that look like old, handmade ones. Take a sample of the original brick with you and compare for size as well as color.

Salvaged brick may also work, but be careful. If reclaimed bricks are soft, they won't last long once exposed to the elements. You can make sure a brick is hard enough to use in repointing by holding it in your hand and tapping it lightly with a hammer. A soft brick will give off a dull thudding sound; a hard one, a metallic ring.

Cleanup—If you've been neat, only a little cleaning will be necessary after repointing. Wait at least two weeks to let the mortar cure completely, then scrape particles and clumps off the wall with a wooden paddle or a chisel. Hose down the wall and give it a good scrub with a stiff-bristled brush.

If stubborn stains remain, or you have splashed mortar down the wall, you may need to scrub it with a solution of 1 part muriatic acid to 20 parts water. Wear hand and eye protection, cover the flowers and shrubs near the house, and hose off the wall when you're done. □

Mixing mortar and repointing—Basic tuck-pointing mortar is composed of portland cement, lime and sand. Mix the dry ingredients in the proper proportions, then dampen until a ball of the mix squeezed in your hand retains its shape (E). This prehydration improves the mortar's workability and reduces its shrinkage. After an hour or two of prehydration, add more water to bring the mix to its final consistency, somewhat dryer than regular masonry mortar. After wetting the joint, press the tuck-pointing mortar firmly into it with a slicker (G), then tool the finish. Once the mortar has set and won't smear, brush off any particles (H).

E

F

G

H

I

J

K

L

Replacing deteriorated brick—Use the plugging chisel to remove the mortar around the bad brick, then the all-purpose chisel to cut out the brick itself (I). Brush away all mortar and brick particles (J). Give both the replacement brick and those still in the wall a thick coating of mortar, and press the brick into place (K). Then point the mortar joints (L).

Rammed-earth walls are built with subsoil that's been excavated from the site. After the topsoil has been cleared away, a foundation is poured, below. Rebar protruding from the foundation will connect the concrete columns, and the ledge will support the forms while the wall is being tamped. At right, a rototiller blends the subsoil with 7% to 15% portland cement. Once the mix is wet, it has about two hours of workability. Facing page: freshly rammed wall sections. The 6-in. gaps between them are for concrete columns. Keyways will lock these columns into the walls. The short wall sections in the foreground will support windows.

Rammed Earth

An ancient building method made easier by new technology

by Magnus Berglund

David Easton likes the earth. Enough that he hangs his hat on walls made of it. Also on these earthen walls hangs his reputation for building comfortable, good-looking houses. An engineer, Easton is one of a handful of people in this country building with rammed earth.

Rammed-earth walls are made by compacting soil. Using plywood forms, a front-loading tractor, a pneumatic backfill tamper and moistened, cement-laden dirt from the site, Easton can build these rock-hard walls quickly and efficiently. As a member of Easton's crew, I build rammed-earth houses, and I think dirt is a superior building material. I also live in a rammed-earth house, and I've had a chance to see it react to its environment. I know from first-hand experience what it's like to live inside those walls, and I'm more enthusiastic about earth buildings as a result. Easton puts it this way: "Rammed-earth walls are dry, fireproof, rotproof, soundproof and termite-proof. They are comfortable, durable and economical.

Rammed-earth houses cost less, use less energy, emit fewer toxins, and will last longer than conventionally built houses of today."

There is nothing new about building with rammed earth. Pliny the Elder refers to it in his *Natural History*, and a 1772 French treatise credits the Romans with introducing rammed-earth construction to England. The Chinese used it in their Great Wall, and the first house erected by Spanish settlers in St. Augustine, Fla., is reputed to have been made of rammed earth. During the 1940s, interest in rammed-earth ran high as prefabricated building materials were diverted to the war effort.

What is new to rammed-earth construction is the use of modern formwork, technique and power tools. A 1938 study by the U.S. Department of Agriculture concluded that three men using shovels, buckets and a hand tamper could complete about 70 sq. ft. of a 14-in. thick wall in one day. Using a tractor, a pneumatic tamper and forms that we designed to assemble and

disassemble in a short time, our crew of five has rammed as many as 23 wall units in one day—about 40 tons of dirt. At that rate, we make walls for a 1,000-sq. ft. house in about 16 working hours.

Preparatory work—A typical project starts with site excavation and the pouring of a 15½-in. wide, reinforced-concrete perimeter foundation (photo above left). The foundation is this wide not only because the 14-in. thick earth walls require the extra width, but also because of the tremendous weight the concrete must carry. A ¾-in. by 1½-in. ledge on the top edges of the foundation supports the form panels as the walls go up.

Dirt from the excavation is stockpiled outside the foundation area into separate piles of topsoil and subsoil. We remove roots and large rocks from the subsoil pile, and add water, preparing the mix right on the ground with a rototiller (photo above right). Because of its organic

content, we don't use topsoil as a building material. At a favorable site, the subsoil needs a quantity of portland cement added to the raw mix as a binder equal to 7% of the compacted volume of the soil. For instance, our average wall unit is 6 ft. long, 7 ft. high and 14 in. wide, and has a volume of about 50 cu. ft. The soil compacts down to about half its volume during ramming, so the average wall section requires 100 cu. ft. of soil mix containing 3½ cu. ft. of cement (7% of 50 cu. ft.) under ideal soil conditions. If the soil is especially sandy or has an abundance of decomposed rock, the amount of cement can go as high as 15%. Not all rammed-earth builders add cement. But Easton believes in its value as a stabilizer to decrease the chance of weather eroding the walls.

Forms—Most historical accounts of rammed-earth construction describe variations on the slip form for building walls. The form is moved around the perimeter of the building, rising in about 2-ft. increments. Our method is different. We build full-height, freestanding boxes, 14 in. wide and from 30 in. to 90 in. long, which correspond to our typical 30-in. rafter spacing. The forms are set up separately for each section of wall and then dismantled as soon as dirt has been rammed to the top. Our crew uses several forms at once; while one is being assembled, a second is being filled and a third is being taken down. This way, the compressor and tractor are always in use.

The forms are built from 1⅛-in. tongue-and-groove plywood panels lined with Masonite and have ¾-in. plywood end-boards (drawing, facing page, left). Holes are drilled in the ends of the panels and pipe clamps are run through them to clamp the 2x10 whalers in place. Each form is squared and plumbed before the ramming begins.

We used to avoid making rammed-earth corners and chose instead to pour concrete columns. But recently we've developed a corner form that's sturdy, yet easily assembled. It relies on opposing pairs of bolted and cross-braced 90° whalers for its strength (photo and drawing, facing page, right).

Soil is dumped into the form box by the front-loading tractor in 6-in. to 8-in. deep increments (photo left) and is pounded down to about half its original volume. If this layer is too thick to begin with, the soil on the bottom won't get properly compacted. Ramming is done with a Thor 33 pneumatic tamper. It has a 5-in. diameter, 33-lb. head. The only real trick to using it is to relax, and listen for the thud, thud, thud that reverberates in the form with a tight, ringing sound, telling you to move on.

As soon as the last course of mix is pounded down, the forms are stripped away, and the wall is finished. No treatment is necessary. If

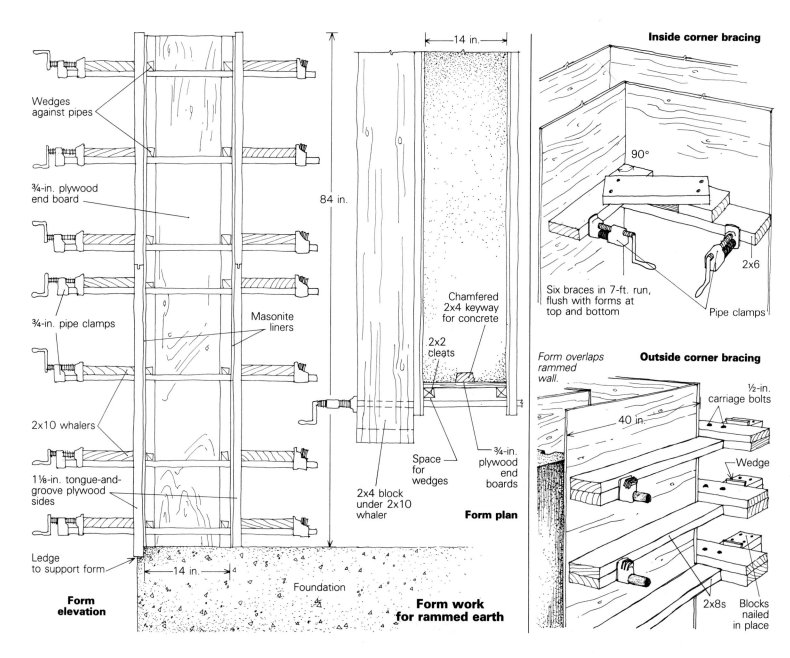

Inside corner bracing

90°

2x6

Six braces in 7-ft. run,
flush with forms at
top and bottom

Pipe clamps

Form overlaps
rammed
wall.

Outside corner bracing

½-in.
carriage bolts

40 in.

Wedge

2x8s   Blocks
nailed
in place

Wedges
against pipes

¾-in. plywood
end board

84 in.

⌐14 in.⌐

Chamfered
2x4 keyway
for concrete

2x2
cleats

¾-in. pipe clamps

Masonite
liners

2x10 whalers

1⅛-in. tongue-and-
groove plywood
sides

Space
for
wedges

¾-in.
plywood
end
boards

2x4 block
under 2x10
whaler

Form plan

Ledge
to support form

14 in.

Foundation

**Form
elevation**

**Form work
for rammed earth**

Illustrations: Frances Boynton

you leave the wall alone, it will be the color of
the mined soil, with a hard, textured surface
like tempered Masonite.

The walls should be left alone for two weeks
while their outer edges harden enough to resist
construction wear and tear. Treat them like
drying concrete during this period, spraying
them lightly from time to time, and sheltering
them from the sun with plastic sheeting. Don't
let them get wet if there's a danger of a
freeze—wet walls that freeze will slough off
their smooth surface and require cosmetic
plastering. Keep the walls covered when it
rains until the roof is raised.

It's pretty hard to do anything so wrong that
it will have an adverse effect on the walls'
strength, but you do want them to look good so
that people who come to see your dirt building
will tell their friends how clean it is.

Interior finish—Once cured, rammed-earth
walls can be left as they are, or treated any way
you like—painted, plastered, drywalled, stuccoed or even wallpapered. Most of our customers prefer the natural look, with nothing more
than a clear sealer applied. The sealer prevents
dusting and superficial flaking, and provides

extra protection from moisture. We have used
Thompson's Water Seal, linseed oil, Varathane
and Shur Bond. So far, each of these products
has worked well, but none of them seems clearly superior.

When a surface finish is needed, Easton prefers dagga, a thin plaster made with two parts
fine sand and one part soil from the site sifted
through a window screen, plus 8% plastic cement. We add enough water to bring the mix to
the consistency of soft ice cream, and then
trowel it onto a moistened rammed-earth wall.
There's no need for a wire base.

In addition to being a compatible color, the
dagga closely matches the coefficient of expansion of the earth wall, lessening the chance that
it may crack and flake off when the wall moves
with temperature changes. This problem of in-
terface, the point at which materials with different qualities touch, is a serious design consideration in earth houses. The best solution is to
limit the number of different materials that go
into a house.

Windows and doors trim out easily with
standard finish lumber because of the ability of
rammed earth to accept and hold galvanized
nails. Windows are trimmed in a conventional

A corner form. Bolted and blocked 2x8
whalers on the outside oppose diagonal 2x6
bracing on the inside (drawing above).

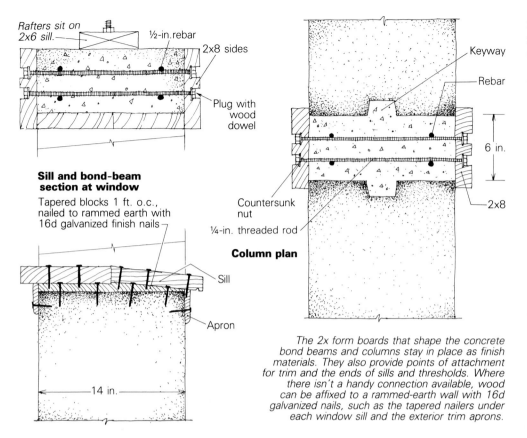

Rafters sit on 2x6 sill.

½-in. rebar

2x8 sides

Plug with wood dowel

Sill and bond-beam section at window

Tapered blocks 1 ft. o.c., nailed to rammed earth with 16d galvanized finish nails

Sill

Apron

— 14 in. —

Keyway

Rebar

6 in.

2x8

Countersunk nut

¼-in. threaded rod

Column plan

The 2x form boards that shape the concrete bond beams and columns stay in place as finish materials. They also provide points of attachment for trim and the ends of sills and thresholds. Where there isn't a handy connection available, wood can be affixed to a rammed-earth wall with 16d galvanized nails, such as the tapered nailers under each window sill and the exterior trim aprons.

manner with a sill nailed over tapered blocks, finished with an apron below, as shown in the drawing at left.

Concrete components—Rammed-earth walls can be built adjacent to one another, but without reinforcing they won't meet the stringent California earthquake-conscious codes. To comply with the demands of our building department, Easton has developed a concrete post-and-beam structure to carry the weight of the oversize roof beams, wide overhangs and sod roof. He erects the wall sections 6 in. apart. A vertical keyway is formed at the end of each section, and four ½-in. rebar studs protrude from the foundation into each gap for a vertical tie to the bond beam. Then the gap between the walls is covered on both sides with 2x8 cedar or redwood. We clamp the boards to the openings between the walls with ¼-in. threaded rod bolted to both sides. Finally, we fill the void with concrete. The boards remain as part of the finish. The ends of the threaded rods are snapped off with a length of ⅜-in. I.D. water pipe. The pipe slips over the end of the rod, and after a few back-and-forth bends the rod snaps off, even with the recessed nut. The hole in the 2x8 is capped with a dowel plug.

Electrical boxes and conduit are attached to the inside face of the 2x8s before the pour. The wiring is then run along the top of the bond beam to junction boxes that feed the post-mounted outlets.

The bond beam resembles a topside perimeter foundation. It joins the wall sections and spans doors and windows. It is formed by more 2x8s, which run horizontally along the top of the walls and are held together with threaded rod and 1x2 spreaders at the top. It's poured at the same time as the columns, and anchor bolts are set in the wet concrete for a 2x6 sill plate that will connect the rafters to the walls.

With the sod roof, the rafters on our houses carry loads of up to 100 lb. per sq. ft. This calls for some big beams. Our favorites are local pines and firs that we mill ourselves. These are usually left exposed on the interior. The roof deck is then built up with a hot-mop process or a vinyl swimming-pool liner to receive a load of topsoil and rolls of sod. Overhangs are usually about 3 ft., to provide shade and to protect the walls from rains.

Soil test—When someone comes to David Easton to ask about building a rammed-earth home, one of the first questions is, "Will my soil work?" Easton's answer is that he can usually make it work if it's not composed entirely of solid rock. Even though the ideal soil for rammed earth is a mix of 30% clay and 70% sand, we've found that a 25% to 40% clay constituent will still make a good wall. If the soil

Columns and bond beams are filled with pumper-delivered concrete. Quarter-inch threaded rod holds the form and finish boards together during the pour. After the concrete sets up, the threaded rod will be recessed into the 2x8s and plugged with dowels. The concrete end columns in this house required forms with extensive diagonal bracing.

has too much clay, we bring in some sand. If there's too much sand or decomposed rock in the soil, we add more portland cement.

To test for your sand-to-clay ratio, fill a quart jar halfway with your subsoil and top it off with water. Shake it up well and let it settle overnight. This doesn't make you a soils engineer but you should be able to see the layer of sand on the bottom, the clay next and the lighter organic materials on top.

You can do a quick field test for the moisture content of your mix. Ball some dirt up in your hand and squeeze it tight. If it sticks together, drop it from a height of 5 ft. onto a hard surface. If it breaks into loose dirt, it's ready to be made into a house. If it doesn't, it's too wet.

If you want to take on a rammed-earth project, my advice is to jump in and give it a try. You'll learn more by tamping your soil for a day than you will from a month of reading about it. Make a small practice wall right on the ground. If you get cracks in the walls, you've used too much clay, so add some sand. If the corners crumble when you brush against them lightly, you've mixed in too much sand; add some clay or cement. If the walls are honeycombed, marked by loose areas that fall out along compaction lines, then the mix is too dry or you tried to ram too much at a time.

Inside the house—Easton doesn't see any reason to install conventional heating and cooling in his rammed-earth houses. Orienting the house to take full advantage of sun and shade, along with lots of south-side glazing, is always a design priority. The 14-in. thick walls function like 12-hour thermal clocks; it takes half of the day/night cycle to transfer heat through the rock-hard thicknesses, making rammed-earth houses especially well suited to climates marked by wide temperature swings.

Under the patterned floor slabs, each house includes a passive-solar convection loop of 12-in. sub-grade culvert connecting the cool, north interior wall to a south-side greenhouse. The sod roof also contributes to passive cooling by insulating the house, and by acting as an evaporative cooler in the summer. The result is a temperate home which, rather than sealing off its occupants inside a vapor barrier, circulates fresh air and controls inside temperatures without the use of mechanical systems.

The ten houses built so far by David Easton's Rammed Earth Works retain much of the flavor of the traditional California adobe, while their exposed rafter ends, wide eaves and natural wood interiors recall the Craftsman style (photos at right). With their earthen walls rising from the same soil, topped by native sod, each one looks as if it just grew there. □

Magnus Berglund lives in Wilseyville, Calif.

Site-cut pines and firs are transformed into rafters, beams, cabinets and countertops in Easton's rammed-earth houses, above. These earthen walls are finished with a coat of dagga, a thin plaster made with local soil. Right, a sod roof keeps this 1,300 sq. ft. house cool on summer afternoons while ridge-mounted sprinklers water the grass.

Surface-Bonded Block

A strong, fast and inexpensive alternative to poured-concrete or block-in-mortar walls

by Paul Hanke

Pouring concrete walls is a difficult and risky business, and I don't recommend it for the inexperienced. Even professionals sometimes have forms let go, creating various degrees of disaster and pandemonium on the site. Laying up block with mortar has drawbacks, too. It is time-consuming, it takes practice, and the result isn't especially strong.

Surface-bonded block, on the other hand, suits owner-builders to a tee, and can be a less expensive alternative for professionals. It is a method of laying concrete blocks without mortar, then troweling both wall surfaces with a

Paul Hanke is a designer and draftsman at Northern Owner Builder, in Plainfield, Vt.

portland-cement coating laced with chopped fiberglass for strength. Built on standard footings, surface-bonded block walls can be used below and above grade, for foundation walls and for finished living spaces. The method is fast and reliable. It requires no particular skill, and the finished wall is stronger than a block-in-mortar wall.

Surface bonding was originally developed as a low-cost construction technique for self-help housing. A USDA booklet on the subject (Information Bulletin No. 374, now out of print) shows a 12-year-old boy doing a successful job after 15 minutes of practice. Even professional masons are reported to be 70% more productive using this method than laying up block in

the conventional way. The USDA estimates that stacking and bonding 100 blocks would take a person an average of 7.4 hours. Several years ago, two friends of mine, Chapin and Donna Kaynor, built an earth-sheltered house using this technique. It took their crew of four inexperienced people, some of whom worked only part time, less than five days to stack and bond about 1,200 blocks.

Strength and cost—Stacked blocks coated with bonding mix have an average tensile strength (ability to withstand longitudinal stress) of from 300 psi to 500 psi, according to lab tests conducted by the USDA and the University of Georgia. This is about equal to the strength of unreinforced concrete, and is six times stronger than block laid up with ordinary mortar joints. Mortar has very little adhesive power, and virtually no tensile strength. Its main purpose is to level blocks between courses. Because the weakest part of conventional block walls is the bond between block and mortar, these joints tend to crack, making water seepage a problem. A surface-bonded wall, with its seamless outer coating, is much more watertight (though the coating alone should not be relied upon below grade).

Having a block-in-mortar wall built costs about twice the price of the materials, plus footing and reinforcing. In our area, concrete foundation walls currently cost around $95 to $105 per cubic yard poured in place, including formwork and labor. A typical full basement accounts for about 5% of the cost of a house, or over $3,200 for the average $65,000 home. The builder using surface-bonded block can save as much as 35% to 40% of this figure.

Estimating materials—To build surface-bonded walls, you will need standard hollow-core concrete block, surface-bonding mix, galvanized corrugated brick-ties for shims (the Kaynors used about 250 for 1,200 blocks), threaded steel rod and connectors, a few sacks of mortar, and the rebar and concrete for footings. Order 8x8x16 block (about 65¢ each) for walls above grade or foundation walls that will extend less than 5 ft. below grade. Order 8x12x16 block (about $1.05 each) for a foundation wall deeper than 5 ft. Use the USDA table at right to estimate the number of blocks you need. Be sure to add extra block for reinforcing pilasters, the column-like buttresses used to strengthen the walls (discussed below), and half-length block, which you may need for the door and window openings. Order 5% to 10% extra to make up for waste and breakage. Have your block delivered to the center of your work area if at all possible, or deposited in strategic piles around the perimeter. To save time and effort, don't carry those heavy blocks any farther than you have to.

As the table shows, nominal 8x8x16 block is actually 7⅝ in. by 7⅝ in. by 15⅝ in. to allow for ⅜-in. mortar joints; so you can't figure in exact 16-in. modules when laying block dry. Having to calculate with fractional numbers would be a real headache, but estimating tables supplied by the USDA or surface-bonding mix manufacturers greatly simplify the task.

Bonding mix, which comes dry and includes the chopped fiberglass strands, is sold in bags of various sizes. A 50-lb. sack will cover about 50 sq. ft. of wall. Check the exact coverage when you order, and allow about 10% for waste from broken sacks, mixing and troweling. A 50-lb. sack of grey-colored mix currently costs about $14 in Vermont. White is about $17 per sack. You can also mix your own, as explained at right. In addition to the bonding mix, get enough sacks of mortar to lay the first course of block, plus a few extra sacks to use in spots where you need to shim more than ⅛ in.

I recommend using ⅜-in. threaded steel rod to connect the sill or top plate at the top of the

wall to the footing below. It is available in 2-ft. and 3-ft. lengths at hardware stores. Threaded rod is expensive, but it makes a secure connection. You'll also need connectors to join the lengths of rod, and nuts and washers to secure the wood sill to the top of the wall. The rod isn't for concrete reinforcement, but to tie footing, foundation and framing together to resist uplift forces. The block cores that contain the rod don't require filling with concrete.

The alternative to running threaded rod all the way up through the wall is to fill the cores at the top of the wall with concrete two or three courses deep every 4 ft., and embed standard ½-in. J-bolts. Use screening to keep the grout from falling all the way to the footing, or stuff fiberglass insulation down the block core. This method works, but it will not provide a continuous connection from footing to sill, and will not resist uplift.

Footings—As a general rule, footings should be twice as wide as the wall above, and as deep as the wall will be thick. A standard 8-in. thick wall calls for a 16-in. wide footing. Pour 24-in. wide footings for either a 12-in. thick wall or a two-story house.

The bottom of the footings should be at least 12 in. below the frost line, and almost anyone can safely pour them. You can pour into shallow forms or directly into trenches of the proper size, provided that their sides and bottoms are of firm, undisturbed soil. Place two No. 4 (½-in. dia.) lengths of rebar near the bottom of a 16-in. wide footing. A 24-in. wide footing will require three lengths of rebar. Check codes for the rebar requirements in your area. Remember to widen the footings for pilasters.

Although you can mix your own concrete for footings, ready-mix concrete delivered to your site is best. Insert the lengths of threaded rod vertically into the concrete at the corners and pilaster locations, on both sides of all the door and window openings, and every 4 ft. to 6 ft. along the wall, as shown in the drawing on the bottom of the next page.

Stacking block—After your footings are poured and have been allowed to cure, you can begin on the walls. Using your batter boards and strings (see pages 10 through 12), drop plumb lines to establish the outside corners. Use the table to determine exact wall lengths, and allow an extra ¼ in. per 10 ft. for irregularities in the blocks. Measure the diagonals to be sure that your corners are square, and adjust if you need to. Snap chalklines from corner to corner as guides, and then lay and level the first course of block in a bed of thick mortar. Check the top of the first course with a 4-ft. level as you go. If a block is too high, tap it down with the butt end of your trowel; if it's too low, remove the block, add more mortar and reset. Don't put mortar in the vertical joints between blocks; just butt them tightly against each other. Some skill is required here. Take your time and do a good job.

The rest of the wall is simply stacked dry in a standard running bond—each block overlapping half the block beneath. Begin by

Mixing your own

Here is the USDA formula for preparing 25 lb. of your own bonding mix. A friend who investigated this option concluded that it costs about 65% as much as a comparable commercial mix.

19½ lb. portland cement (78% by weight), white or type I grey, which is more common. This is the glue that holds things together. It comes in 94-lb. sacks.

3¾ lb. hydrated lime (15%) for increased workability. It comes in 50-lb. sacks.

1 lb. glass-fiber filament (4%), chopped into ½-in. lengths. Use type E fiber or, better yet alkali-resistant type K fiber, available from plastic and chemical-supply dealers, building-material dealers or boatyards.

½ lb. calcium chloride flakes or crystals (2%), to speed setup time and harden the mix. It's available from agricultural-chemical supply houses. Calcium chloride is also used for salting roads.

¼ lb. calcium stearate (1%), wettable technical grade, makes the mix more waterproof. You can obtain it from chemical distributors.

Since the bonding mix sets rapidly after water and calcium chloride have been added, do not make more than a 25-lb. batch at one time (dry weight).

Begin by mixing the powdered ingredients, except for the calcium chloride. Add the glass fiber, and remix only enough to distribute the fibers well. Overmixing breaks the fibers into individual filaments, which makes application difficult. Be sure to wear a proper respirator. The chemicals are very corrosive, and you don't want to breathe fiberglass, either.

Mix the calcium chloride with 1 gal. of water, and slowly add this solution to the dry ingredients. Mix thoroughly. Add about ½ gal. more water, until the mix is the right consistency—creamy, yet thick enough for troweling. A mix that's too thick is hard to apply and may not bond properly. —*P.H.*

Wall and opening dimensions for surface bonding

Number of blocks	Length of wall or width of openings		Number of courses	Height of wall or openings	
1	1 ft.	3⅝ in.	1	0 ft.	7⅝ in.
2	2	7¼	2	1	3¼
3	3	10⅞	3	1	10⅞
4	5	2½	4	2	6½
5	6	6⅛	5	3	2⅛
6	7	9¾	6	3	9¾
7	9	1⅜	7	4	5⅜
8	10	5	8	5	1
9	11	8⅝	9	5	8⅝
10	13	0¼	10	6	4¼
11	14	3⅞	11	6	11⅞
12	15	7½	12	7	7½
13	16	11⅛	13	8	3⅛
14	18	2¾	14	8	10¾
15	19	6⅜	15	9	6⅜

Blocks sold as 8x16 are actually 7⅝ in. by 15⅝ in. to allow for the size of mortar joints in standard block construction. Remember that cement blocks are not uniform. Add ¼ in. to every 10 ft. of wall length to take this into account, and before beginning to build, make a trial stack to measure the precise height your wall will be.

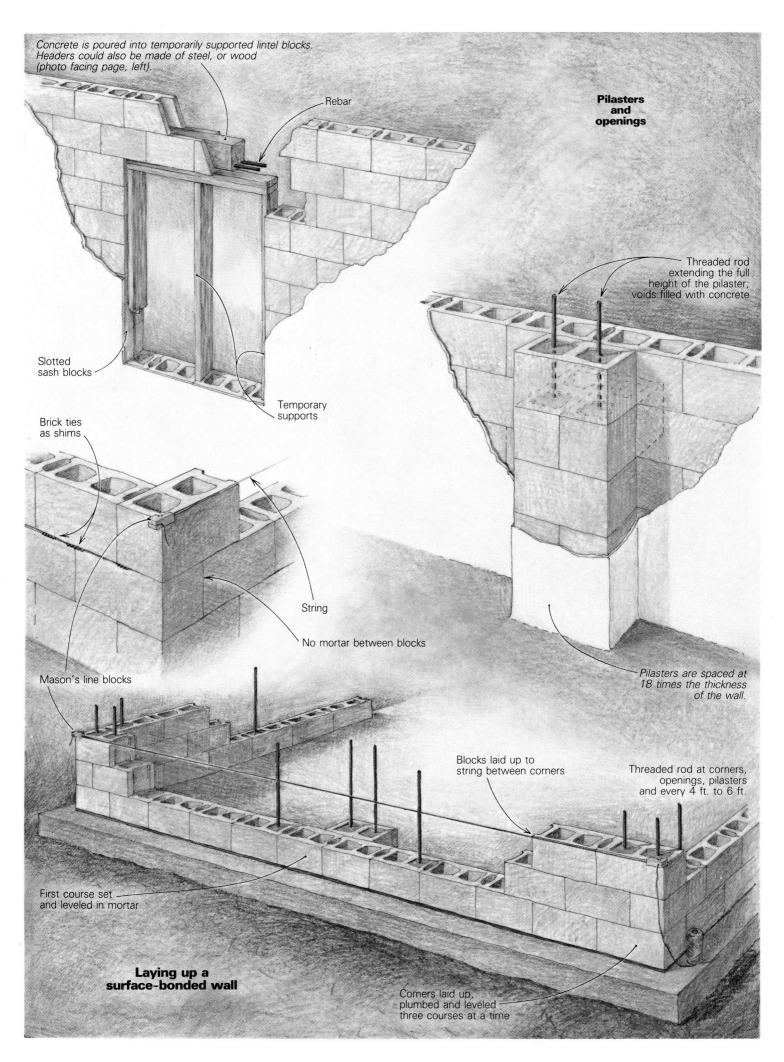

Concrete is poured into temporarily supported lintel blocks. Headers could also be made of steel, or wood (photo facing page, left).

Rebar

Pilasters and openings

Threaded rod extending the full height of the pilaster; voids filled with concrete

Slotted sash blocks

Temporary supports

Brick ties as shims

String

No mortar between blocks

Mason's line blocks

Pilasters are spaced at 18 times the thickness of the wall.

Blocks laid up to string between corners

Threaded rod at corners, openings, pilasters and every 4 ft. to 6 ft.

First course set and leveled in mortar

Laying up a surface-bonded wall

Corners laid up, plumbed and leveled three courses at a time

Illustration: Christopher Clapp; Photos: Chapin and Donna Kaynor

The first course of block is laid and leveled in mortar, as at right. The corners are built up three courses, then a level line is strung between corners. Dry block fills in up to it. Shims are used where necessary to keep blocks aligned. A mason's corner block, which holds a level line that is also the correct plane for the face of the blocks, is visible on the second course. Once the walls have been built up around openings, headers must be installed. These can be steel, concrete or wood, as above.

building up the corners three courses high. Check them for plumb with the 4-ft. level held vertically, and for level with a water tube. Then stretch a taut string between the top outside edges of each course of the built-up corners. Use mason's line blocks (available where you buy concrete blocks) to secure the line at each end. Fill in the length of the wall up to the string, and repeat the process every three courses, inserting metal shims as necessary to keep the wall level and plumb. If more than ⅛ in. of shimming is required, use mortar instead. Check the wall for plumb at least every three courses. Connect new segments of threaded rod as you go.

Pilasters—These are engaged columns that reinforce the wall against lateral forces and keep it from buckling under heavy loading. For basement walls, pilasters should be on the inside to resist the pressure of the surrounding earth. The Kaynors put theirs outside to get them out of the living area. They are tied in by rotating the blocks of every other course 90° so that they become a part of the wall itself (drawing, facing page). Threaded rod or rebar should extend through the block cores the full height of all pilasters. After the wall is laid up, fill the voids of the pilasters with concrete.

For above-grade construction, pilasters are usually spaced along the wall at a distance equal to 18 times the thickness of the wall (for example, every 18 ft. for a 12-in. wall, or every 12 ft. for an 8-in. wall), or on a shorter wall, at midspan. The pilasters on the house shown here are on 8-ft. centers for earth-bermed walls, which is probably a good precaution for any below-grade construction.

Weight-carrying beams should also be supported by pilasters at each end. Be sure that the beam pockets extend into the wall at least

3 in. to get good bearing surfaces. Once the beam is in place, you can continue dry-stacking blocks in the usual manner.

Openings—For doors and windows, just omit blocks in the proper locations. This is where half-blocks come in handy. With these, you don't need to cut standard block down to size. The blocks at each side should be slotted sash-blocks, which accept a metal or wood spline that attaches to specially made framing.

Headers are required above openings, as in any other type of construction, and they should be properly sized for their span and load. Consult standard tables, codes, or an engineer if necessary. Headers can be made of wood (photo above left), steel angles or U-shaped bond-beam blocks. You support the blocks temporarily over the opening you want to span, then fill their cores with rebar and concrete. Once the concrete has cured, remove the supports, and you have a solid beam.

Coating the wall—Once all the blocks are stacked and the pilasters are filled with concrete, the walls are coated with surface-bonding mix (see sources of supply, below). Commercial mix consists primarily of mortar and strands of fiberglass chopped into ½-in. lengths. Add water according to the instructions, and mix with either a garden cultivator or a mason's hoe (the kind with two large holes in the blade). The mix will cure in about an hour and a half, so don't whip up too much at one time. Hose down the block wall so that the mix won't dry out too quickly, then trowel on a 1/16-in. to ⅛-in. coat of the paste.

Both sides of the wall get surface-bonded. Use a hawk to hold a comfortable amount of the mix while you work, and press its edge against the wall to limit slop and spilling. Use a

plasterer's trowel, a steel trowel about 12 in. long, and work from the top of the wall down so you can moisten the block as you go if it begins to dry out in hot weather.

The USDA breaks the procedure down into four steps. First, with a series of sweeps of the trowel, spread the mix 2 ft. or 3 ft. upward from the hawk over a section about 5 ft. wide. Then even out the surface by going over the area lightly with your trowel slightly angled. Repeat these two steps over the area just below the block you've just covered, and cover as much area as you can in 15 or 20 minutes. Lastly, clean the trowel in water and retrowel the plastered area with long, firm, arced strokes to achieve a final, smooth surface.

The glass fibers bridge the joints between blocks, and the tensile strength of the wall increases as the concoction cures. Because the fibers are so short, the system won't work if you lay the block in mortar before coating the walls with the bonding mix. The fibers would be spanning almost their entire length, and this would destroy their effectiveness. The interior surfaces of walls can be textured with a light pass with a stiff brush. Mortar pigment can be added during mixing to color the wall. Surface-bonded walls can also be stuccoed or furred out, if you prefer something other than the bonding mix as a finished surface. □

Manufacturers of surface-bonding mix

Fiberbond Surface Bonding Cement: Stone Mountain Mfg. Co., Box 7320, Norfolk, Va. 23509.

Q-Bond: Q-Bond Corp. of America, 3323 Moline St., Aurora, Colo. 80010.

Stack & Bond: Conproco, Box 368, Hooksett, N.H. 03106.

Surewall: W.R. Bonsal Co., Box 241148, Charlotte, N.C. 28224.

Quick Wall: Quikcrete, 1790 Century Circle, Atlanta, Ga. 30345.

Building a Block Foundation

How to pour the footing and lay the concrete block, and what to do about waterproofing, drainage and insulation

by Dick Kreh

Foundation walls are often the most neglected part of a structure. But they are actually the most structurally important element of a house. They support the weight of the building by distributing its entire load over a large area. Apart from structural requirements, foundations have to be waterproofed, insulated and properly drained.

Although the depth of a foundation wall may vary according to the specific needs of the site or building, the footings must always be below the frost line. If they're not, the foundation will heave in cold weather as the frozen earth swells, and then settle in warm weather when the ground softens. This shift-

ing can crack foundations, rack framing, and make for wavy floors and sagging roofs.

Concrete blocks are composed of portland cement, a fine aggregate and water. They have been a popular choice for foundations because they're not too expensive, they go up in a straightforward way, and they're available everywhere. Block foundations provide adequate compressive strength and resistance to fire and moisture. They don't require formwork, and they're not expensive to maintain.

All standard blocks are 8 in. high and 16 in. long—including the usual ⅜-in. thick mortar head and bed joints. But they come in different widths. The size given for a block always

refers to its width. The size you need depends on the vertical loads and lateral stresses that the wall will have to withstand, but as a rule, most concrete-block foundations are built of 10-in. or 12-in. block.

Footings—After the foundation area has been laid out and excavated, the concrete footings are poured. Footings should be about twice as wide as the block wall they will support. A 12-in. concrete-block foundation wall, for example, should have a 24-in. wide footing. The average depth for the footing, unless there is a special problem, is 8 in.

Concrete footings for homes or small struc-

Chalklines snapped on the cured footing guide the masons in laying up the first course of concrete block for a foundation wall. The blocks are stacked around the site to minimize legwork, yet allow the masons enough room to work comfortably.

tures need a compressive strength of 2,500 pounds per square inch (psi). You can order a footing mix either by specifying a five-bag mix, which means that there are five bags of portland cement to each cubic yard, or by asking for a prescription mix—one that is ordered by giving a psi rating. Some architects and local building codes require you to state the prescription mix when you order. Either way, footing concrete is a little less expensive than regular finishing concrete, which usually contains at least six bags of portland cement to the cubic yard. The six-bag mix is richer and easier to trowel, but isn't needed for most footings. For more on mixing and ordering concrete, see pp. 136-143.

There are two types of footings—trench footings and formed footings. If the area where the walls are to be built is relatively free of rock, the simplest solution is to dig a trench, and use it as a form. Keep the top of the concrete footings level by driving short lengths of rebar to the proper elevation. Don't use wooden stakes because later they'll rot and leave voids in your footing. You'll need a transit level or water level to get the rods at the right height. After you install the level rods even with the top of the proposed footing, pour concrete in the trench, and trowel it flush with the tops of the rods. Some building codes require that these stakes be removed before the concrete sets up.

If the ground is rocky, you may have to set up wooden forms and brace them for the pour. I've saved some money in this situation by ordering the floor joists for the first floor and using them to build the forms. This won't damage the joists and will save you a lot of money. When the concrete has set, I remove the boards and clean them off with a wire brush and water. The sooner you remove the forms, the easier it will be to clean them.

After the footings have cured for at least 24 hours, drive nails at the corners of the foundation. To find the corner points, use a transit level or drop a plumb line from the layout lines that are strung to your batter boards at the top of the foundation (for more on laying out a foundation, see pp. 10 through 12). Next, snap a chalkline between the corner nails on the footings to mark the wall lines. Stack the blocks around the inside of the foundation. Leave at least 2 ft. of working space between the footing and your stacks of block. Also, allow room for a traffic lane so the workers can get back and forth with mortar and scaffolding.

Mortar mix for block—For the average block foundation, use masonry cement, which is sold in 70-lb. bags. You have to supply sand and water. Masonry cement is made by many companies. Brand name doesn't matter much,

Illustrations: Christopher Clapp

but you will need to choose between mixes of different strength. The average strength, for general masonry work, is universally classified as Type N. Unless you ask for a special type, you'll always get Type N. I get Type N masonry cement unless there is a severe moisture condition or stress, in which case I would use Type M, which is much stronger. The correct proportions of sand and water are important to get full-strength mortar. Like concrete, mortar reaches testing strength in 28 days, under normal weather conditions.

To mix the mortar, use one part masonry cement to three parts sand, with enough water to blend the ingredients into a workable mixture. Mortar for concrete block should be a little stiffer than for brickwork, because of the greater weight of the blocks. You will have to experiment a little to get it right. The mortar must be able to support the weight of the block without sinking.

The mixing water should be reasonably clean and free from mud, silt or organic matter. Drinking water makes good mortar. Order washed building sand from your supplier. It's sold by the ton.

The following will help you estimate the amount of mortar you'll need: One bag of masonry cement when mixed with sand and water will lay about 28 concrete blocks. Eight bags of masonry cement, on the average, will require one ton of building sand. Remember that if you have the sand dumped on the ground, some will be lost since you can't pick it all up with the shovel. For each three tons, allow about a half-ton for waste.

Laying out the first course—Assuming the footing is level, begin by troweling down a bed of mortar and laying one block on the corner. Tap it down until it is the correct height (8 in.), level and plumb.

A block wall built of either 10-in. or 12-in. block requires a special L-shaped corner block, which will bond half over the one beneath. The point is to avoid a continuous vertical mortar joint at the corner. Now lay the adjoining block. It will fit against the L-shaped corner block, forming the correct half-bond, as shown in the drawing and photo at right.

When the second L-shaped corner block is laid over the one beneath in the opposite direction, the bond of the wall is established. On each succeeding course the L corner block will be reversed.

Once the first corner is laid out, measure the first course out to the opposite corner. It's best for the entire course to be laid in whole blocks. You can do this simply by using a steel tape, marking off increments of 48 in., which is three blocks including their mortar head joints. Or you can slide a 4-ft. level along

the footing and mark off 48-in. lengths. In some cases, of course, dimensions will require your using a partial block in each course, but it's best to avoid this wherever possible. If a piece of block must be used, lay it in the center of the wall or where a window or partition will be, so it is not as noticeable. After the bond is marked on the footing, a block is laid on the opposite corner and also lined up. Then you attach a mason's line to the outside corner and run it to the opposite corner point. This is called "ranging" the wall.

Sometimes there are steps in the footings because of a changing grade line. The lowest areas should always be built up first to a point

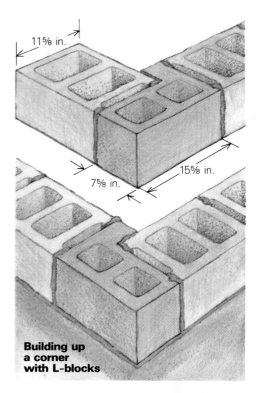

Building up a corner with L-blocks

where a level, continuous course of block runs through from one corner to the other. Steps in footings should be in increments of 8 in. so that courses of block work out evenly.

After one course of block is laid completely around the foundation to establish the bond and wall lines, it's time to build up the four corners. But before you begin laying block, you should make a story pole, sometimes called a course rod. Do this by selecting a fairly straight wooden pole and marking it off every 8 in. from the bottom to the height of the top of the foundation wall.

Any special elevations or features, such as window heads, door heads, sills and beam pockets, should be marked on the pole to coincide with the 8-in. increments wherever possible. After checking all your pencil marks, make them permanent by kerfing the pole lightly with a saw. Then cut the pole off even with the top of the foundation wall and number the courses of block from the bottom to the top so you don't find yourself using it upside down.

Now you can start laying up the corners so that you end up with only one block at the level of the top course. Successive courses are racked back half a block shorter than the previous ones (photo top left), so trowel on only enough mortar to bed the blocks in a given course. If the local code or your specifications call for using wire reinforcement in the joints, leave at least 6 in. of wire extending over the block. At the corners, cut one strand of the wire, and bend the other at 90°, rather than butting two sections together and having a break in the reinforcement.

Check the height of the blockwork periodically with the story pole. The courses of block should line up even with the kerfs. Once the corners are laid up, you can begin to fill in the wall between. Keep the courses level by laying them to a line stretched between the corners. Keep the corners plumb by checking every course with a spirit level.

Using manufactured corner poles—So far, I've described laying a foundation using the traditional method of leveling and plumbing. But in recent years, manufactured metal corner-pole guides have become popular with builders. They guide the laying up of each course and require less skill than the old way. They work like this. The corner poles are set on the wall once the first course of block is laid out. They are plumbed, then braced in position. Each pole has course heights engraved on it. Line blocks are attached to the poles on opposite corners at the desired course height, and the wall is laid to the line. There is no doubt that the use of manufac-

tured corner-pole guides has increased the mason's productivity without adversely affecting the quality of the work.

If you have to tie a porch or garage wall into a main foundation at a higher elevation, lay a concrete-block lintel in mortar from the corner of the wall being built to the footing at the higher elevation (photo bottom left). Then lay blocks on the lintel to form the wall. This saves time and materials in an area that doesn't require a full-basement foundation.

Stepping the wall at grade line—As you build up to the natural grade line of the earth, you can set the front of the wall back about 4 in. to form a shelf for a brick veneer, if the plans call for it. This is done by switching to narrower block—from 12-in. block to 8-in. block, or from 10-in. block to 6-in. block. The inside of the wall stays in the same plane.

Making the last course solid—On some jobs, specifications require that the last course of block be solid to help distribute the weight of the structure above and to close off the holes. You need only grout the voids in the top course of block. Broken bits of block wedged into the voids in the course below will keep the concrete from falling through. The sill plates will rest on this top course, and the floor joists on top of the plate.

The sill plate has to be bolted down to the top of the foundation wall. So you have to grout anchor bolts into the top of the wall every 4 ft. or 5 ft. These bolts should have an L-bend on the bottom and be mortared in fully so they don't pull out when the nut is tightened against the sill plate. They should extend about 2 in. out of the top of the wall. In some parts of the country, building codes require that the walls include a steel-reinforced, poured-in-place concrete bond beam in every fourth course.

Waterproofing the foundation—The traditional method of waterproofing a concrete block foundation is to parge (stucco) on two coats of mortar and then to apply a tar compound on top of that. This double protection works well, unless there is a severe drainage problem, and the soil is liable to hold a lot of water for a long time.

There are various mortar mixes you can use to parge the foundation. I recommend using a mix of one part portland cement to one-half part hydrated lime to three parts washed sand. This is a little richer than standard masonry cement and is known as type S mortar. The mix should be plastic or workable enough to trowel on the wall freely. Many mortars on the market that have waterproofers in them are all right to use. However, no two builders I know seem to agree on a mix, and most have worked out their own formulas.

Prepare the foundation wall for parging by scraping off mortar drips left on the block. Next, dampen the wall with a fine spray of water from a garden hose or a tank-type garden sprayer. Don't soak the wall, just moisten it. This prevents the parging mortar from drying

After the first course is laid, the corners are built up to the topmost course. Above left, a mason checks course heights against a story pole, which is graduated in 8-in. increments.

Reinforced concrete lintels, left, are used to tie the main foundation to walls that are laid at a higher level, such as porch foundations or garage walls.

Photos: Dick Kreh

Troweling technique

Laying up concrete blocks with speed and precision takes a lot of practice. But it's chiefly a matter of learning several tricks, developing trowel skills and performing repetitive motions for several days. A journeyman mason can lay an average of 200 10-in. or 12-in. blocks in eight hours. A non-professional, working carefully and after practicing the techniques shown here, ought to be able to lay half that many. If you've never laid block before, what follows will show you the basic steps involved in laying up a block wall.

1. First, mix the mortar to the correct stiffness to support the weight of the block. Then apply mortar for the bed joints by picking up a trowelful from the mortarboard and setting it on the trowel with a downward jar of the wrist. Then swipe the mortar onto the outside edges of the top of the block with a quick downward motion, as shown.

2. Apply the mortar head joint pretty much the same way. Set the block on its end, pick up some mortar on the trowel, set it on the trowel with a downward jerk and then swipe it on the top edges of the block (both sides).

3. After buttering both edges of the block with mortar, press the inside edge of the mortar in the head joints down at an angle. This prevents the mortar from falling off when the block is picked up and laid in the wall.

4. Lay the block on the mortar bed close to the line, tapping down with the blade of the trowel until the block is level with the top edge of the line. Tap the block in the center so you won't chip and smear the face with mortar. Use a hammer if the block does not settle easily into place.

5. The mortar in the head joint should squeeze out to form a full joint at the edge if you've buttered it right. The face of the block should be laid about 1/16 in. back from the line to keep the wall from bowing out. You can judge this by eyeballing a little light between the line and the block.

6. Remove the excess mortar that's oozing out of the joint with the trowel held slightly at an angle so you don't smear the face of the block with mud. Return the excess mortar to the mortarboard.

Check the height of the blockwork by holding the story pole on the base and reading the figure to the top of the block. Courses should be increments of 8 in.

Finishing the joints—Different types of joint finishes can be achieved with different tools. The most popular by far is the concave or half-round joint, which you make by running the jointing tool through the head joints first, and then through the bed joints to form a straight, continuous horizontal joint. If you buy this jointing tool, be sure that you get a convex jointer. These are available in sled-runner type or in a smaller pocket size. I like the sled runner because it makes a straighter joint.

After the mortar has dried enough so it won't smear (about a half-hour), brush the joints lightly to remove any remaining particles of mortar. —D.K.

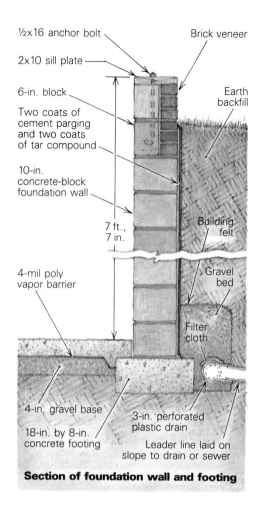

½×16 anchor bolt

2×10 sill plate

6-in. block

Two coats of
cement parging
and two coats
of tar compound

10-in.
concrete-block
foundation wall

7 ft.,
7 in.

4-mil poly
vapor barrier

4-in. gravel base

18-in. by 8-in.
concrete footing

3-in. perforated
plastic drain

Leader line laid on
slope to drain or sewer

Brick veneer

Earth
backfill

Building
felt

Gravel
bed

Filter
cloth

Section of foundation wall and footing

The completed foundation has been sealed with two ¼-in. thick parging coats and topped with an application of tar compound, which finishes the waterproofing. Backfilling should happen only after the first floor is framed and the walls framed up, so the added weight of the structure will stiffen the walls and make them less liable to bulge from the pressure of the earth.

out too quickly and allows it to cure slowly and create a better bond with the wall.

Start parging on the first coat from the bottom of the wall to the top, about ¼ in. thick. A plastering or cement-finishing trowel is excellent for this. After troweling on the parging, scratch the surface with an old broom or a tool made for this purpose. Let the mortar dry for about 24 hours or until the next day, and repeat the process for the second coat. Dampen the wall between coats for a good bond. Trowel the final coat smooth, and let it dry for another 24 hours.

To complete the waterproofing job, spread on two coats of tar compound (photo, top). You can do this with a brush or roller if the weather is warm. Many builders in my area use a product called Hydrocide 700B (Sonneborn Building Products, 57-46 Flushing Ave., Maspeth, N.Y. 11378). It comes in 5-gal. containers and is available from most building-supply dealers. I like it because it stays a little tacky and seals the wall very well. It's gooey, though, so wear old clothes and gloves when you're applying it. Kerosene will get it off your hands and tools when the job is done.

Drain tile—Most codes require some type of drain tile or pipe around the foundation to divert water build-up and to help keep the basement dry. The design of the drain-tile system is important. Generally, drain tile or pipe is installed around the exterior wall of the foundation, below the wall but above the bottom of the footing, as described below.

Begin by spreading a bed of crushed stone or gravel around the foundation next to the wall. Lay the drain tile or perforated plastic pipe on top of this bed. The bottom of the drain pipe should never be lower than the bottom of the footing, or it won't work properly. Lay filter cloth over the drain pipe to keep mud and dirt from blocking the holes. Then place another 4-in. to 6-in. layer of crushed stone or gravel over the pipe, as shown in the drawing above.

The water collected by the drain tile has to flow away from the foundation. One way to make this happen is to drain the water to a natural drain or sewer away from the foundation area by installing a leader line on a slope lower than the drain pipe. The other method is to drain the water under the wall of the foundation and into a sump pit inside the basement. The water that collects is then pumped out through the plumbing to a sewer or septic tank.

A third method, which has worked well for contractors in my area, is to put the drain tile inside the foundation on a bed of crushed stone, just beneath the finished concrete floor, which will be poured after the drain tile is in place. One-inch plastic pipe is installed about 6 ft. o.c. through the wall at the bottom of the head joints in the first course of block. When the foundation wall is done, crushed stone is spread around the exterior edges as before, but no drain pipes are needed.

The idea is that any water that builds up outside the foundation wall will drain through and into the drain tiles. In addition, the water inside the foundation area will also flow into the drain tile and into the sump pit in the basement floor, where it can be pumped out into the septic or sewer system.

Insulating the foundation—In recent years, the use of rigid insulation applied to the exterior of the foundation wall has helped to reduce dampness and heat loss. This is especially important in the construction of earth-sheltered homes. There are a number of products that will do a good job. Generally, rigid insulation is applied to the waterproofed wall with a mastic adhesive that's spread on the back of the foam board. Use a mastic or caulking that does not have an asphalt base. Most panel adhesives will work. The building-supply dealer who sells the insulation will know the proper adhesive to use for a specific type of insulation board. Also, there are granular and other types of insulation that can be poured into the block cells.

After all of the foundation work has been completed, the backfilling of earth should be done with great care so that the walls don't get pushed out of plumb. It is always better to wait until the first floor is framed up before backfilling. This weight resting on the foundation helps prevent cracking of the walls, and the framing material will brace the block wall and make it more rigid. If the walls are cracked or pushed in from backfilling, the only cure is very expensive—excavate again and replace the walls. □

Dick Kreh is an author, mason and industrial-arts teacher in Frederick, Md.

Facing a Block Wall With Stone

A good rock supply, tight joints and hidden mortar are the secret to the solid, structural look

by Tim Snyder

Building with fieldstone and building with concrete block represent two extremes in masonry construction. Concrete blocks aren't especially interesting to look at, but they go up fast, and it's easy to build a sturdy wall with them. Stone construction demands patience, skill, and above all, lots of rocks. Even with these ingredients, the different shapes and sizes of the material make it tough to keep a wall of stone plumb and strong. Given these considerations, it's easy to appreciate a construction technique that combines the beauty of stone with the strength and practicality of concrete block.

Larry Neufeld laid up his first stone face to cover a block chimney in a house that he and his brother were building. He had never worked with stone before, but as a general contractor he knew enough about masonry to take on the project. By the time work began on the solar addition shown here, he had developed a technique and style that take the best from both building materials. The finished wall—20 tons of mass facing the windows and skylights on the south side of the addition—shows little mortar at all, and unless you examine the joints carefully, they seem to be dry-fit.

A flexible system—Neufeld's method uses found stone, and thanks to the New England countryside, he can usually gather what he needs from the fields and stone walls on the owner's property. Working against a 6-in. thick block wall, he lays up a face 8 in. thick, using odd-sized stones from 2 in. to 7½ in. thick. The void behind the stone is filled with mortar, which sets up around the masonry ties set in the block's joints.

Neufeld's system can work just as well with a poured wall or a bearing wood-frame wall, as long as the footing is beefed up to hold the extra load, and there is a mechanical connection between wall and face.

It's good to begin the job with plenty of

Hiding the block. With a depth of 8 in., the stone face that covers this block wall doesn't require rocks of uniform thickness. Careful fitting is still important, though. At left, Neufeld works against temporary grounds that frame an opening in the wall. These boards were later replaced by the oak casing shown on the facing page. The finished wall faces a bank of windows in a solar addition, and looks like a solid stone wall.

stones. As you look for rocks, pick out natural corners, base-course stones with especially flat, broad faces and pieces with unusual colors or mineral formations. Toss these in separate piles before you start building, and each time you sort through your rock, take stock of the sizes and shapes you've got. Cataloging like this can make the job go a lot more quickly. But even with a good collection of stone, you can expect to be missing a few key pieces. In the middle of a job, Neufeld often finds himself driving more slowly past stone walls after work, seeking out an elusive corner or curved face.

Before laying up the face, be sure that any wall or ceiling surfaces that will be adjacent to the stones are finished. This means drywalling, paneling, plastering and painting earlier than you normally would, but it's far easier to do this work before the face goes up.

Laying up the wall—For bonding stone to block, Neufeld uses a mix of three parts sand to one part portland cement. Working his mix in a wheelbarrow, he adds just enough water to make a very stiff mortar. Then at the front of the barrow he adds a bit more water and trowels up a small section of wet mix. The stiff mix is used between the stones so that no mortar will flow out of the joints onto the exposed rock face. The wet mix fills voids closer

to the wall, and bonds the back sides of the stones solidly to the block.

Neufeld uses a tape to check the thickness of the face as he lays it up. A level isn't much help because of the irregularities in the stone, so he uses it only for rough checking. Working against a plumb block wall is pretty good insurance that the stone face will be plumb, but with many rough facets to account for, Neufeld does plenty of adjusting by eye. Fortunately on this job, a temporary post had to be nailed up to support the second-floor overhang where the circular stair would go. By plumbing the post both ways, Neufeld was able to use it to align the face of the wall as he laid it up. He also measured against the post to check the arc of the curved wall section.

The secret to achieving the dry-stack look is to test-fit all stones carefully before laying them in place, and then to keep the mortar away from the face edge of the joint. Test-fitting the stones is the first step in building the face, and it's like working on a big jigsaw puzzle. Working horizontally across the wall, you have to find stones that fit well together. A good fit means not only that the joints are tight, but also that they are staggered vertically, just as they would be in a solid, structural wall (see Neufeld's finished wall, above).

In many instances, you have to do some coaxing to get a joint right. And it's always a

good idea to flatten the top edges of your stones slightly before casting them in the wall. This ensures a stable surface for the next course of stones to rest on.

Neufeld uses a mason's hammer to knock off leading edges, and a cold chisel to fracture thick stones. Sometimes you can split a sedimentary rock along its bedding lines, but more often than not you'll end up with random fragments. This is one reason why Neufeld prefers to trim off as little as possible, using smaller pieces to fill gaps rather than trying for an ideal fit between two stones. He doesn't like to chip into the exposed face of a stone if it can be avoided, explaining that a split face has a harsh look that will never be lost inside a house.

Your test-fit stones should be able to rest on the previous course without falling off. They don't have to be exactly plumb at this stage, but you're looking for a gravity fit. Once you're satisfied that a group of stones fits well into the wall section, memorize their relative positions and remove them from the wall. Then prepare a bed of mortar by packing some stiff mix on top of the previous course. Work out from the block wall and leave the inch of joint area closest to the outside face bare of mortar. Lay down just enough mortar so that each stone will seat securely in its preassigned position. After pressing the

Balance and alignment. At right, stone chips inserted along joint lines to serve as temporary wedges prevent tall, thin stones from leaning out of plumb. They're removed after the mortar has set. The facing is thick enough to conceal a heating duct in the wall. Below, corner and curve construction depends on a good selection of shapes and sizes. A mixture of small and large stones also makes it easier to stagger the joint lines.

Hiding the mortar. The key to the dry-fit appearance is to keep mortar away from the face edge of each joint. Neufeld packs the mortar close to the block, left, then presses the stone in place and seats it in its mortar bed with several hammer blows, as shown above.

stones into their mortar bed, Neufeld sometimes uses a hammer to help seat them.

As you seat the stones, check to make sure they're plumb. Broad, narrow stones that don't extend the full depth of the face tend to lean out farther than they should. To make minute adjustments in orienting these stones, Neufeld inserts small rock fragments that serve as temporary wedges. They hold the stones in alignment until another course is laid up and the mortar sets; then Neufeld removes them.

It's best not to pack mortar behind a course until the mortar between the stones has set. Then trowel in the wetter mix to fill the space behind the stone, and you're ready to test-fit another course. This way the wetter mix can't ooze out of the joints and dribble down the face of the stone.

At the end of the day when you're using up

the last of your mortar, don't fill all the way up behind the last course of stones. It's better to leave a slight depression because this forms a keyway for the mix you trowel in the next day. Another important practice at the end of the day is cleaning the stone you've laid up. Go over the joints with a pointing chisel or a sharp piece of wood, and rake them back so that there's little or no mortar showing. Then use a broom to sweep down the face of the wall so that any drops of mortar are removed before they adhere.

Curves, corners and openings—To build the curved section of the face, Neufeld traced the clearance arc for the circular stair onto the concrete floor. Then he fit and laid up the stones as if he were working on a straight section. The only differences were that he had to use smaller stones to get smoothly around the arc, and that he could no longer sight off the temporary post to check for plumb. He used a level instead.

Successful corners are mostly a matter of having a good variety of cornerstones to choose from. Your first inclination may be to overlook small, right-angled rocks in favor of large, squarish stones. But what you actually want is a mixture of large and small; this creates overlapping joints and integrates the corner with the rest of the wall, as shown in the photo, bottom left.

You don't need an exact 90° angle to make a cornerstone. The secret is to aim for a right-angle average over several courses. Stones that come within 10° of 90° should work in a corner, as long as you get a good combination of large and small, acute and obtuse.

At door or window openings, both the stone face and the block wall are exposed. On this job, Neufeld hid this joint with trim. Before constructing the face, he erected temporary grounds from 2x stock along the trim lines. Once the face had been built into these plumb and square housings, they were removed and replaced by finish trim.

Fine points—Neufeld admits that it takes time to develop technique and style, a consistent choice of stones that will look nice in the finished wall. He likes to play one shape off against another, but stresses that the joint lines should give an impression of horizontality. Using a mixture of small and large stones is important to the overall composition, and also makes it easier to stagger the joints. But with this method of laying up stone, you've got the flexibility to try out your own ideas. Neufeld says that on his next job, he'd like to do a rock pattern in relief.

Building this 32-ft. wall took Neufeld about 400 hours. Since he was the general contractor for the entire solar addition, working on the wall kept him at the site through the arrival and departure of most of the subcontractors. Because the structural part of the wall—the concrete block—was finished at an early stage, there was no need to rush in laying up the stone. Having the time to find and fit the rock is an important advantage. □

Concrete splint

On a trip to England years ago, I saw an old fence whose posts had rotted off at ground level. The rest of the fence was still in good shape, so its owner had replaced the below-grade post sections with concrete splints. I've since mended my fences using the same solution, with excellent results.

I built a form about 4 ft. long, with sloping ends and one sloping side, as shown in the drawing. For each splint, three short pieces of ⅜-in. galvanized pipe are mounted on the form's axis to secure two lengths of ⅜-in. rebar. Two of the pipe sections are in the above-

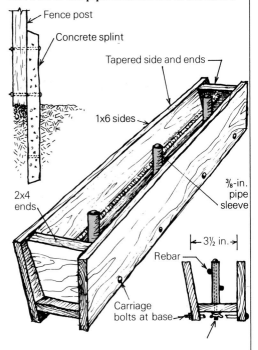

Fence post

Concrete splint

Tapered side and ends

1x6 sides

2x4 ends

⅜-in. pipe sleeve

3½ in.

Rebar

Carriage bolts at base

grade half of the splint, and act as sleeves for ¹⁵⁄₁₆ in. bolts during installation.

Before filling the form, I lubricate the sides with a little axle grease to aid in removal. Nearly every splint I've cast has been with the leftover concrete from some small household job. —Sidney McGaw, Albany, Calif.

Stabilizing plumb bobs

As a construction millwright, I frequently need to establish very accurate plumb lines for machinery alignment in places where a transit is impractical. I rely on my 24-oz. brass plumb bob. When it's not necessary for the end of the bob to be visible, I immerse it in a bucket of oil to reduce motion from vibration and wind.

To avoid the problem of winding and unwinding line, I replaced the string in an empty chalk box with 50-lb. test braided nylon fishing line and tied the bob to the end; it makes a convenient reel. —David Walter, Rossville, Ill.

Blocked-up chimney

Prefabricated metal chimneys (whether insulated or not) have become almost universal for use in woodburning appliances. When we installed one, we thought, like most people, that our $20-a-foot expenditure was a lifetime investment. It hasn't been for us—after two years we have already had to paint the exposed pipe, which was badly corroded (in an admittedly severe coastal environment). We have heard that metal chimneys can be badly damaged by undetected chimney fires. With many builders installing them in hard-to-reach places, it seems that the alternative of terra-cotta chimney liners set in standard 16-in. by 16-in. by 8-in. concrete chimney blocks is a safer and more permanent choice. The cost is comparable or less, the units self-aligning, and the blocks can be tied together with rebar for earthquake resistance.

Flue liners and chimney blocks vary in size from location to location, but the local sources probably make compatible blocks and liners. The blocks around here are as shown in the drawing below. Liners are usually started 4 in. above the bottom so they align each chimney block as it is laid.

Depending on local codes, earthquake and wind conditions, and the height of the chimney, it's usually wise to run ½-in. rebars up from the footings to the top of the chimney on opposite diagonal corners and grout them in place. Usually 4-ft. to 5-ft. rods are as long as is comfortable to work with. Be sure to over-

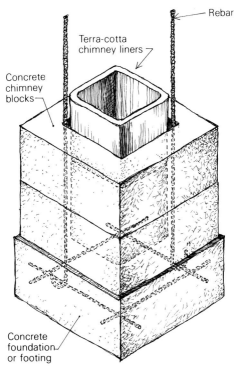

Rebar

Terra-cotta chimney liners

Concrete chimney blocks

Concrete foundation or footing

lap them 20 diameters (10 in.). It is also advisable to tie the chimney to the ceiling and roof structure with steel straps, leaving at least a 2-in. clearance to combustible materials. —Tom Bender, Nehalem, Ore.

Grout gun

We bought a house with a nicely remodeled bathroom. All it lacked to make it look really finished was a tile baseboard.

With the sink, toilet and other fixtures still in place, I was able to spread the mastic to hold the tiles to the bathroom wall with a fair amount of contortion but no real trouble. The only problem I foresaw was with the unwieldy rubber grout trowel—there simply wasn't enough room to maneuver it in some of the tight places.

Casting about for an alternative grouting method led me to the kitchen drawer and the seldom-used cake-decorating device. It looks like a caulking gun, with several interchangeable metal tips and a piston plunger at one end to control the flow of icing, or whatever. The cap unscrews for loading. I filled the thing with grout and went to work.

The device worked remarkably well for this application. I was able to reach remote crevices with ease and fill them with a high degree of accuracy. The only problem came when I failed to screw the cap completely on after reloading: it popped off, firing a messy blob onto the floor.

My new tool was easy to clean when I was done with it, and I returned it to its kitchen duties none the worse for wear.
—Chuck Gomez, Ponca City, Okla.

Gypboard concrete forms

We recently did a foundation job in very crumbly, sandy soil. The first task was to set 27 pier blocks in pier holes that were 18 in. on a side and 12 in. deep. But by the time we had dug down a foot, we often had a hole that was more than 2 ft. across at the top and growing. Faced with filling these craters with concrete, we calculated that we would waste more than a cubic yard.

Instead of ordering the extra concrete, we transformed four sheets of gypboard into form boxes. We cut the sheets into 6-ft. lengths, and scored them along their length at 18 in. o.c., leaving the face paper intact. Perpendicular to these scored lines, we cut the board into

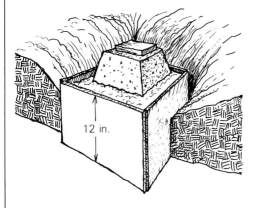

12 in.

12-in. wide strips. These strips were then folded into square boxes, placed in the over-size holes and backfilled. The forms not only saved concrete, but also gave us an accurate way to calculate our ready-mix order.
—Sunrise Builders, Santa Cruz, Calif.

Extruded flashing

Bending your own flashing on site usually produces wavy, crinkled pieces of sheet metal that make you wish you'd had them made up

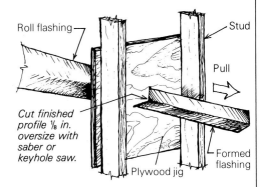

Roll flashing

Stud

Pull

Cut finished profile ⅛ in. oversize with saber or keyhole saw.

Formed flashing

Plywood jig

at a shop. Here's a fixture you can make on the site that produces near-perfect results.

Using a jigsaw or keyhole saw, cut the full-size cross section of the flashing you want to make into a plywood scrap that's at least 18 in. across. Nail the plywood across the studs of an unsheathed partition at chest level. Then cut a piece of roll flashing to length and insert it into one side of the cut in the plywood. With one person pulling and another supporting and feeding the flashing, this fixture makes the job easy.

—Mick Cappelletti, Newcastle, Maine

Built-up roof blade

I've installed a lot of skylights in tar and gravel roofs. As a consequence, I've gummed up plenty of circular-saw blades trying to burn through the built-up roofing. Since this kind of cutting ruins the blade anyway, I decided to modify a blade so that it would work better.

I selected my dullest combination blade, and set the teeth way out—¼ in. or more. This radical set allowed the blade to clear itself with each revolution, and worked even better than I'd hoped.

—Bob Whiteley, San Rafael, Calif.

Snapping lines

When snapping a series of chalk lines, as on roofs or siding, two people can hook their chalk box lines at the clip by inserting one line through the other, as shown in the draw-

ing. Worker A reels in the hooks to his end and they snap lines until worker B's line goes dry. Then worker B cranks his line while exposing A's for more snapping. There is no walking back and forth, and no time wasted rechalking the lines.

—Jackson Clark, Lawrence, Kans.

Soldering water pipes

An old plumber's trick for resoldering copper water lines when making repairs is to take a piece of white bread, make two dough balls and then stuff them up the pipe in both directions. The dough balls will block the water while the soldering is being done, and then dissolve easily in the water system.

—Chris Valenzuela, Millboro, Va.

Adjusting doors

Sometimes a new door will hang cockeyed in an old frame, despite the fact that the door and frame are properly sized, square and plumb. This can happen if the depth of the hinge mortises is slightly off, if there's a twist in the jamb, or if there's any variation in the butts themselves.

To remedy this situation, most carpentry books advise shimming the hinges to throw the hinge pins closer to or farther from the jamb. This trial-and-error method takes quite a lot of time. Instead, simply close the door and remove the hinge pins from either the top and middle hinges, or bottom and middle hinges, and temporarily shim the bottom of the door so that the gap is even along both edges of the door. Now take a 6-in. adjustable wrench or smooth-jawed pliers and carefully bend the hinge knuckles until they once again align, and replace the pins. I know that this approach sounds brutal, but it does work. If necessary, you can pull a door as close as ⅟₃₂ in. of the jamb without making it hinge-bound. In fact, making minor adjustments with this method will automatically relieve the stiff action of bound hinges without ever having to loosen a screw or cut a cardboard shim.

—David Walter, Oakwood, Ill.

Tub drops in

Using materials and fixtures in a way that wasn't originally intended is a mainstay of remodeling work. This can even be true of bathtubs. My project last summer was a tub installation that called for wood paneling around the walls and included a low bench at the side. What I needed was a drop-in tub. However, these are very expensive, hard to color-match to existing fixtures, and take a long time to get after ordering.

I bought a standard tub. By installing the

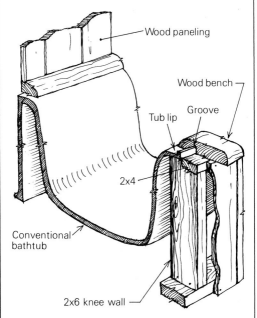

Wood paneling

Wood bench

Groove

Tub lip

2x4

Conventional bathtub

2x6 knee wall

finished skirt against the back wall of the tub area instead of in front as usual, I was left with a completely open outer side for framing the knee wall and applying the seat, which I grooved to fit over the lip of the tub.

—James B. French, Portsmouth, R.I.

Scissor framing

The scissor-like tool illustrated below can be useful in plumbing and aligning stud walls during framing. With it, a worker can exert and maintain great pressure against a wall while another worker nails bracing to hold the wall in its correct position.

The scissor can be assembled quickly on the job site from a pair of 8-ft. 2x4s and a ⅜-

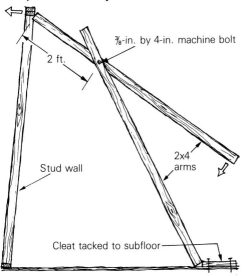

⅜-in. by 4-in. machine bolt

2 ft.

Stud wall

2x4 arms

Cleat tacked to subfloor

in. by 4-in. machine bolt. The proportions shown in the illustration above are only approximate; I usually trim off one end or the other until the tool feels right and gives the best leverage.

To use the scissor for plumbing a wall, tack a cleat to the subfloor about 6 ft. from the bottom plate and place the foot of the tool against it. Next, open the arms of the scissor until its raised end can be firmly wedged against the top plate of the stud wall. By pressing down on the tool's handle, you will increase its span and force the top of the wall away from you. Fine adjustments can be made by altering the pressure on the handle.

—Malcolm McDaniel, Berkeley, Calif.

Form-fit vapor barrier

When I install a plastic vapor barrier over a wall or ceiling, I ignore the electric outlet boxes until I finish hanging the sheet. Then, rather than cut the plastic on the outside of the box, I make the cut ¼ in. to ½ in. inside the perimeter of the box. I stretch the plastic around the outside, which gives me a nice tight fit that cuts down on cold-air infiltration.

This trick works for other utilities that poke through the vapor barrier, such as ductwork or plumbing. A bead of caulk or a duct-tape collar can make the seal even tighter.

—Susan Caust Farrell, Searsport, Maine

Plugged-up plumbing

If you're working on an old galvanized water supply system (replacing a section with copper pipe to improve water pressure, for example) and nothing comes out of the faucet when you turn it on, the faucet screen is probably packed with rust particles and flux from the soldering operation.

—Chas. Mills, Leadville, Colo.

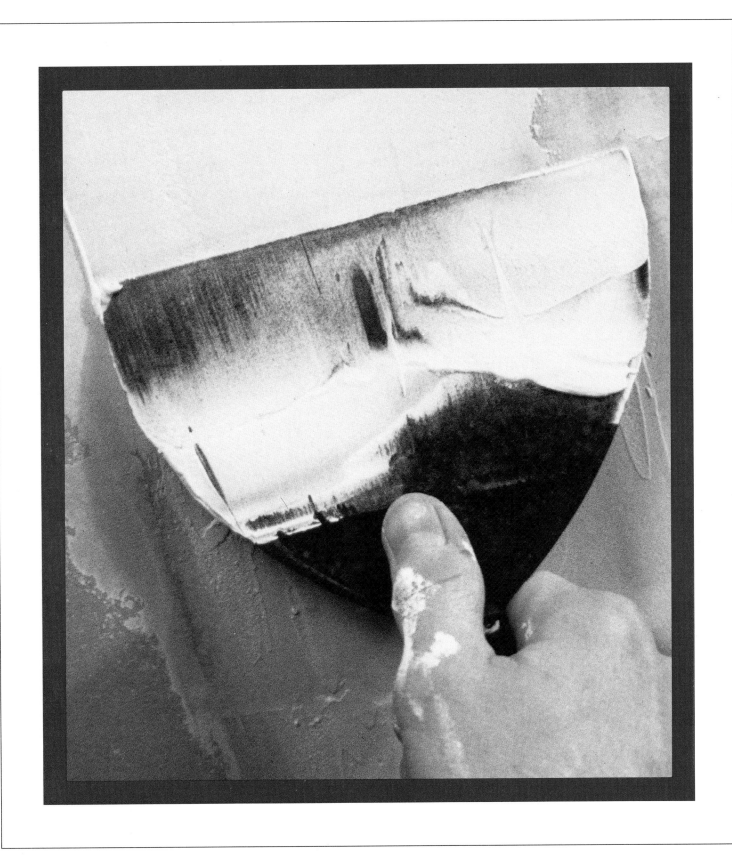

On-Site Shop

A portable table saw and jointer and several useful jigs

by Sam Clark

Given enough big-shop tools, it's fairly easy to do accurate work quickly; given more time and skills, accurate work can be done by hand. But often there is a need to do accurate work quickly when shop tools are not available. In my work, remodeling, many jobs require accurate cutting but are too small to justify bringing in the 10-in. table saw. For such jobs I have developed a traveling shop consisting of light, portable jigs and setups that extend the uses of portable circular saws and routers. All these gadgets are cheap; convenient to make, move and set up, and precise.

Plywood-cutting jig—I learned about this gadget from John Borden, a designer from Cambridge, Mass. Carpenters commonly make long

cuts in plywood, counter stock and other large panels by clamping a straightedge to the work and running a circular saw against it. Setting up is tedious, particularly if many such cuts must be made, and usually the cuts are off a bit. My plywood-cutting jig has a fence, but it also has a base that aligns exactly on the cutting line so you don't have to measure.

You can make this jig in five or ten minutes. First cut a strip about 5 in. wide off the long edge of a piece of ½-in. or ¾-in. fir plywood. I use A/B, A/C or A/D grade. Flip the strip over so the factory-cut edge ends up as the working side of the fence, and screw it down firmly to the remainder of the sheet with eight or ten screws. Leave enough of the sheet sticking out to the left of the fence so that you can clamp it to your

work surface without obstructing the saw. Run the saw down the track once to cut the jig free from the plywood. To use the jig, line up this edge with the line you want to cut, clamp the jig down, then saw.

Homemade table saw—The heart of the traveling shop is the homemade table saw. George Carson of Bloomfield, Conn., rigged up a table saw in his basement by suspending a circular saw upside down underneath a table. My gadget is based on his idea. The table is simply a piece of ¾-in. plywood with a hole for the saw and a routed slot for the crosscut fence to slide in. It rests on a pair of folding sawhorses, braced to prevent sagging. I used A/B, A/C or A/D grade Douglas fir, which will stay flat for years if it is

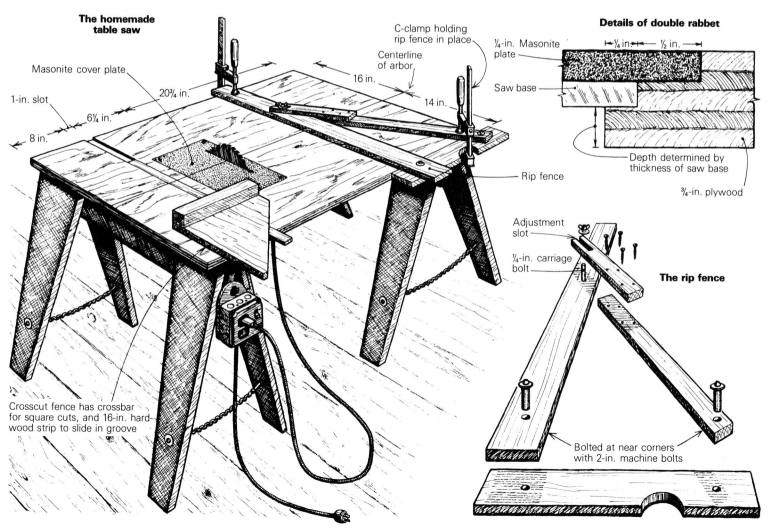

The homemade table saw

Masonite cover plate

1-in. slot

6¼ in.

8 in.

20¾ in.

C-clamp holding rip fence in place

Centerline of arbor

16 in.

14 in.

Rip fence

Crosscut fence has crossbar for square cuts, and 16-in. hardwood strip to slide in groove

Details of double rabbet

¼ in.   ½ in.

¼-in. Masonite plate

Saw base

Depth determined by thickness of saw base

¾-in. plywood

The rip fence

Adjustment slot

¼-in. carriage bolt

Bolted at near corners with 2-in. machine bolts

stored in a dry place. Cut the plywood table using the plywood cutting jig; the drawing gives table dimensions.

The accuracy of this homemade table saw depends on aligning the rip fence precisely parallel to the rabbets around the saw hole and the groove in which the miter gauge slides. The simplest way to accomplish this is to use the rip fence itself when you rout the saw hole and the miter gauge slot. Therefore, it's best to make the rip fence first.

The rip fence is simply an L-square with a two-piece diagonal brace for rigidity. It is bolted at the two near corners with 1-in. machine bolts. At the apex, a $\frac{1}{4}$-in. carriage bolt and wing nut ride in a slot cut in the diagonal brace to allow for adjustment. In operation, the fence slides side to side against the near edge of the table, much like a drafting T-square. Once set up for a cut, the fence is simply C-clamped in place. Use a rafter square to square up the rip fence, then tighten the wing nut. A wood screw installed at the apex, next to the wing nut, will serve as a setscrew to make the adjustment permanent.

Use the rip fence to draw the location of the saw hole on the table. This hole is double-rabbeted. The lower shelf supports the saw base and the upper one holds a slotted $\frac{1}{4}$-in. Masonite cover plate, which should be fitted flush with the table top. This cover deflects sawdust and keeps small scraps from falling through. The saw is sandwiched between the lower rabbet and the cover plate, which is screwed down to the table. No other fastening is necessary. Rout the top rabbet first, fencing your router with your new rip fence. After completing both rabbets, cut the hole out with a saber saw. The precise dimensions of these cuts will be determined by the size of your portable circular saw. Now make the crosscut fence, a 45° triangle of plywood with a crossbar screwed to its top for square cuts, and a $\frac{5}{16}$-in. by 1-in. by 16-in. hardwood strip screwed to its bottom. Rout a groove, using the rip fence. Make the fit snug, and lubricate with wax or silicone spray.

If you fasten the hardwood strip perfectly square to the plywood triangle, the guide will stay accurate. I've been using mine for five years now and it's still true. It would not be hard to make an adjustable miter gauge, but when I need precise angled cuts, I sometimes make a jig similar to the plywood cutting jig, but only 16 in. long for convenience. This jig is particularly useful when less experienced people are on the job, because anyone can make precise cuts easily with it.

Now you're ready to insert the saw. Use a good saw with a rigid base. I prefer the Milwaukee. Hold the saw trigger down with a clip or tape, and plug the saw into an extension cord fitted with a switch, so you can turn it on and off like a regular table saw.

The most annoying aspect of the homemade table saw is that blade height and angle adjustments are awkward. I'm not sure this problem can be solved.

Router table—The homemade table saw cannot cut dadoes, but you can solve this problem by hanging a router underneath the table. Replace the saw cover plate with one of $\frac{1}{4}$-in. Masonite drilled to receive the router bit. Remove the plastic disc from the router, and bolt its metal base to the plate. For greater rigidity, use $\frac{1}{2}$-in. plywood, which will necessitate rabbeting the edges of the bottom side of the plate. The router table can be used for dadoing, sliding dovetail joints, rounding over and similar operations using the crosscut and rip fences devised for the circular saw.

With a special fence, you can also use this setup to join the edges of boards with reasonable accuracy and speed. On a power jointer, the outfeed table lines up exactly with the jointer knives. The infeed table is slightly lower but parallel, so the work travels over the cutters in a straight line even while part of the wood is being cut away. To make a jointing fence for your router, take a straight board at least 4-in. wide and rip between $\frac{1}{32}$ in. and $\frac{1}{16}$ in. off half of its length. You can use the homemade table saw for this cut if the blade is sharp. The full-width part of the board will be the outfeed fence and the narrower part will be the infeed fence. Cut a hole for the bit. Clamp the fence to the plywood table so that the outfeed fence lines up perfectly with the router bit. Use a sharp $\frac{1}{4}$-in. straight bit. Because the jig's capacity is determined by the length of the bit, make sure the one you use is at least $\frac{3}{4}$-in. long.

Dadoing jig—I have found this jig indispensable. It consists of two 1x2s screwed with drywall screws to a base of $\frac{1}{4}$-in. Masonite. The 1x2s are spaced a distance equal to the diameter of the router base. They form a track for the router. The jig in the drawing is for $\frac{3}{4}$-in. dadoes. When the strips were in place, I put a $\frac{3}{4}$-in. carbide bit in the router, lowered the router into the track, and made the slot in the base. To make a dado, clamp the slot directly over where the cut is to be made. Because there is no measuring, this jig is precise. For angled cuts or cuts in slightly warped pieces, it is often more convenient than a table saw.

On deep cuts, make several passes with the router. Set it for the maximum cut, but make the first pass with two thicknesses of $\frac{1}{4}$-in. Masonite between the router base and the base of the jig. Make the second cut with one thickness, and the final pass with none. □

Plywood cutting jig

Fence is screwed to sheet, then jig is cut free

Flip 5-in. piece so factory-cut edge becomes work side of fence

Overhang for clamping

← 3 in. → ← 5 in. →

Detail of jointing fence

Outfeed

Router bit

Infeed portion of jointing fence is $\frac{1}{32}$ to $\frac{1}{16}$ in. narrower than outfeed

Dadoing jig

Use different sized slots for different sized dadoes

$\frac{1}{4}$-in. Masonite

1x2

Sam Clark, of Cambridge, Massachusetts, is the author of Designing and Building Your Own House Your Own Way *(Houghton Mifflin Co., Boston, 1978).*

Working with Green Wood

Getting the most out of unseasoned native lumber sawn at a local mill

by Paul Hanke

Green, roughsawn lumber from a sawmill is both cheaper than commercially dried wood and stronger in the long run. Its price can be as much as 40% to 50% lower—which could amount to a 15% savings on a typical house-building project—and sawmill lumber is ultimately stronger than kiln-dried because it is cut to full dimension (a lumberyard 2x4 is only $1\frac{1}{2}$x$3\frac{1}{2}$). I also think it makes more sense to use native wood than to import commercial lumber from far away on gas-guzzling transcontinental trucks. There are some disadvantages, though.

The high moisture content of sawmill lumber and the shrinkage that will occur as it dries usually cause distortions that have to be compensated for. Commercial lumber has customarily been dried to a moisture content of 19%, but a green 2x4 will sometimes be so full of water that it will squirt you in the eye when you drive a nail into it.

Problems also occur during construction because the dimensions of green framing lumber vary by as much as $\frac{3}{8}$ in. Such problems are not insurmountable, but you need some practical advice to minimize them.

Buying lumber from a sawmill—You may not be able to find local mills simply by looking in the phone book. In our area of Vermont there are some good-sized ones, a number of small operations, and some specialty mills that saw only clapboards or cedar shingles. Ask around and you'll probably find several leads. Then shop and compare. Mills and sawyers differ in their reliability, capabilities, prices and quality.

Be sure to allow enough time for the sawyer to do your job. A single house may be a pretty small undertaking for a large mill, and in any case the operator will want to give priority to work for regular customers. Depending on the location and size of the mill, the sawyer may also have to find and buy special logs for really large pieces, like big timbers for post-and-beam construction, or long 2x12 rafters. Order early to give yourself and the sawyer plenty of lead time.

In shopping for lumber, make a list of the sizes, lengths and species you want. Then begin inquiring whether the mills can supply them from inventory, or whether they will have to be custom-sawn. Other questions to ask: How far in advance should you order? Does the mill de-

liver? At what cost? Does the mill have a thickness planer, or can you arrange for special milling like tongue-and-groove or shiplap if you need it? How much extra will planing cost? Is there any inventory that has been around long enough to be pretty well air-dried? If so, you're in luck, but give it a test lift to see how dry it is.

Ask for references from other builders if you haven't already gotten one from someone you know and trust. The two most important things to find out are if the sawyer delivered on time and the quality of the lumber. Some experienced sawyers are especially good at making straight cuts and reducing dimensional variation, and their work is worth more money.

Once you have shopped around and are ready to place your order, set delivery dates a week or two in advance of when you need the materials.

Grading and drying—Kiln-dried lumber is inspected and graded based on its species, the existence of defects (see the box on the facing page) and mechanical testing of its strength characteristics (primarily fiber stress in bending, f, and modulus of elasticity, E). Such lumber will have a

---

Buying green lumber... *by Paul Fuge*

When you're buying wood from a sawmill, remember that straight lumber comes from straight trees. Understanding this, you can talk to your sawyer about what trees he'll be cutting to fill your order. He may already have the logs in his yard. If he does, be sure they're straight, and that they haven't been there too long. Check to see the bark is still tight, that no rot has set in, and that the wood is not infested with ants.

The best boards and trim stock come from the part of the log some distance from its center. Lumber sawn from near the pith is subject to stresses that can lead to warping and twisting. One good approach is to tell your sawyer that you want him to box the hearts of a certain number of the logs. This way, you'll wind up with 4x4 or 6x6 posts from the center of the logs, and lots of high-quality stock from the outside—perfect if you're building post and beam. If you've ordered hardwood, this also forces the sawyer to use butt logs (from the lowest and straightest part of the trunk) rather than industrial-quality tie logs (cut from the more twisted, upper portion of the trunk). This isn't as important when you're ordering softwood, which tends to be more uniform. As long as the lumber is straight and the knots are tight and less than 2 in. across, it doesn't matter if it was cut from a tie log or a butt log. Boxing the hearts, though, is always a good idea if you need posts.

Different parts of the country, of course, have different native woods to work with. Around much of New England, the most common are oak, white pine and hemlock. Oak is great for structural posts and beams, and is a safer bet than any softwood where water is liable to stand. White pine makes fine siding and trim, but doesn't hold nails well, and yields lousy studs and beams. Hemlock is denser, heavier and stiffer than white

pine, and makes much better structural members. It's decent for framing, too, though no one will ever mistake it for Douglas fir. Hemlock tends to be a bit brittle. Watch out for shakes (see the box on the facing page).

A sawmill might have some logs with subtle flaws—seams or stains or little branches—that keep it from being sawn into high-grade lumber. It's liable to be pretty wood, and you can often buy it at a bargain by offering a few cents a board foot over the price of industrial lumber.

When your order is delivered, stack and sticker it immediately. I recommend shading the pile with scrap or waste wood so the sun doesn't dry out the top boards too quickly. I don't believe in covering stacked lumber with plastic, which I feel results in higher humidity and slower drying, even if the ends are left open in the air.

It's worthwhile to take the trouble to stack imperfect boards with their bends, bows and cups opposing each other. They will straighten somewhat as they dry. This is particularly important for hardwoods.

The irony of all this is that most commercial, milled framing lumber these days is stamped S GRN, which means *surfaced green*. It's not kiln-dried, and probably has a moisture content significantly higher than the standard 19%. On most production jobs it is nailed in place and restrained before it dries, so it doesn't twist or bend badly. If you were to let it dry unrestrained, though, you could have problems.

One more thing. When you order from a sawmill, the sawyer is custom-sawing for you and has no alternative market. Figure carefully, find a sawyer you think will do a good job, and then consider a deal a deal.

Paul Fuge operates a lumberyard and millwork shop in Shelton, Conn.

grade mark stamped on it that gives you some assurance of quality. At a sawmill, you are on your own in determining the quality of what you buy. A conservative rule-of-thumb for sizing joist and rafters (or for using span tables) is to assume an f-value of 900 psi to 1,000 psi (pounds per square inch) and an E of 1,000,000 psi for ungraded lumber. (For more on sizing roughsawn lumber, see p. 160).

If you order your lumber far enough in advance, you can dry it yourself. Drying your own lumber will reduce shrinkage, checking and warping, increase strength, relieve residual stresses, and minimize the possibility of decay and stains. According to the USDA, 1-in. boards will dry to 15% to 20% moisture content in about two months of warm, dry weather. Two-inch stock will take two to three months. Double that in cool weather. Here's how to do it properly: First stack your lumber pile at least 6 in. above the ground. Put the thicker material on the bottom of the stack, as it is better able to resist the weight of the boards above. (It's harder to get at when you need it, though.) Place stickers (1-in. spacers) between layers, perpendicular to the boards at 12-in. to 18-in. intervals. Be sure they are directly over each other, and that the ends of the pile are supported. Cover the top and sides, but not the ends of the pile (as shown in the drawing at right); this way you protect it from rain yet also allow air to circulate. You may want to paint the ends of the boards with aluminum paint to slow down the rate of drying and reduce the chance that end checks will develop.

A simple polyethylene plastic "greenhouse" with a small fan to circulate air can speed up the drying process. Make it black on the north side, clear on the other sides, and slope the south side to face the sun. The USDA Forest Products Lab (Box 5130, Madison, Wis. 53705) has detailed information on the solar drying of lumber.

Financing—Because sawmill lumber is not planed smooth or stress-graded, many building inspectors, codes and lending agencies prohibit its use. Banks naturally want to protect their investment, and the final condition of a green-wood house frame is impossible to predict, since many defects won't appear until after a year's drying. Of course you the owner can find, evaluate and repair any defects that develop, but this isn't likely to satisfy other people who are in a position to make decisions about your project. However, if you are building without a loan or in an area without a lot of building regulations, you will probably have no trouble.

Working with green wood—Luckily, given the inconveniences and problems you can run into working with green lumber, there are approaches and techniques that will help you do so with every expectation of success. Here are a few tips.

Always use threaded nails. Threaded nails have 160% to 180% more holding power than equivalent straight-shanked nails, because wood fibers expand back into the grooves after the nail is driven. They come in various sizes and in two types: spiral-groove (thread is cut into the

As green wood dries

Most construction lumber will shrink from 5% to 10% as it dries (about $\frac{1}{16}$ in. per inch across the grain) and about half that amount radially across the growth rings. Lengthwise shrinkage is insignificant (about 0.2%), since wood cells reduce mostly in diameter, not length, as they dehydrate. A board that starts out 6 in. wide may be 5$\frac{3}{4}$ in. or less after it dries. That could mean $\frac{1}{4}$-in. to $\frac{3}{8}$-in. gaps between all your beautiful vertical siding. Dense woods like oak shrink most, but they are also very difficult to work if dry, because they are so hard. All wood, green or dry, will shrink or swell to some extent with changes in temperature and humidity. You can usually live with these small changes if you let the wood acclimate itself to the temperature and humidity of your site for a few days before it is used. It is the defects that develop as a result of drying that you must be sure to compensate for, not the small changes. These defects include:

Cup—curling of edges across the grain. This can pop the heads of flooring or siding nails fast if you don't use threaded nails;

Bow—a lengthwise curve along the flat surface of a board;

Crook—a lengthwise curve along the edge of a board. In boards used on edge, such as joists, this is called a crown. It is a good practice to place crowns up when installing joists or rafters because this helps them to resist

downward deflection under load;

Twist—a curve in two dimensions;

Checks—minor separations of the wood along the grain, harmless if not too severe;

Splitting—longer and deeper separation along the grain (often through the board). This may seriously weaken beams or joists. End splits measuring more than one-third the depth of the piece of wood can result in horizontal shear under load. If such splits develop during the first year, repair or replace the board;

Shakes—similar to checks and splits, but the wood is separated between the growth rings. Shakes can be serious if they are over $\frac{1}{8}$ in. or greater than one-third the board thickness.

Other defects, not caused by drying, are:

Rot—a serious defect, caused by moisture and fungus. Never use for a structural purpose any piece of wood that appears punky or soft;

Knots—If there are any large knots near the edge of a piece of wood being used for a joist or rafter, place them up, since compression forces in the wood under load will then clamp the knot tightly in place. This is more important advice than the crown-up rule, but if there is a conflict, or if the knots are large or numerous, play it safe and use another piece;

Wane—a beveled section along the edge of a board where the bark used to be. This is a visual defect, and you can use such a piece where it isn't fully exposed. —*Paul Hanke*

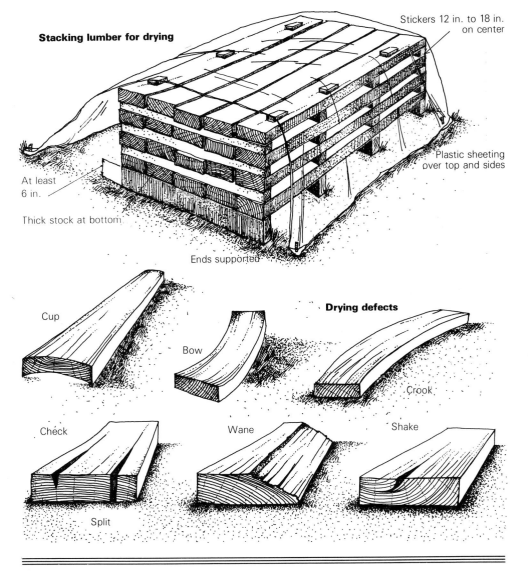

Stacking lumber for drying

Stickers 12 in. to 18 in. on center

At least 6 in.

Thick stock at bottom

Ends supported

Plastic sheeting over top and sides

Cup

Bow

Crook

Drying defects

Check

Wane

Shake

Split

Nailing board and batten

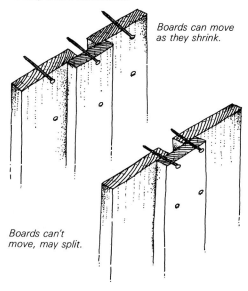

Boards can move as they shrink.

Boards can't move, may split.

In board-and-batten construction, don't nail the batten to the boards in an attempt to prevent cupping. As the boards dry, they will shrink and move. If they are nailed to the battens, they will probably split.

Green wood for joists and studs

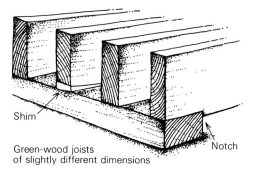

Shim

Green-wood joists of slightly different dimensions

Notch

Because green wood isn't cut to precise dimension at the mill, joists often must be shimmed or notched. Choose the widest stock to be plates, then set studs flush against the interior side of the plate.

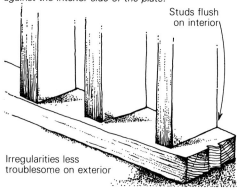

Studs flush on interior

Irregularities less troublesome on exterior

Threaded nails have a lot more holding power than straight-shanked nails. They are less likely to pop as green wood dries and shrinks.

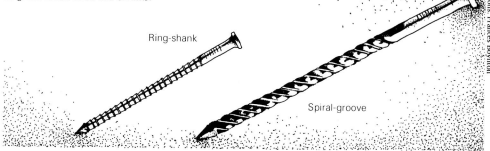

Ring-shank

Spiral-groove

shank) and ring-shank (looks like a screw, but doesn't spiral (see the drawing, bottom).

I once used 9-in. threaded spikes to attach dormers to the roof of an old Vermont farmhouse. Occasionally one would bend after being driven an inch or two, and we couldn't pull it out with a wrecking bar! It had to be hacksawn off, and a new nail tried elsewhere.

Threaded nails are the best insurance that your joints will stay together. They cost about 50% more than the straight-shanked type, but they are worth it, especially when you consider that the cost of nails is typically only about 1% of the cost of a house. Except for 6d flooring nails, threaded nails may not be readily available at building supply dealers, so inquire in advance, and order if necessary.

Larry Hackenberg, who wrote *The Green Wood House*, suggests nailing battens into boards when applying siding to reduce cupping (drawing, top left). This is the opposite of conventional construction practice. I think you'd get extra splitting using Hackenberg's method. A better solution is to pay attention to the Heart Side Out rule, explained at the end of this article.

Always nail at close intervals into green wood, but don't nail within about 1½ in. of the ends of 1-in. boards or you'll split the wood. Driving nails at a slight angle makes a stronger connection than flat nailing.

Get lumber planed on two edges. The dimensions of roughsawn lumber will quite likely vary slightly from one piece to another. Sawmills are not precision machines, and the experience and finesse of sawyers vary. A batch of 2x6s from the same mill might vary in width from 5⅞ in. to 6¼ in., which can be quite a headache when it comes to doing finish work.

In floors and roofs you can usually cope with some variation in widths by either notching or shimming (drawing, center left) or by sorting and laying joists or rafters in descending order. This second method will result in a floor that is slightly out of level, though, and the first method is preferable. For walls, it's best to line up all your pieces in order of thickness, use the widest ones for plates, and align interior edges for a smooth surface, as shown in the drawing at left. Sheetrock is a much less forgiving material than exterior sheathing and siding.

The problem can be avoided altogether by having your lumber planed on two edges, so that all pieces are a uniform width. This is what I recommend, especially for wall framing.

Temporarily brace some beams. Certain rafters or joists may carry especially heavy loads

Further reading

Designing and Building Your Own House Your Own Way by Sam Clark ($8.95 from Houghton Mifflin Co., 2 Park St., Boston, Mass 02108). Includes air-drying, defects of lumber, sawmill vs. lumberyard, and step-by-step procedures. By an experienced green-wood carpenter.

The Green Wood House by Larry Hackenberg ($4.95 from University Press of Virginia, Box 3608, University Station, Charlottesville, Va. 22903).

The Owner Built Home by Ken Kern ($7.95 from Charles Scribner's Sons, 597 5th Ave., New York, N.Y. 10017). Chapter 24 discusses wood and various threaded fasteners.

Low-Cost Green Lumber Construction by Leigh Seddon ($8.95 from Garden Way, Charlotte, Vt. 05445). Includes practical advice, case studies and plans for a solar drier.

From the Ground Up by Charles Wing and John Cole ($9.95 from Little, Brown & Co., 34 Beacon St., Boston, Mass. 02106). Grading and seasoning lumber, plus span tables for roughsawn stock.

in deflection. They'll be 30% to 100% stronger when they're dry than when they were green. A temporary brace will keep them from bending out of shape while they season.

Do finish work only after the first year. Most shrinkage occurs during the first year after construction, especially if the house has gone through a heating season. By leaving the house unfinished on the inside during that time, you'll be able to see and correct any problems that occur. Insulate and install your vapor barrier, but resist the urge to hang that Sheetrock or nail down your hardwood floor. You will immediately notice any splitting that occurs, and eliminate nail-popping in finish materials.

Never use green wood for finish work. Wood will warp and shrink as it dries, leaving gaps where you don't want them. Also, using green wood for window frames may result in cracked glass later.

Use solid blocking. Use solid blocking between joists, studs and rafters at mid-height, midspan, or every 6 ft. to resist the inevitable twisting forces that occur when green wood dries. The only exception to this rule is in the airspace of ventilated roofs. Here, use blocking 2 in. smaller than the rafter size to leave an airspace.

Use narrow pieces. To minimize gaps caused by shrinkage, use narrow boards wherever possible. A rule-of-thumb is to use a maximum width of 8 in. for 1-in. thick boards. Using shiplap or tongue-and-groove decking and siding makes gaps less apparent.

Put heart side out. The heart side of a board is the side that had been closest to the center of the tree—the side to the inside of the circumference described by the growth rings. As the board dries, it will tend to cup so that the heart side is convex. If it has been put in place facing out (siding) or up (flooring), such cupping won't pop nails, as it would if edges were allowed to curl away from the nailing surface. □

Paul Hanke is a designer and architectural draftsman at Northern Owner Builder in Plainfield, Vt.

Illustrations: Frances Boynton

On-Site Carpentry with a Circular Saw

Cutting in place saves time and trouble

by Jud Peake

The portable electric circular saw isn't just a labor-saving device. It's a tool that can do things which would be impossible with a handsaw. The circular saw is hazardous, and its electric cord tethers you to an outlet, so you've got to organize your work differently than you would if you were using a handsaw. But if you can develop the right habits, tricks and sequences for moving materials, nailing and cutting, you can work safely and also save time and energy.

I think worm-drive saws are best. They are better balanced, harder to stall and, most important, the blade is on the left, so a right-handed person can see the cut in progress. If I were left-handed, I would investigate sidewinder models with the blade on the other side.

Whenever wood binds on the blade, a circular saw will kick back toward the operator. Always support the workpiece in such a way that one part will fall away after the cut. Cut framing lumber near the floor, supported on your foot, as shown below, not on a sawhorse. A right-hander

Supporting Lumber During a Cut

Right-handers brace wood with left hand, with wood supported on right foot.

should hold the wood with the left hand, supporting the end closest to the cut on the right foot, with the right hand a little to the right of the body. Make sure at least part of the foot of the saw is always resting on the work.

Cutting in place—If you set out to build a house you must realize you're going to have to move its entire weight from wherever the material is

dropped to the site. This consideration should dictate how the carpenter organizes his or her work. One of the first things you realize is that the circular saw is best used for cutting in place. In addition to reducing physical labor, cutting in place can reduce the necessity for measuring and marking on site and at the lumber pile.

Before you move anything, consider the stack of lumber as a convenient place to cut many pieces to the same length—say, 30 studs for a low wall. First square up one end of the stack by beating on the ends with a hammer and check-

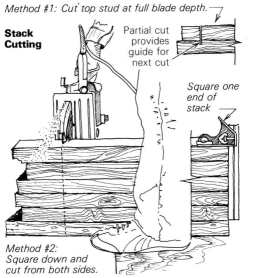

Stack Cutting

Method #1: Cut top stud at full blade depth.

Partial cut provides guide for next cut

Square one end of stack

Method #2: Square down and cut from both sides.

ing with a square, then cut across the top or along the side at the uneven end (method #1, left). If you cut across the top of the stack, the depth of the blade will exceed the thickness of the wood enough to score a guide for cutting the next layer. Be sure your cut is square.

Another way (method #2, left) is to square down the side of the stack across the edges of the lumber. You must be able to get at both edges to complete the cut. This method means you'll have to move more material, but it is usually more accurate and faster in the long run.

Joists are usually lapped, but if they do require cutting (to fit between hangers, for example), lay them between the beams with one end resting on top of the hanger, as in the drawing below. Eyeball a cut along the top of the overlapped beam, allowing the blade to graze the corner of the beam. If the tool is new to you, set the blade depth $\frac{1}{8}$ in. deeper than the joist is thick. With a little practice, you'll be able to rest the toe of the

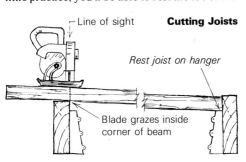

Line of sight **Cutting Joists**

Rest joist on hanger

Blade grazes inside corner of beam

COMMON BUILDING TERMS

Blocking—Short pieces cut to fit between framing members. Blocking is used structurally to prevent framing members from twisting, or to provide nailing surfaces for wall panels. Blocking is also used as a draft stop to keep fire from spreading within a building's concealed spaces.

Cripples—Framing members that support the ends of a header, also called trimmers.

Double plate—The plate at the top of a stud wall, doubled to allow overlapping plates at intersecting walls.

Eyeball—To gauge a cut by sighting, unaided by a line, string or square.

Hangover—The portion of the material that will be cut off in place later.

Kerf—The material lost to sawdust. You can't cut a piece $80\frac{1}{2}$ in. long into two pieces $40\frac{1}{4}$ in. because you'll lose about $\frac{1}{8}$ in. in the saw kerf.

Sidewinder—A saw in which the shaft of the motor is parallel to the spindle of the blade.

Skip-sheathing—1x material laid with large spaces between courses to allow the shingles or shakes they support to breathe.

Snap a line—To lay out a cut by making a line with a chalk box.

Staggers—Intersections of structural plywood sheathing should form a T, not a cross. In other words, three pieces of plywood—not four—comprise a correct intersection. This practice is called staggering the joints and requires some portion of a full sheet (usually half) to begin alternating courses of sheathing. These partial pieces are called staggers.

Tack—To nail only enough to hold in place. Used when the material might be removed, or nailed off later.

Tear-out—The splintering of wood fibers caused by a blade as it leaves the cut. Backing the workpiece with a scrap board will prevent tear-out.

Titch—A small cut or notch made as a guide for the main cut.

saw on the work and adjust the depth of your cut by pivoting the saw.

The subfloor provides another excellent opportunity to use the technique of cutting in place. Tack the plywood in position for half the area to be covered, allowing the ends to run wild over the perimeter of the building. Snap lines around the perimeter and cut. Use the large scraps as staggers for the second half, snap the rest of the perimeters and cut. If you are going to have a lot of hangovers 4 ft. long or more, it is a good idea to pre-cut them first on the plywood stack before spreading.

When you are framing walls, you should cut various cripples, jacks, blocking and double plates in place. Spread your pre-cut studs and plates and nail them together. Now spread your double plates across the studs near the top of the walls and position cripples with one end resting on the floor tight against the header; the long end extends over and beyond the bottom plate. Also spread lengths of material for head jacks, sill jacks, and atypical blocking. Cut all these components to length at one time. Don't interrupt the process to nail, measure or possibly snag the saw's cord on uncut material.

Cut double plates by eyeballing a square cut in line with the appropriate channel mark on the top plate. For cripples, eyeball along the top of the bottom plate, as in the drawing below. The

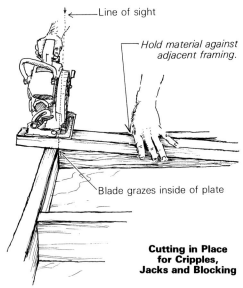

Line of sight

Hold material against adjacent framing.

Blade grazes inside of plate

Cutting in Place for Cripples, Jacks and Blocking

tight cut you need here will cause the blade to scratch the top flat side of the bottom plate and can cause the saw to kick, if you don't hold the cripple firmly enough. Keep your knee behind your elbow and your foot out of the line of the cut. Cut jacks and blocks in the same manner. If the material can't be well braced during a cut, make a slight titch, or nick, with the sawblade, as a reference point, then cut nearby with adequate support.

Pocket cuts—This trick has a wide range of applications. Start with only the toe of the saw's foot on the wood. Instead of pushing the saw along the line of cut, hinge the saw on its toe, hold the blade guard back with one hand (drawing, top of next column) while slowly lowering the blade into the work. This method allows cuts directly into the middle of the work, without access cuts from the side. It can be used to cut holes

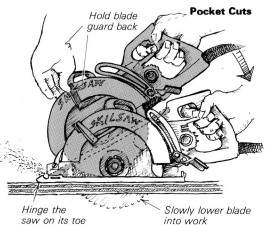

Hold blade guard back

Pocket Cuts

Hinge the saw on its toe

Slowly lower blade into work

in subfloors for toilet flanges, for example.

The diagonal let-in brace installation provides a test for both cutting in place and pocket cutting. Lay the 1x brace (a brace 1 in. by some other dimension) in position diagonally across the wall, carefully avoiding nails in its path. Cut off the bottom of the 1x at the angle formed by the bottom plate, as in the drawing below. The

Letting in a Brace

Hold diagonal brace with your foot. Make ³/₄-in. deep pocket cuts using edge as guide.

Turn saw sideways and hold the blade guard back. Release trigger before cut is complete.

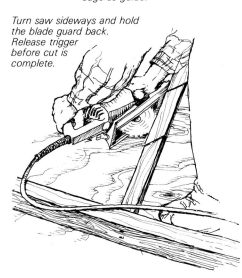

brace needn't be tacked; hold it in place with your foot as you work. While resting the foot of the saw on the 1x and guiding the blade along its edge, lower the saw blade into the studs and plates the ³/₄ in. necessary to let in the brace. Next, kick the 1x out of the way, and turn the

saw sideways, so the plane of the blade is parallel to the plane of the floor. Hold the trigger handle with one hand and, with the other, grasp the blade guard flange and the top handle. Now lower the blade into the wood and keep your feet out of the way. Because you're holding the guard back, it's important to let off the trigger just before the end of the cut; the inertia of the blade will finish the cut and the blade will be dead by the time you're ready to take it out.

You can also use the pocket cut in conjunction with an improvised guide for trimming corners on existing shingled walls. Tack a length of scrap 1x over the shingles to coincide with the trim. Set the saw depth so the blade doesn't quite reach the tarpaper below the shingles and run the saw along the 1x guide. Break the shingles off the rest of the way and apply the trim. If the tarpaper should accidentally be cut, cover the slice with asphalt-base hydroseal before trimming. This method is an inexpensive alternative to weaving the shingles at both the inside and outside corners.

Cutting two at once—If you can cut single pieces in place, why not cut two? To cut skip-sheathing, lap both pieces over the rafter, making sure that each piece reaches center. Don't tack the material in your line of cut. After all the pieces are tacked in place, cut all the laps in one pass each, as shown. Plywood sheathing can be

Cutting Two at Once

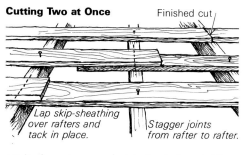

Finished cut

Lap skip-sheathing over rafters and tack in place.

Stagger joints from rafter to rafter.

cut in the same way, but instead of eyeballing, you will have to snap a line.

You can cut two boards at once to bisect any angle. For rough work, such as the intersection of plates, lap one piece over the other and eyeball a titch on the outside corner. Cut through both pieces, heading for the titch from the inside corner. The kerf left in the bottom piece will indicate the angle for completing the intersection. This trick is useful when the walls don't meet at 90° angles (drawing below). For more careful

Intersecting Plates

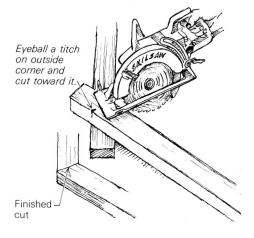

Eyeball a titch on outside corner and cut toward it.

Finished cut

work, you can square up on both of these corners and draw a connecting line. In this case, you will need to move the material together after the cut to make up for the saw kerf.

If, for some reason, you are unable to move the material ends together, you can always find a bevel at which to set the saw to hide the kerf. As you can see in the drawing, it is impractical to

Hiding the Kerf

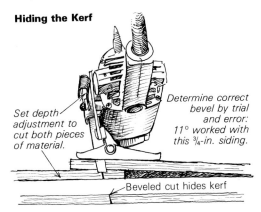

Set depth adjustment to cut both pieces of material.

Determine correct bevel by trial and error: 11° worked with this ¾-in. siding.

Beveled cut hides kerf

try to predict the exact angle. Trial and error works best. The beveled lap cut can sometimes be used when doing exterior trim or siding. Start from the bottom and cut laps in place as you go. Don't wait until it's all up to cut.

Cutting a bevel in thicker material can be difficult. The 45° bevel is easy to visualize; it bisects a right angle. But with a circular saw it can be hard to execute. Let's take an example from rough work. You've cantilevered some joists for a bay window that require 45° cuts, and snapped your lines. Square down the face of the joist, set the bevel on your saw, and cut. You find out soon you can't cut one side of the bay window from above; you usually can't reach it from below, either. Instead, square down from the long point and make a square cut, as in the drawing below;

Cutting a Bevel in Thick Material

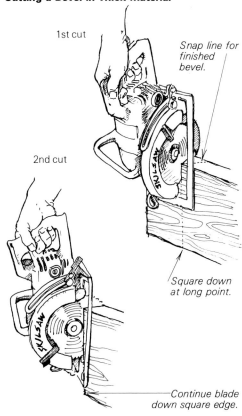

1st cut

Snap line for finished bevel.

2nd cut

Square down at long point.

Continue blade down square edge.

then cut into the top edge of the joist along the snap-line mark until the blade has reached its full depth. Continue the cut by leading the front of the blade down the square end edge of the joist. I have found this method to be consistently more accurate than relying on the saw's beveling capabilities.

Trim work—It can be hard to use the circular saw here, but often it's the only power saw on hand. In this case, the best place to cut is on sawhorses. Cut from the backside of the material, so the tear-out is hidden. For outside miters you just have to cut carefully. Tilt the square up toward the saw, so that, instead of running over it, the saw runs against the edge of the square, which acts as a guide. If the material is wide and tends to get forced onto the untoothed portion of the blade where it binds, make relief cuts on the waste side. Don't try to make a pocket cut on a beveled miter.

For inside corners on trim of rectangular cross section, I think the butt joint is best. It's easiest, and, if the material cups or shrinks, it's much more difficult to see any gap in the joint. If a miter joint is used, any gap will be obvious to people as they walk by.

Sometimes you can get stuck cutting molded material for interior trim without a miter box; either you don't have one on hand or your material won't fit in the one you do have. You can make perfectly acceptable inside corners by first making a 45° bevel with a circular saw and then back-cutting the joint with a coping saw. A back-cut joint is cut at slightly less than 90° to allow the trim's visible edge complete contact with the adjacent trim. This is useful when joining trim with complex contours (drawing below). To in-

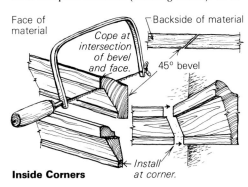

Face of material

Cope at intersection of bevel and face.

Backside of material

45° bevel

Install at corner.

Inside Corners

stall picture molding, for example, butt the material between the walls on one end of the room. Measure and cut the material for the adjoining side just as if you were going to miter the inside corner. Make this cut from the backside of the material with a circular saw and follow the line formed by the intersection of the face of the material and the end-grain bevel made by your circular saw cut with a coping saw. Make the piece ¹⁄₁₆ in. long, and the back-cut trim will dig into the adjoining piece and hide any errors.

Dimensions and notching—If you want to crosscut or rip something at 1½ in. or 3½ in., you don't have to measure and mark. The circular saw has these dimensions built in. From one side of the foot to the far side of the blade is about 3½ in.; from the other side of the foot to the near side of the blade is about 1½ in. Eyeball the ap-

propriate side of the foot along the edge of the material to get your cut. If you have some odd-sized rip to make that is not greater than 3½ in., and a rip fence isn't handy, make a mark on the toe of the foot at the correct distance from the blade and lead this mark down the edge of the material to be cut.

Notches around posts for exterior decking can be accomplished quickly by cutting from the backside. Start a saw cut into a scrap of the same material and stop just as the blade touches the lower corner, as in the drawing. Make a pencil

Notching Posts

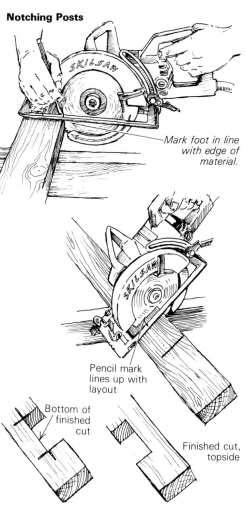

Mark foot in line with edge of material.

Pencil mark lines up with layout

Bottom of finished cut

Finished cut, topside

mark on the foot of the saw in line with the square edge of the wood (this mark won't be accurate if you change the depth adjustment of the saw or work with varying thicknesses of material). Lay out the notch on the backside of the work and cut past the intersecting lines until the pencil mark lines up with the layout; this hides tear-out and saves handsawing.

If you use the circular saw enough, someday it's going to kick back on you when you really don't expect it. That's why it's important to make your safety precautions habitual. Support the wood well, stand to the side of the line of cut, release the trigger before the end of the cut, and back up your elbow with your knee when possible. Eyeball your cuts square—the more you do it the better you'll get. Cut in place whenever practical and work your material—moving and nailing into a pattern with your cutting to save unnecessary movement. Don't use your tape measure unless you have to. □

Jud Peake is a contractor in Oakland, Calif..

Drywall

Hanging and finishing gypboard can be an aggravating mess. A veteran builder shows how to do it right the first time.

by Bob Syvanen

Gypsum board, gypboard, Sheetrock (a brand name), drywall—whatever it's called, it's probably the most disliked and misunderstood stuff in house building. The sheets are heavy, the sanding seems endless and the dust is unpleasant. But a drywall job well done is very satisfying. There are many ways to do it, but I have settled on this system after years of trial and error, watching others and asking questions.

Drywall comes in sheets from 4x6 to 4x16 and in ¼-in., ⅜-in., ½-in., and ⅝-in. thicknesses. Quarter-inch drywall is used over old plaster walls so it is always fully backed; ⅜-in. is for cheap construction, and the framing must be no greater than 16 in. o.c. You can use ⅜-in. material double, so you end up with a ¾-in. thick wall. The most commonly used drywall is ½ in. thick; ⅝-in. is for extra good work. There is also a ⅝-in. fire-rated sheet, usually required on a garage wall shared with the house. Use water-resistant sheets in bathrooms, especially around the tub and lavatory.

Most lumberyards carry drywall sheets, but I prefer to buy from a drywall supplier who has a truck with a hydraulic arm. This is the best thing to come down the pike since drywall itself. Loaded with sheets of gypboard, the arm will reach right in the door or window, and sometimes even to the second floor. You might have to take out a window sash, but it's a small price to pay for such a glorious convenience. For a small load delivery, you may have to wait until the truck has a full load and is heading your way. It's advisable to put the sheets into the rooms where they'll be hung and to keep the sheets to be used first on top of the pile.

Editor's note: Bob Syvanen, a draftsman and carpenter in Brewster, Mass., is consulting editor to Fine Homebuilding. *This article is an expanded version of a section in his book* Interior Finish *($7.95 from East Woods Press, 820 E. Boulevard, Charlotte, N.C. 28203).*

Preparatory work

Before you begin hanging Sheetrock, it's best to clear the floors of tools and clutter. The sheets are just too heavy and cumbersome to carry through an obstacle course. A clean sweep-up is nice too, even though you'll shortly be up to your ears in debris again.

New England is the only place I know of where 1x3 strapping is used as a base for drywall ceilings. Although it's a lot of work, it's an excellent way to keep things flat with today's varying lumber sizes. Eyeball the ceiling for any bad joists **(1)** and trim with a Skilsaw before nailing up the strapping. You can also work to a stretched string with spacers at each end. Make minor adjustments with shims of wood shingle tips between the strapping and the joists **(2).** I like to double-nail for more holding power, but single-nailing works too. When a ceiling needs joints in the strapping, be sure to stagger them from one course to the next. Check the walls for bowed studs by holding a long, straight 2x4 against the wall **(3)**. A badly bowed stud in a finished wall really shows, so replace or straighten it. To straighten a bowed stud, saw well into it on the side opposite the hump **(4)**. The stud can then be forced into alignment. Wood shingle tips driven into the kerf will keep the stud straight while cleats are nailed on each side, just like splinting a broken bone together. It might take two such cuts if the bow is bad.

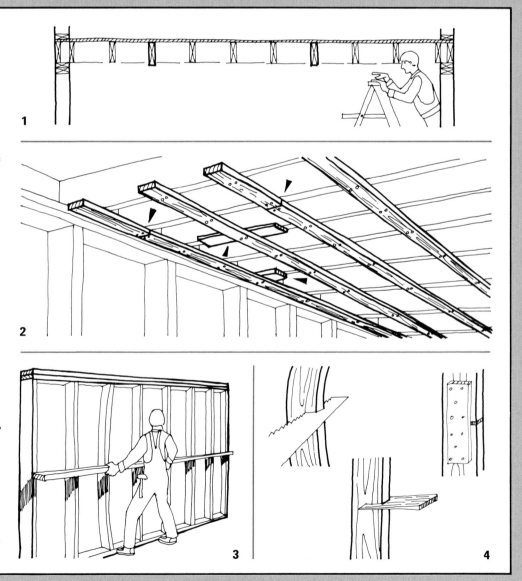

Ceiling joints

The fewer joints, the better. Ceilings are best done with a minimum of joints where the untapered edges of the sheets come together (5). So where possible, span the width with a single sheet. Untapered edges make for bad joints (6), but there are several ways to overcome this. Where the ceiling requires an untapered edge joint, it should fall between the joists or straps. The first solution is to nail up blocking 16 in. o.c. across a joist space and then a 2x4 or 1x3 strap parallel to the joists (7), so that when the sheets are nailed they will be depressed about $\frac{1}{8}$ in. Another way is to cut four 12-in. by 12-in. squares of Sheetrock. Butter these squares with joint compound and slip them in on the backside of the panel already in place (8). Nail the next sheet up, place a piece of strapping along the seam and hold it in place with crosspieces of strapping (9). These crosspieces will depress the joint, and the buttered squares will dry, holding everything in place like glue. Then remove the strapping braces, and tape and spackle the joint like any other. Although these methods are not commonly used by professionals, either one will result in a smoother surface, and require less work when these joints are taped and spackled.

I like to make a sketch of all the walls and ceilings to decide on sheet size and placement (10). Sometimes I make my sketch on a stud or on a header or each wall, so I don't lose it. If you do the ceiling and butt the ends together, figure on using full sheets, plus the pieces cut to fit (11). When working around the doors and windows, put the sheets in place first, and then cut out the openings. This is extra work, but if pieces are patched in over the doors and windows, they will crack at the seams as the framing shrinks and the house settles. The bumps at these joints won't let the casing lie flat at the corners without a lot of extra work (12). If economy is your goal, figure the wall sheets any way you want, but if quality is what you want, run the sheets over the openings and then cut out.

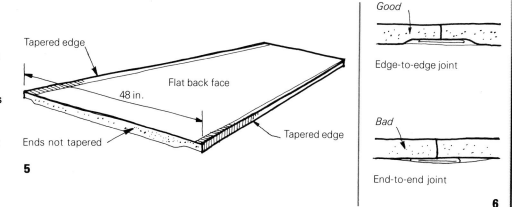

Tapered edge

Flat back face

48 in.

Ends not tapered

Tapered edge

5

Good

Edge-to-edge joint

Bad

End-to-end joint

6

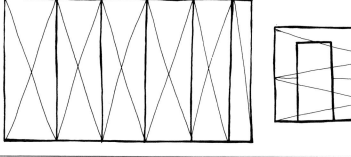

7

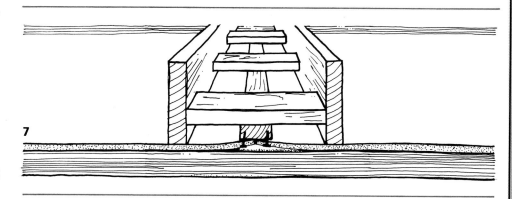

8

9

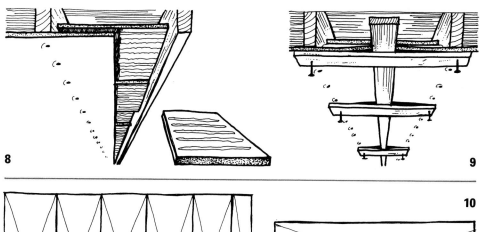

10

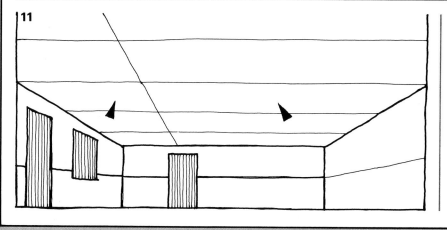

11

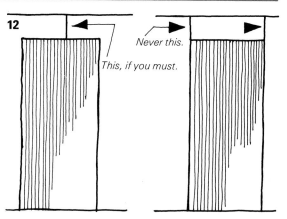

12

Never this.

This, if you must.

Tools

For cutting drywall, a utility knife is useful. A 48-in. aluminum T-square is a must for laying out and guiding the cuts. A 12-in. keyhole saw **(13)** works best for cutting holes in the sheets. A Stanley Surform **(14)** is good for shaving a wallboard edge. A claw hammer works fine, but a drywall hammer **(15)** is lighter, and has a rounded head with a waffle pattern that leaves a dimple without tearing the paper. Use the blade to trim framing or as a jacking wedge. An aluminum hawk can be bought for about $7. A homemade plywood hawk works almost as well **(16)**. A galvanized metal or

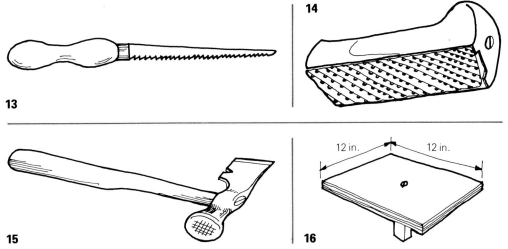

Fitting and nailing

The ceiling is the place to start, and you need two people to do it. I have done it alone, but don't recommend it at all. A ½-in. 4x12 sheet weighs almost 100 lb. A couple of T-braces will make the job easier. Be sure you can reach both of them and still control the sheet overhead. It also helps to stick a few nails in the panel where you'll be nailing. Holding the sheet and fishing for nails can be tough. The T-brace should be a little longer than the floor-to-ceiling height **(20)**, so that there is a slight wedging action. If it is too long, the brace won't stay in place, and if it's too short, the whole business comes down on top of you. The easiest way to hold a sheet against the ceiling is with your head. A sponge inside an old cap can make this a little less painful.

When nailing, be sure the sheet is pushed hard against the joist or strapping before driving the nail home. If it isn't, the nail will pull through the surface of the drywall sheet **(21)**. Once the paper surface is broken, the nail will not hold the sheet in place. If a nail misses a joist or stud, pull it out. Hit the hole with a hammer hard enough to depress the surface without breaking the paper. These dents, or dimples, can easily be filled with joint compound **(22)**. Just spot-nail the sheets enough to hold them in place at first. Secure them later. Cement-coated and annular-ring nails are okay for walls, but screws do a better fastening job on ceilings. Tool-rental shops have both the screws you'll need, and the driver. It's the size of a ¼-in. electric drill with a magnetized Phillips bit and an automatic depth stop. If you must use nails, a good way to get ceiling sheets to pull up is by double nailing **(23)**. First, single-nail 12 in. o.c., starting from the middle to prevent bulges from forming. Then come back with a second set of nails about 2 in. from the first set. You might have to drive

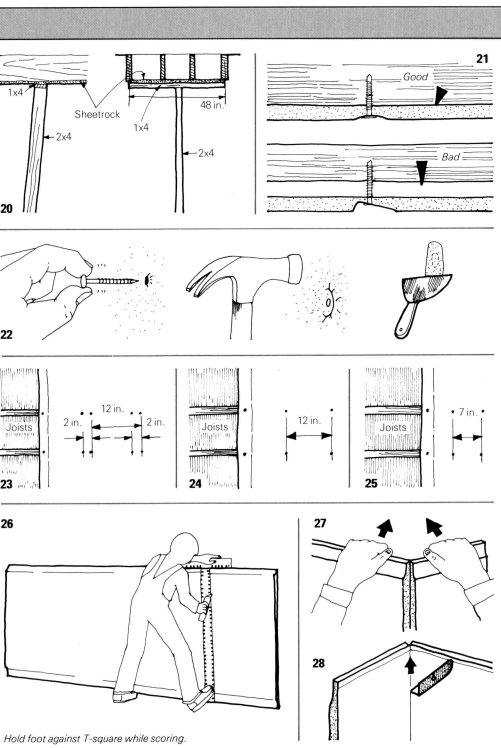

Hold foot against T-square while scoring.

114

plastic mud pan has the advantage of high sides, and runs about $4. At least four trowels are required—a 6-in., a 10-in., a 12-in. and a corner trowel **(17)**. The 12-in. trowel is much stiffer than the others, and it has a curved bottom **(18)**, so it takes a good bit of pressure to smooth and feather out the final coat. I use fiberglass tape for the ceiling joints, and paper tape for the corners and walls. Paper tape costs about $2 a roll, and a fiberglass roll **(19)** about $6. Fiberglass tape eliminates the first step of putting on a layer of compound. It has a sticky surface, and you just stick it on, and then apply a coat of compound—a big help when you're working overhead.

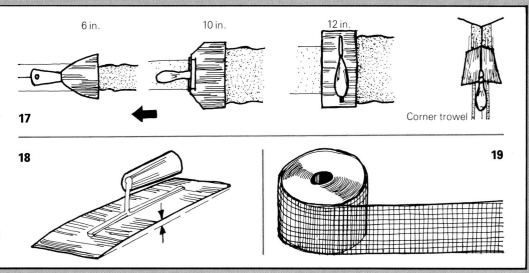

the first set of nails again as the sheet snugs up. Every nail should be driven just below the surface of the sheet, without tearing the paper, creating a dimple for the joint compound. Screwing is 12 in. o.c. **(24)**. Single nailing is 7 in. o.c. **(25)**. Cutting wallboard is probably the easiest part of the job. Just line up the T-square on the sheet and score **(26)**. One stroke with a sharp knife is usually enough. Snap it by folding the sheet back quickly **(27)**. Then cut along the backside **(28)**. If the edge needs cleaning, use the Surform.

Drywall walls are really simple. The panel that touches the ceiling should go in first to ensure a good joint there **(29)**. A few nails stuck into the sheet, on line with a stud, will make it easy to hold the sheet up snug and nail at the same time **(30)**. Leave about a ½-in. gap at the floor on the bottom sheet, so that a lever can be slipped under to push it up tight **(31)**. The bottom of the bottom sheet will be the cut edge, a good surface for the baseboard to butt against **(32)**. If you use corner clips or blocks at inside corners, install the sheet that runs parallel to the first so you don't have to nail into the blocks. The adjacent sheet is nailed to the stud in the corner and holds the first sheet in place **(33)**. Tape and spackle the whole business in place. The thing to remember about outside corners is that the drywall must extend far enough so that the corner bead has adequate backing **(34)**. Some professional drywallers cover the wall and then cut the outlet and switch holes. A safer way is to rub the outlet box with block chalk, hold the sheet in place, smack it with an open palm, and cut along the resulting lines left on the back of the sheet. Another way is to measure the location of the box and mark it on the sheet. This way isn't bad, but expect a few mistakes. Don't forget to mark all the openings, like medicine cabinets, recessed shelving and fans, as the sheets go up. They are easily lost.

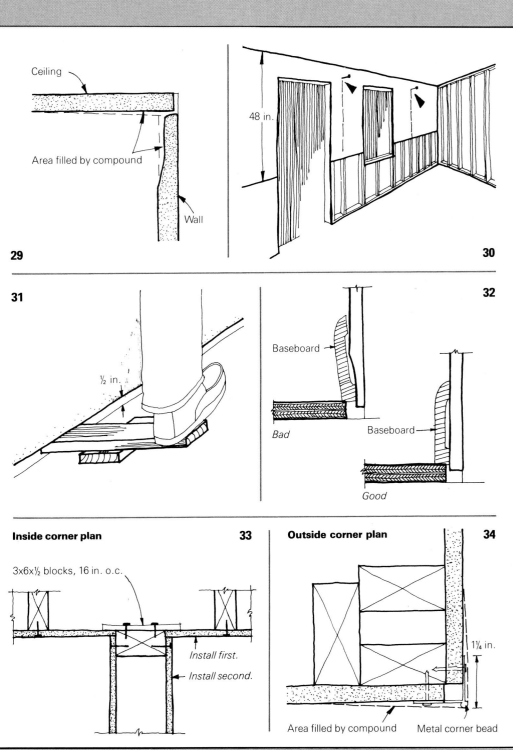

Spackling

Now we come to the spackling, and it's not as hard as most people think. With the proper technique and a little care, anyone can do a professional job. Cleanliness is a must. Any hard lumps or pieces of dirt in the spackling compound will mess up a joint (35). Load the hawk by scooping compound with a wood shingle or a 3-in. by 16-in. stick. Keep the scoop in the bucket, and the cover on so the compound stays moist and clean. Scrape the trowel clean as you work (36). If the scrapings are soft and clean, mix them with the stuff on the hawk. Try to keep the compound in a single gob so it won't dry out so fast. If it gets dirty or lumpy, dump it out. To do a joint, start with a 6-in. or 8-in. trowel and run a layer of compound as wide as the tape, the whole length of the joint (37). Holding the trowel at a low angle will leave a nice bed of compound (38). Lay the tape on the buttered joint and push it flat with the 6-in. trowel (39). Get all the bubbles out from under the tape. Spread a thin layer of compound over the tape. Keep it smooth; there's less sanding that way. Raise the trowel towards the perpendicular for smoothing (40). Try different angles until you get the best results. Each coat must be sanded when it's dry with 80-grit sandpaper. Silicon carbide works the best. The next coat is done the same way, but with the 10-in. trowel (41): Lay on a coat, wipe the trowel clean, then run it the full length of the joint to smooth it out. Press hard. If the trowel loads up, scrape the excess compound on the hawk and continue the run down the joint (42). Every time the trowel is picked up from the joint, it pulls some compound with it, leaving a ridge (43). Further, every bump

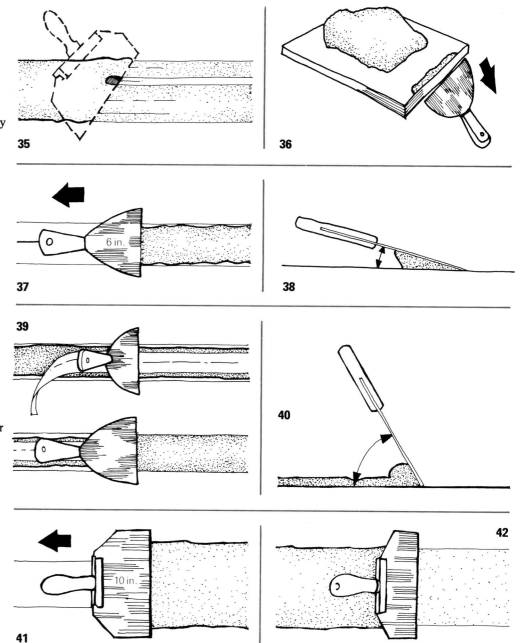

Finishing

Use the drying compound still on the hawk to fill the nail holes and dents. Run the trowel at a low angle to leave a layer of compound in the dimple (48). Then hold the trowel almost perpendicular to scrape the surface clean (49). Go to the next dimple; deposit, scrape, and so on. It will take a few coats, sanding between each coat. Always wear a mask when sanding. There is no asbestos in the compound, but breathing all that dust is not too healthy. Eye goggles are helpful, but you have to keep wiping them off.

I make a pad of sandpaper by cutting a sheet in half (50). Drag the sandpaper across a corner to break the glue (51). This will make the sandpaper more supple, and will prevent its forming a sharp ridge that will dig into the sanded area. Fold one

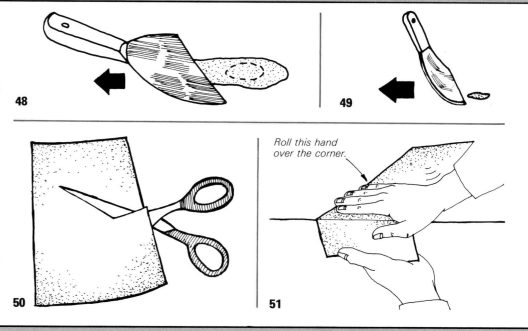

Roll this hand over the corner.

the trowel hits telegraphs onto the surface of the joint. Smoothing out may take a few strokes to get the excess off. Make that last stroke a nice long one, since it forms the base for the next coat. The last coat is done the same way, but with the 12-in. trowel **(44)**. Try to make a long smooth run with this coat. If there's a long ceiling joint, it's best to have a long unobstructed run on a plank raised high enough to allow some good arm pressure against the joint. The 12-in. trowel is much stiffer than the others, so it takes some real force to smooth and feather out the final coat **(45)**. This is the last coat, so make it a good one, and sand it with 120-grit paper. The better the joints are feathered, the less sanding you'll have to do. Sanding is dusty work, and I try to do as little as possible.

Inside corners are done in a similar fashion, but here the paper tape must be prefolded to fit the corner. Drywall tape has a crease down the middle, so folding is easy. The procedure is the same, but use the corner trowel to lay in the first coat of compound **(46)**. Lay in the tape and smooth out the bubbles. A coat of compound is next; when it dries, sand it and then smooth on a final coat of compound. Experiment with the trowel angle for the best results. A few tries and you'll have it. Use a flat 6-in. trowel to scrape off the compound that squeezes past the edge of the corner trowel. Don't disturb the corner you just troweled.

Exterior corners are the easiest of all. Nail on the corner bead and spackle it with a 6-in. trowel **(47)**. Do the last coat with the 10-in. trowel. There will be excess compound running around the corner as you trowel, but it's easy to scrape off because the corner bead gives you an edge surface to trowel against.

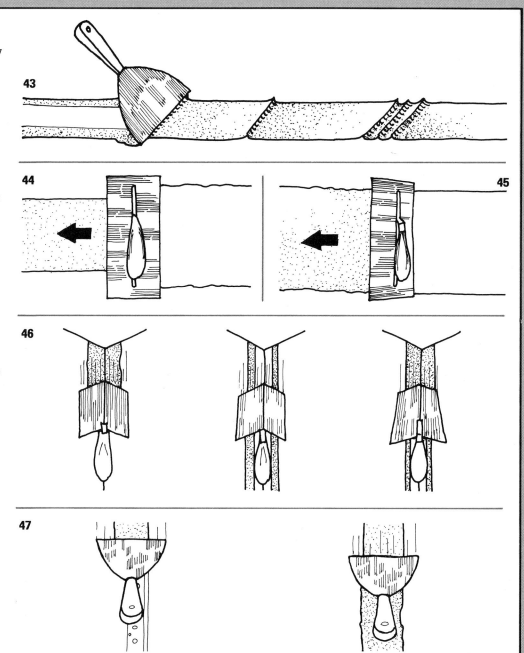

sheet into thirds **(52)**, and then fold the other sheet in with it. This makes a nice soft pad. Sand using your fingertips **(53)**, and stroke perpendicular to your fingers. The control is better, and the sanding surface is broader. Especially on ceilings and in corners, a pole sander **(54)** can save a lot of stretching and bending. It's worth the $12 purchase price. Sandpaper is easy to change, and the swivel head allows flat sanding from any angle of approach.

Check for spots you missed in sanding or spackling, and after you have admired the job, paint it. It's easy to paint without the trim in the way. Use a sealer first, or tiny filaments of paper from the sanding will show through successive coats. Pros spray-paint, and if the polyethylene vapor barrier is uncut and left over the windows, spraying is a breeze. ☐

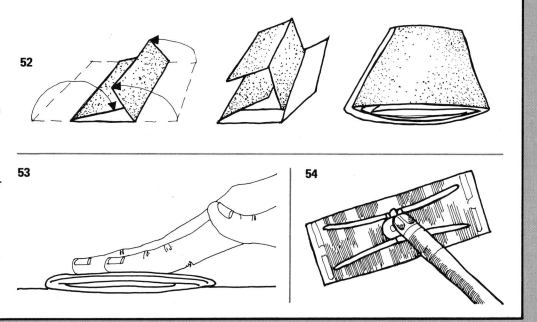

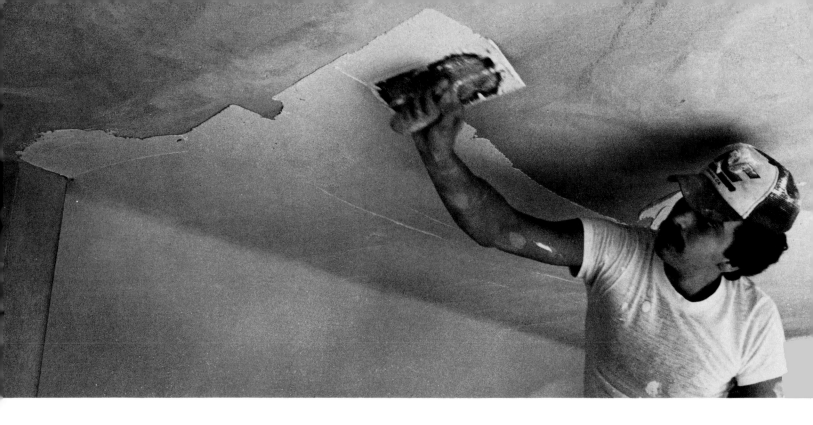

Veneer Plaster

Getting the look and texture of plaster on lath with gypboard and mud

by Tim Snyder

Finishing with plaster isn't the awesome job that it once was. Modern plastering systems have replaced the traditional wood or metal lath with gypboard sheets, textured on one side so they hold the plaster. After hanging and taping this substrate, you trowel on a thin (3/32-in.), seamless surface that's far stronger (up to 3,000 psi) and harder than the drywall itself. It can be painted or left as is, with no sanding or other treatment required.

Called Thincoat, Skimcoat, Kalcoat, Imperial Plaster, or simply veneer plaster, depending on what manufacturer or contractor you're talking to, the finish costs a few cents more per square foot than a conventional drywall finish. It takes time to learn how to handle the mud, though, and you'll need at least two people to do the job right. The plaster sets quickly, and it's meant to be worked fast.

Materials—Veneer plaster isn't a new material, but it's only in the last several years that manufacturers have put together complete systems based on its use. The systems consist of gypsum-core backing board (which is nothing more than regular gypboard with a bluish, textured paper surface), more commonly called blueboard; high-strength plasters for one or two-coat finishes; a retarder compound to extend the setting time if you're mixing big batches or working slowly; fiberglass-mesh

joint tape; metal or plastic corner beads; and plastic edge terminals for transitions between plaster and other surfaces. You'll need all these items for most jobs. Manufacturers have their own product lines and encourage you to use their stuff only, but in practice, everything but the plaster pre-mix is interchangeable.

Veneer plaster can be either a one or a two-coat finish. The one-coat system is quicker, but if you want a slightly stronger, smoother wall surface, use the two-coat system. With the two-coat system, the base, or scratch coat is a grey, coarse plaster that bonds well with the substrate and with the white finish coat that covers it. This final coat can be floated to a satiny-smooth finish, or you can texture it by adding washed sand to the mix, or by going over the plaster roughly with your float. The finish coat can also be tinted with powdered pigment, though this is risky, since color may vary slightly from batch to batch.

Apart from sand, pigment and retarder, all you add to the plaster is clean water. The retarder gives about 15 minutes more working time—a blessing if you're troweling in corners or around contoured areas that demand more attention than flat work. A retarder works best when it's pre-mixed in water and then added to the batch as it's being mixed. Use the water-to-weight ratios recommended on the bag, both for plaster and retarder.

Veneer plaster can be applied as a one or two-coat finish over specially textured gyprock or concrete. The finish is far stronger than drywall and needs no sanding or painting once the finish coat has been floated smooth. Facing page, top, the white finish coat goes on over an equally thin base coat. Facing page, bottom, base-coat application. Stilts are a must for ceilings and high sidewalls.

Tools—Here you'll need a combination of drywall and plastering equipment. A straight-edge and utility knife work fine for cutting the blueboard, just as they do with gypboard. Drywall screws aren't usually necessary, so all you'll need to hang the backer board is a hammer and ring-shank drywall nails.

The plaster has to be mixed by machine. Most pros use a heavy-duty commercial mixer (photo below right), but a drill-driven blade will also work, if the drill can take the strain of heavy mixing. This isn't paint. With either rig, you'll need a couple of clean, strong containers to mix in. Plastic drywall-compound buckets work well, but they won't hold an entire bag's worth of mix. If you can get a couple of 30 or 40-gal. steel drums, use them instead. It's important to wash out the container completely with every other batch you mix. Otherwise, old plaster can contaminate a new batch and cause it to set too soon.

You'll need a hawk to hold the mud, and a plastering trowel to apply it. These can be either wood or metal. Once the mud is on the wall, the trowel becomes a smoothing tool. To produce an extra-smooth finish, you'll need a sponge-surfaced float and a handy source of water to keep the sponge clean and wet as you work the plaster. A garden hose with a misting nozzle does a good job. Some plasterers also use a blister brush for smoothing. It's similar to a paintbrush, but with thick felt strips rather than bristles. Like the sponge float, the felt has to be kept wet. Corners are best finished with a corner plow, which is simply a specialized float that's bent into a right angle so that both sides of a corner can be smoothed at once.

Get plasterer's stilts for working on ceilings. Speed and mobility are important parts of the job, so it's worth renting or borrowing a pair. With a two-man crew working on stilts, the mortarboard has to be set up at tabletop height, so that both mudslingers can get to the mix without dismounting.

Preparation—First, hang the blueboard, just as you'd hang drywall. Since the total thickness of the plaster finish should be only $\frac{3}{32}$ in. or so, get the substrate plumb, with as few voids and dips as possible. Wherever the blueboard meets wood, brick or any surface that won't be plastered, install metal or plastic edging along the transition line (photo above right). These special edge terminals act as grounds so that you plaster up to the dissimilar material but not against it, allowing each surface to expand and contract at its own rate. Outside corners require metal or plastic beads, and inside corners should be taped. Reinforce all other joints with fiberglass-mesh

Preparing the surface. Before the first coat of mud goes on, corners, above, are beaded with metal or plastic grounds. All joints are reinforced with fiberglass mesh-tape. The plaster compound can be mixed using either a commercial rig, below, or a powerful electric drill and spade. Water must be clean enough to drink, and solid should always be added to liquid.

Applying the plaster. Above, the last of a batch of finish coat is about to be hawked and applied to the scratch-coated wall in the background. Learning how to handle the fluid mix takes practice. Speed is essential, since the plaster sets in less than an hour. Once applied, each coat is either floated smooth or textured. Using an overlapping circular motion, below, and keeping the float clean and wet help speed the work. Lighting fixtures and trim are installed after the finish dries.

tape. The self-stick kind and the staple-down kind work equally well.

Veneer plaster can be applied over materials other than blueboard. It will stick to a concrete-block wall as long as you've got a clean surface. Remove dirt and grease by scrubbing down the wall with a weak solution of muriatic acid and water. Poured-concrete walls have to be treated with a bonding agent before you start plastering. This is a brush-on preparation, available from masonry suppliers, that creates a surface the plaster will adhere to.

Using the mud—Mixing, applying and floating techniques are the same whether you're plastering against blueboard or concrete. To mix, always add solid to liquid, except if you need to add retarder to a batch that's in danger of setting too soon. Wet-to-dry proportions can depart slightly from the recommendations on the bag (usually 8 gal. per 80-lb. bag). In hot weather, for example, a quart more water will keep the mud workable longer. It's important to blend the ingredients thoroughly. But don't overmix them, because this accelerates the setting time.

If there are just two people working on the job, the best strategy is to spill your batch on a mortarboard so that both workers come and go with hawks, trowels and floats (photo above left). As the batch runs out, one remixes and the other starts to level the plaster just applied. A batch may set faster than one person can apply it. Working time is about an hour, less if the weather is warm. With three workers, two can apply the mud while the third supplies and mixes, so no mortarboard is necessary.

Do the ceilings first (photo below left). For a one-coat job, pre-hit all the joints with mud as you work section by section. This helps to keep the fiberglass-mesh tape in place as you coat the rest of the surface. Veneer-plaster mud is more fluid than drywall compound, so you have to load your trowel, then move it from hawk to ceiling or wall in a single, continuous motion. This particular stroke takes some time to master. Wear a hat.

Follow your application stroke with a reverse stroke that passes back over the plaster. This double-back technique is the fastest way to get mud on the wall in relatively smooth fashion. The object is to move a lot of material as quickly as you can; then go back over the plaster while it's still workable and float it smooth. A sponge-surface float will produce a smoother finish than a plain metal or wood one. With either tool, use a sweeping circular motion, and keep the float wet. You can use a blister brush or a sponged float to create a completely smooth surface, or just let the roughly leveled plaster set for a more rustic appearance. With either approach, plaster that will be covered by casing needs to be flat so that there won't be gaps between wood and plaster after the trim is applied.

Don't apply more plaster than you can float in 45 minutes. As the plaster begins to set, it darkens noticeably, and once this happens, it's folly to try to work it further. □

Nail Guns

Pneumatic nailers and staplers give even the small builder affordable speed and precision not possible with hand nailing

by Angelo Margolis

Nail guns used to be considered the exclusive domain of tract builders—quick and dirty production tools for huge projects. These days, they are also found in cabinet shops and on building sites where pride in craft is still the dominant influence. Pneumatic fastening tools use compressed air to drive nails, staples, corrugated fasteners and brads. The air is delivered to the tool by a portable compressor through small-diameter flexible hose. The guns, which are usually aluminum-alloy castings, range in weight from 2¼ lb. to 10 lb. and are balanced for one-handed use.

The primary virtue of nail guns and staple guns is the speed you can develop building with them. A subfloor that would take half a day to nail off with a hammer can be completed in an hour with a nail gun or stapler. The time that you save can be spent on more creative work, which requires the judgment and experience a machine can't provide. Even more important, the time saved can mean building a high-quality house for less money. I saved about 15 days of labor on my own house by using pneumatic fasteners.

I do new construction and remodeling on a modest scale, and would no more build a house exclusively with a hammer than I would with a handsaw, though there is a place on every job for the limited use of each.

I first used a nail gun in 1964, when, as an apprentice, I put down a subfloor in an apartment complex. The work was still hard, but it was undeniably faster than nailing off with a hammer. In my mind, however, nail guns were relegated to use by large crews on big projects. It took me ten years to rediscover them.

In the meantime I built a number of spec houses by myself. Because of the hours of tedium that come with siding a house or nailing off the roof, building became a seige, and I often felt I was barely gaining ground. I still work by myself much of the time, but nail guns and staplers have reduced the drudgery.

Guns and staplers—Pneumatic fastening tools were originally developed for the assembly-line production of wood pallets, but residential construction has been the biggest market in recent times. Though nail guns and staplers are most often thought of for nailing off structural work like plywood subfloors, roof sheathing and wall framing, they can also be used on a wide variety of materials including asphalt and wood roofing, drywall, ply-

The biggest advantage of nail guns and staplers is the time you save when nailing off sheathing, siding or subflooring. However, new generations of pneumatic tools can fasten roofing, drywall, and even fine finish work without any sacrifice in quality.

wood and board siding, interior wood paneling, trim and other millwork items. Some guns even shoot a fastener that will penetrate concrete, something that used to require powder-actuated fastening tools.

Most guns operate on an air-pressure setting between 60 psi (pounds per square inch) and 110 psi. The nose of the gun is fitted with a spring-loaded safety, often in the form of a wire bail, that has to be depressed against the work before the trigger will release. The gun fires when a charge of compressed air fills a cylinder and forces a piston forward. The piston strikes a nail or staple in the guide track, and drives it out of the gun and into the work at high speed. Fasteners can be driven in one blow as fast as you can trigger the tool.

Most nail guns can also be used for bounce nailing, or bottom tripping, when you nail off large areas. Keeping your finger on the trigger will fire the gun as fast as you can depress the safety by tapping the surface of the work.

Nail guns and staplers vary slightly according to brand, but all the major manufacturers (listed at the end of this article) produce a variety of guns for different uses. Staples and nails are typically not interchangeable between brands. The trend seems to be toward specialized tools that can handle one type of fastener in a range of sizes appropriate to the task. Paslode, for instance, sells a gun designed for hanging drywall. It weighs just under 6 lb. and drives a ring-shank nail at the same time it dimples the gypboard, something drywall tapers like because the nail is always centered in the dimple.

Guns designed for framing, siding and nailing subfloor and roof underlayment are the most versatile. They generally shoot headed nails from 16d down to 8d or 6d. The magazines of these framing guns are angled away from the surface of the work so that the tool can be used in tight spots for toenailing, fireblocking and the like. Guns that will be used strictly for nailing off large areas have extra-long magazines so that the tool doesn't have to be loaded often.

Nail guns also come in smaller sizes. Some of these shoot finish nails or brads for finish work. These guns are light enough to be held vertically or overhead.

One of the guns I own is a T-nailer for cabinet assembly and finish work. The fastener is shaped like a T, and the gun sets it just below the surface. Lining up the long part of the

Framing guns. Nail guns have a place in small-scale remodeling because they make the work faster and easier. A large framing gun like the one being used below to nail off sheathing is usually the first pneumatic tool bought. It can shoot a range of nails—smooth shank, screw shank, and ring shank—in sizes from 16d down to 8d or less. The angled magazine gives these guns surprising maneuverability. Toenailing studs for a grade-beam pony wall, left, shows off this advantage. Unlike the blows of a hammer, a nail from a gun moves the material very little when it penetrates.

Staple guns and T-nailers. Two-legged staples hold better than nails of the same length, and aren't as likely to split the material. The staple gun being used to join two cabinet partitions, above, can also be used for shingles and shakes, and for nailing off shear walls, subfloor, and roof diaphragms. At left, Margolis uses a T-nailer to fasten roughsawn hemlock near a skylight opening. These guns get their name from the fastener they shoot. It's driven with the grain, just below the surface. Brad-drivers are also used for overhead finish work because they are light, and their fasteners are inconspicuous. Photos: Ron Davis.

T-head with the grain makes the fastener fairly inconspicuous. There are also some good small finish-nail guns. One of these will be the next air tool I buy.

Pneumatic staple guns also come in different sizes based on the range of staples they can handle. The larger guns use staples with legs 2 in. or longer for sheathing and framing. Medium-weight staplers can be used for cabinet assembly. Guns that shoot wide-crown staples are used by roofers for putting down felt and asphalt shingles. These staplers are often equipped with an adjustable shingle gauge. Unlike framing guns, the magazines on staplers are usually at right angles to the head of the gun because the stapler is held flat on the work when it's being used.

Fasteners—All pneumatic fasteners are collated either in coils, clips or strips. Headed nails, which come in lengths from 1½ in. (4d) to 3½ in. (16d), are often manufactured with a crescent-shaped piece missing from their heads so that they can be nested with their shanks touching. This economy of space yields between 60 and 80 nails per strip, depending on the size of the nail.

Pneumatic nails have smaller-diameter shanks than hand-driven nails. For example, a standard 16d box nail is .160 in. in diameter. The nail-gun equivalent, which meets all of the required code standards, is just .131 in. in diameter, because it doesn't need to stand up to repeated hammer blows.

Headed nails also come with smooth shanks, ring shanks and screw shanks (drawing, above right). Some companies manufacture a hardened nail for fastening wood to concrete. It penetrates up to ¾ in. into the concrete. This seems like a good way to attach furring strips and mud sills for nonbearing partitions, but I haven't tried it yet. Most nails are available with a galvanized coating for weather, or a resinous coating that heats up when the nail is driven, acting as a lubricant and improving adhesion when it sets—something like hot-melt glue. Senco has a moisture-resistant coating called Weatherex, which the company claims is as good as a hot-dipped galvanized coating used on nails driven by hand. I've had good luck with them, and they don't seem to leave black marks on siding the way galvanized nails do. Stainless-steel nails are also manufactured for some guns, but they're very expensive, and you can't always get them at a moment's notice.

The wide distribution of pneumatic fastening tools has brought with it a real increase in the use of staples. They are classified by their crown width (the distance across the top of the staple), leg length and gauge (metal thickness). Crown widths range from ³⁄₁₆ in. to 1 in., the wider ones for asphalt roofing and tacking up building paper. Staple leg lengths vary from ⁵⁄₃₂ in. to 3½ in., and gauges vary depending on the shear value needed.

Staples are said to have greater holding power in straight withdrawal tests than nails of equal length. I have tried to remove plywood sheathing after it was stapled down and

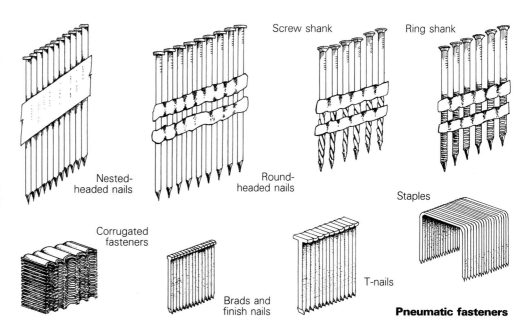

Screw shank   Ring shank

Nested-headed nails   Round-headed nails

Staples

Corrugated fasteners

Brads and finish nails   T-nails

Pneumatic fasteners

destroyed the wood in the attempt because the staples held and the wood tore. Nails have more strength than staples in resisting horizontal movement, but you can use more staples in a sheathed wall to increase its shear strength. Sometimes I feel as if I'm actually stitching a building together. If you tried to use the same quantity of nails, you would just split the studs.

Staples are even being used for 2x wall framing. A 14-ga. staple 3½ in. long can now be used as a 1:1 equivalent for a 16d nail with code approval in many areas—a significant fact considering that staples cost a lot less than nails. Staples are also much less liable to split the wood than a nail is, particularly when driven near the end of a board or when several fasteners are needed in a small area. If you have questions about whether staples can be substituted for nails in a certain application, check with your building department, or write to the Industrial Stapling and Nailing Technical Association (ISANTA) or one of the manufacturers listed on p. 53.

Most manufacturers offer a tool that drives corrugated fasteners. As the fastener is driven, it pulls the two pieces together for a tight fit. A friend of mine built some machines for joining mitered picture frames using one of these guns. The frame is held against a fixture, and the gun is lowered and fired with a pedal. The machine works beautifully.

Buying a gun—The first decision to make when you're buying a pneumatic tool is not which brand, but which type of gun. Consider your needs carefully, and buy the tool that best covers the range of work you do the most. If you are buying a gun to nail off plywood for roofs, floors, sheathing and shear wall—a good place to start with nail guns—consider buying one of the larger pneumatic staplers instead of a nail gun.

The best way to get started is to buy a package. Pneumatic-gun manufacturers and their dealers make their profit primarily from the sale of the ammunition, so they can be very competitive with the price of compressors,

hoses, and guns. It is in their interest to have their equipment in the field. In areas where there is a lot of new construction going on, some companies offer a nail gun free if you buy ten cases of nails at the list price. It pays to shop around, but when you compare price, consider the cost of the entire package you want, and not just one component.

I have used three makes of guns—Bostitch, Paslode and Senco—and they have all performed well; so differences in quality will probably not be the deciding factor in which brand to buy. Instead, consider the price, and the quality and speed of service you can get on the tool you buy. These things will vary according to where you live. Some of the manufacturers will repair a nail gun or stapler right on the job site within a day for a minimal charge. Check with other builders to see if they have gotten the attention they needed once they bought the tool. The service that will be most important to you is being able to get fasteners when you need them. An empty nail gun is useless.

For my first gun, I made a choice based principally on price and bought a package from a Senco dealer. It consisted of a SN IV gun, which shoots nails from 16d to 6d, a Rolair compressor, a hose, a regulator and some nails. My most recent catalog (winter 1983) lists the following prices on these tools: nail gun, $518; 50 ft. of ⅜-in. hose with connectors, $49.94; filter and regulator mounted in a frame, $82; and compressor, $689. You'll probably be able to do better than list.

Compressors—The compressor I bought is quite different from the small, top-heavy, single-tank compressors that you see in shops. It is designed to be moved from one location to another on a job site, or to ride easily in the back of a van or pickup truck at the end of the day. With one wheel mounted between two low tanks that support the pump and motor, it maneuvers like a wheelbarrow. This keeps the center of gravity low, so it doesn't tip over easily. Emglow (Johnstown Industrial Park, Johnstown, Pa. 15904) and

This compressor has twin tanks, a front wheel and a low center of gravity; it won't tip when moved around a site or transported. The small steel frame on the left contains a filter to trap moisture, and a regulator that adjusts the air pressure delivered to the tool. Photo: Ron Davis.

Using a nail gun can give a carpenter working by himself the extra hand needed for stubborn material. Only one hand is necessary to drive a 16d nail into this decking, leaving the other one free to lever the board into line with a framing chisel. Photo: Angelo Margolis.

Finish nailers that shoot brads or finish nails can be used both in the shop and at the site. Adjusting the pressure of the air delivered to the gun determines whether the nails are driven flush, or countersunk. Some of these guns are fitted with a removable rubber nose to protect the work from being marked. Where trim or paneling is to be stained or left natural, the heads of the nails can be painted to match the material.

Dayton Speed Air (W.W. Grainger, 5959 West Howard, Niles, Ill. 60648) sell similarly designed compressors.

The motor on my compressor has a rating of 1½ hp. It was sized to produce enough compressed air to operate the large framing gun I bought with the package. This motor also allows me some versatility in where I can get power. When a compressor switches on under load it will draw over twice the current than when cycling without a load. Many compressors use 220 volts, but this can be a problem when remodeling or adding on. My 1½-hp motor allows me to plug in to any 110-volt, 20-amp circuit.

The output of a compressor is measured in cubic feet of air per minute (cfm). There are several methods of rating this flow, and you should check with the manufacturer of your gun about the size compressor you'll need for the guns you'll be using. I bought a compressor that gives me the reserve power to bounce-nail 16d nails or run more than one gun at a time. Roofers and cabinetmakers can use smaller compressors because their finish

nailers and staplers require much less air than a framing setup.

The difference between home-owner and industrial equipment is usually obvious, and this is as true with compressors as with other tools. Lower-quality compressors can be bought cheaply on sale, and I would encourage you to buy one of these, rather than miss out on using a nail gun because you can't afford a better unit. The compressors that Sears sells seem to fall into this category.

Hose, regulator and filter—You will need at least 50 ft. of hose to get started. Although auto-parts stores sell 150 psi, ⅜-in. hose, it is sometimes not as well reinforced or as flexible as the hose you can buy from the dealer who sells you the nail gun or stapler. Make sure the hose has the correct end fittings, or couplers, to fit your compressor, regulator and gun. Unlike electricity, which suffers a serious drop in voltage on long runs, compressed air can be efficiently delivered over distances, allowing you to leave the compressor at the power source and use more air

hose. You can also run more than one tool off a single compressor by making up a simple T-shaped adapter.

Down line from the compressor is a filter and regulator. The filter reduces the amount of moisture in the air coming from the compressor. A small cock at the bottom of the filter lets you drain the water that's collected.

From the filter, the air goes to a regulator—essentially a valve with a gauge on it. This is where you can adjust the amount of air pressure the gun receives. Many companies sell a filter and gauge that are combined into a single frame that is separated from the compressor by just a few feet of hose. This is a helpful combination because it's compact and protects the regulator and filter from the usual beating that tools take on a job site.

The operating pressures of pneumatic tools are usually stamped right on them, with final adjustments left up to the operator. Keep in mind that the pressure gauge on the compressor will always give higher readings than the regulator, because the compressor will be cycling on and off to keep pressure up in the tank. My compressor cycles between 125 psi and 140 psi. The regulator settings most often used on a building site range from 80 psi to 110 psi. Hardened concrete nails, large framing nails and screw shanks require settings at the upper end of the scale. Small brads take less pressure. You can adjust the regulator to countersink them or not.

Maintenance and safety—Maintenance is simple, and should be a part of setting up and shutting down each day. Most failures in guns and compressors result from dirt or moisture in the system, and from insufficient lubrication. Clean the compressor's air filter regularly and watch the oil level. I set up my compressor away from where I am working to minimize how much sawdust it inhales.

The setup procedure is simple, too. Make sure the drain cocks are closed on the compressor; then connect the filter, regulator and hoses, and start it up. I usually leave the drain cock on the filter open just briefly to blow out any residual moisture in the system. Once you close it, the gauge on the compressor should begin to register increasing pressure. Set the regulator for the psi you'll need at the gun. While you're waiting for the system to come up to pressure, put a few drops of oil—I use mineral oil kept in a small squeeze-bottle—in the air inlet of the gun. If it blows an oil mist after you've used it for a while, give it a little less the next day. Only highly refined oils should be used, and check with the field representative for your tool if you have a question. Some manufacturers recommend an automatic oiling device that attaches to the air inlet of the gun, but I've had no trouble with any of my pneumatic tools when following a strict regimen of adding oil every time I hook up the gun. This lubrication is for the inner workings of the gun—the piston and cylinder walls. Be sure to lubricate the nail track of the magazine on the outside of the gun as well, with a non-sticky product like WD-40.

When you connect the gun to the hose, which is now under pressure, make sure it's pointing at the ground. Do the same when you are loading. If the gun jams while you are using it, disconnect it from the air before you work on it. Some guns have a quick-clearing device that doesn't require you to take the nose of the gun apart to get at a crumpled staple or nail. If you want to test-fire the gun after clearing it, check to see if there are some fasteners in it. Dry-firing the gun will make it wear out more quickly.

Safety—Most safety rules for pneumatic tools are common sense. Don't use bottled gases to power the tool—oxygen will cause an explosion, and carbon dioxide will freeze the regulator. Keep an eye on the compressor gauge to make sure that it's cycling on and off properly. Lastly, don't ever pull the safety mechanism on the nose of the gun back manually, or fire the gun at something to see if it will penetrate. Consider the forces involved, and you won't be tempted to fool around.

The compressor should be emptied at the end of the day. This makes it safer to transport, and lets the moisture that has collected during the day evaporate. Open the cocks at the bottom of the compressor tank or tanks, and the filter. Keep them open until the system is drained of air. Treat the hoses with care. Coil or chain them like utility cords and keep the couplers out of the mud. Keep the magazine and the nose of the gun out of the dirt as well, or some of these particles may end up on the cylinder walls of your gun.

More advantages—If you are bidding a job, the labor savings possible with pneumatic tools can mean the difference between making a profit or not. They reduce the time you'll have to spend nailing off to the point that it's not much of a factor in the bid. Pneumatic fasteners can cost as much as three times more than hand-driven nails. But nails are a negligible expense when compared to labor costs for driving them with a hammer.

On time-and-materials jobs, my wage usually includes all my overhead and tools, but my nail guns are an exception. I charge half the local rental-yard price for using them, in addition to my hourly labor figure. Using the guns saves money for my employers.

I do no more than four asphalt-shingle roofs a year, and my roofing stapler has made the difference between my being able to do the roof competitively on a small job, and having to sub it out. An experienced roofer will hold his hammer in one hand, the nail in another hand, and position the shingle with his third hand as he drives the nail home. I have never done enough roofing to develop a third hand. With a roofing stapler I can hold the shingle and staple it down in one easy motion. With the pneumatic tool, I shingle more than twice as fast. I don't think a roofer who worked at it every day would double his time, but the savings would still be significant.

Nailing off exterior plywood sheathing with pneumatic tools is not only faster but safer than leaning way out from scaffolding or ladders and hand-nailing. Instead of having to use two hands, one to hold the nail and the other for the hammer, you can hold on with one hand at all times.

Nail guns can improve the quality of your work. The heads of finish nails and brads can be painted to match the color of thin, solid-wood paneling or trim that has to be face-nailed. This eliminates countersinking and filling, which often looks more conspicuous on a naturally finished wall than the flush-driven heads of these tiny nails. Splitting is seldom a problem, even with narrow trim.

A subfloor will go a lot longer without squeaking or lifting if you drive screw nails or ring shanks. But they require many more blows with a hammer than smooth-shank nails, and this translates to a lot of extra time. They're often not used for this reason. With a nail gun, all you need is a little extra pressure at the regulator.

One of the seldom recognized advantages of pneumatic tools is that they can drive a nail or staple without moving hand-held work significantly. This makes jobs like toenailing and installing forgotten drywall backing blocks much easier. In some cases, this lack of percussion is what makes a job possible. I was involved in building a 12-ft. high single-wall form for a concrete retaining wall up against an excavated hill. Face-nailing the form boards to the 2x4 uprights was impossible because of all the rebar inside the form. The only thing we could do was toenail through the uprights into the back of the form boards. But this just wasn't practical with a hammer because there was no effective way to hold, or buck, the boards while the nails were driven. With the nail gun, there was no problem.

One of the drawbacks of pneumatic nailers is that they won't draw a joint together as repeated blows with a hammer will. This can be a disadvantage in rough-wall framing where headers and sills should be drawn in tightly against the studs and trimmers to prevent problems later on. Another situation where a hammer is a better tool is installing bowed 2x6 decking on a roof. Gaps between boards can be eliminated by toenailing the leading edge and hitting that nail until the board moves into line. A nail gun won't do that. However, all of these boards also have to be face-nailed, and for that part of the job, a nail gun is a better tool than a hammer.

I don't use pneumatic tools everywhere, but I think every serious builder should have at least one. As my collection of nail guns grows, I feel like a golfer with his bag of clubs, using one and then choosing another as the game progresses, reaching for them in the same instinctive way I do my Skilsaw or Sawzall. □

Angelo Margolis lives in Sebastopol, Calif.

Sources of supply

The following companies are the major manufacturers of pneumatic tools and fasteners used in the U.S. They are all members of the Industrial Stapling and Nailing Technical Association (435 North Michigan Ave., Suite 1717, Chicago, Ill. 60611), which acts as the educational and technical arm of this industry.

BeA America Corp., 280 Corporate Center, The Briscoe Building, Roseland, N.J. 07068.
Bostitch, Division of Textron, Inc., 806 Briggs Drive, East Greenwich, R.I. 02818.
Duo-Fast Corp., 3702 North River Rd., Franklin Park, Ill. 60131.
Hilti Fastening Systems, 4115 South 100th East Ave., Tulsa, Okla. 74145.
International Staple & Machine Co., Box 629, Butler, Pa. 16001.
Paslode Co., 8080 N. McCormick Blvd., Skokie, Ill. 60076.
Senco Products, Inc., 8485 Broadwell Rd., Cincinnati, Ohio 45244.
Spotnails, 1100 Hicks Rd., Rolling Meadows, Ill. 60008.

Among Japanese tool manufacturers, Hitachi (Air Nail, 1407 S. Powell, Springdale, Ark. 72764) has entered the pneumatic-fasteners market in the United States.

The Renovator's Tool Kit

A versatile collection that fits in a carpenter's toolbox

by Craig F. Stead

If your business is renovating houses, you are typically faced with a sequence of tasks requiring several skills, many unknowns and a number of different tools. Replacing a water heater, framing a wall and installing new outlet boxes could easily fall into a day's work schedule. The workspace is usually cramped, the work itself unpredictable, and it's easy to spend half the day driving back and forth just picking up tools you forgot to bring. Wasting time on gofer runs can put a renovator out of business.

To work efficiently and to minimize hunting trips, I've organized my tools into a basic core group that goes to every remodeling job. The power tools—a 7¼-in. circular saw, a ⅜-in. variable-speed reversing drill and a reciprocating saw—have their own carrying cases. All the hand tools in the kit can fit into a standard carpenter's toolbox. Mine is 9 in. wide, 9 in. deep and 32 in. long, and has the usual tray on top. Packed with the renovating gear you see above, it weighs just under 50 lb.

One indispensable item that doesn't go in the box is a folding, clamping bench (mine is a

The author with his tool kit. All the tools shown at the top of the page fit into Stead's 9-in. by 9-in. by 32-in. toolbox.

Black & Decker Workmate). There's not much room to set up a work surface on most remodeling jobs. I've seen people use chairs, cardboard boxes or even their knees to support stock for scribing and cutting. This isn't very safe or accurate. The clamping feature of the bench means you don't need a heavy bench vise or C-clamps, and I can set up just about anywhere. I organize the rest of my tools by use.

Layout equipment—You don't need anything longer than a 25-ft. steel tape for renovation work. I use a Stanley Powerlock II. I carry a framing square and a combination square. A sliding bevel is indispensable for fitting angled framing and trim on old houses. My aluminum level is 30 in. long, and just fits into the toolbox. You'll also need a chalk box and refill for snapping layout lines, as well as nylon string, and a plumb bob and sheath. The sheath keeps the bob's point sharp, and gives you a place to wind the string. I find a 6-in. machinist's rule comes in handy for measuring drawings and doing precision layout work. For fitting trim to uneven surfaces, I carry a small compass, or scriber, and I use a 6-in. compass for laying out curves and circles. My tool kit also includes a compass/divider. For layout on concrete or brick, I use red and blue marking crayons. I also carry flat carpenter's pencils, which last longer than round pencils and are easier to sharpen with a utility knife.

Pliers, snips and strippers—I carry a pair of channel-lock pliers with a capacity of 1½ in. for opening sink traps and loosening large bolts. Side cutters come in handy for service entrance and heavy cable work. Vise-grips can often lend you a third hand, and I carry a 5-in. and a 10-in. pair. I also find frequent use for needle-nose and slip-joint pliers, aircraft-type tinsnips and diagonal cutters. My wire stripper and cutter is adjustable for #24 to #12 wire.

Wrenches and sockets—Renovators need two 1⅛-in. capacity adjustable wrenches for sink plumbing. I also carry a 1½-in. pipe wrench. Rigid makes a good aluminum pipe wrench, which keeps the toolbox lighter. Nut drivers (3/16 in. to ½ in.) are for electrical service work and appliances. Deep sockets (3/16 in. to ½ in.) will drive lag screws. A square-to-hex adapter allows you to use your ⅜-in. electric drill to drive nuts and lag screws. I carry nine Allen wrenches (⅝ in. to ¼ in.). Eklind makes a set with an integral holder that keeps all the keys together, an important feature.

Screwdrivers and awls—I carry one square-shank screwdriver with a ¼-in. blade, and drivers with 3/16-in. and 7/32-in. tips. A small, instrument-type screwdriver is good to have, as well as #1 and #2 Phillips heads. I prefer Klein rubber-handled screwdrivers, but Stanley Jobmaster and Sears Craftsman are also good. You can save money by buying sets. Include a couple of awls for removing knockouts on electrical boxes, starting screws and scribing.

Files—My standard selection includes a 10-in. flat bastard, an 8-in. mill bastard, a 6-in. slim taper, a 6-in. round bastard, a 7-in. auger-bit file and a four-in-hand. Buy handles for the first four, and you'll work a lot easier. The four-in-hand is the one to carry around in your pocket. You'll also need a good file brush with metal bristles on one side and fiber ones on the other. Buying files as a set is a good idea because they're less expensive that way, and you can use the pack as a carrying case.

Edge tools—I keep my good paring chisels in the workshop. For the rougher, less precise work that renovation requires, a plastic-handled set of ¼-in., ½-in., ¾-in., 1-in. and 1¼-in. butt chisels is adequate. I keep mine in a leather roll to protect their edges. I use a Stanley pocket Surform for fitting drywall and other rough shaping; I also pack a block plane. I prefer a non-adjustable utility knife, because they are more durable than adjustable knives. You can store your spare blades in the handle of either type. I use a cold chisel for breaking rusted bolts free and for cutting cast-iron pipe.

Bars and putty knives—For pulling nails, you'll need a Stanley wonder bar, or a similar cat's paw. Check the claw before you buy; the quality of this forged part can vary. I like a ship's scraper for heavy scraping and removing molding. Red Devil makes a good one that can be sharpened easily with a file. For puttying, I carry a 1¼-in. flexible putty knife, and I carry a stiff knife the same size for scraping and rough patching that will need a lot of sanding over. For cleaning plaster cracks before patching, a machinist's scribe works well.

Saws—For rough work, I use an inexpensive straight-back crosscut, with 8 points per inch. Stanley has a cheap one with hardened teeth that looks to me like just the ticket for renovators. I like hacksaws with tubular backs that hold extra blades. Both Rigid and Milwaukee make good ones. Try to keep hacksaw blades with different teeth counts on hand. In tight spots, I use a Stanley wallboard saw.

Punches—I carry a center punch for starting drill bits in metal, and a Stanley #1113 self-centering screw punch for mounting hinges and lockplates. If you're hanging even a few doors, a self-centering punch will save plenty of time. I use #1, #2 and #3 nailsets whenever I'm installing casing or interior trim.

Electrical—A Romex ripper allows me to strip plastic-sheathed electrical cable without cutting into the insulation, and also avoid cutting myself with a jackknife. It's also good to have a couple of 3-wire to 2-wire adapters for tool hookup. A low-cost pocket electrical tester could save your life. I also make sure I have electrician's tape and a flashlight (the rubber-handled models hold up well under renovation's rigors). One 50-ft. extension cord should reach anywhere inside the house; two will get you to the yard for siding or trim work. Buy only cords with #12 wire. Lighter wire will actually starve a power-tool motor, causing overheating and insulation failure.

The screw box—The Plano Mini Magnum #3213 plastic organizer (available at fishing-tackle stores or from J.C. Penney) holds a versatile variety of screws: drywall screws (3 in., 2¼ in., 1¼ in.) for hanging gypboard and cabinets; self-tapping, hexhead sheet-metal screws, also good for general woodwork (#6 by ⅜ in., #10 by ¾ in., #8 by 1 in., #8 by 1½ in.); joist-hanger nails; flathead wood screws (#8 by 1¼ in., #12 by 1¼ in.). The #12 screw is perfect for mounting a toilet flange.

Hammers—Choosing a hammer is a subjective decision, so look around until you find one that feels right in your hand. I've found that wooden handles break too easily, while steel handles transmit a lot of shock, so for me fiberglass is a good compromise. My tool kit has one rip-claw framing hammer with waffle head, and one curved-claw finish hammer. □

Craig Stead specializes in house renovation, and lives in Putney, Vt.

Shopping for tools

In general, I buy the best tools I can. Renovation work often pushes a tool to its design limit, and failure can be dangerous and costly. If you're buying new tools, it may pay to wait for the annual January and July sales. You'll probably get discounts of 15% to 30%. If you are an established contractor, get a contractor's account with W.W. Grainger (look in the Yellow Pages under Electric Motor Distributors), which will allow you discounts on many items, particularly Milwaukee power tools.

The used-tool market can offer substantial savings for the knowledgeable buyer. A power or hand tool in good condition is worth 50% to 60% of its new list price. If you are buying a tool privately, 30% to 50% of current list is a fair price. Examine any second-hand tool carefully. Look for clean, high-quality tools with little evidence of wear. Don't buy any worn, off-brand or very old power tools because replacement parts are often unavailable or very expensive. January and February are good times to buy used tools because this is when contractors sell equipment they don't need to raise cash for taxes. Keep a list in your wallet of tools you want to buy, along with approximate retail prices. This allows you to evaluate whether you're getting a fair price.

Auctions can be a good source for tools if you are well prepared. Before the auction, list all the items you'd like to bid on and the price you are willing to pay. Use catalogs such as Sears, Silvo and W.W. Grainger as price references. Test power tools before the bidding, and examine them for wear. Don't bid over your intended price unless you have a very good reason. A single dollar increase bounced back and forth between two eager buyers can quickly add up to much more than you wanted to pay.

Patience and a little homework can occasionally be highly rewarding. I once attended an auction for a large tract-house builder who had filed for bankruptcy. The notice read like a builder's letter to Santa Claus. Saws, drills, sanders and other shop equipment were featured. I was interested in many items, so the morning before the auction I inspected equipment and prepared my wish list. Then I called local distributors to determine market prices.

The day of the auction broke cold and rainy, but a crowd of over 400 people was milling around the tools and equipment. Bidding was hot, and most of the tools were selling for close to list price, far in excess of any maximum price on my want sheet. I had my eye on a 14-in., 5-hp radial arm saw for my cabinet shop. It looked to be in excellent condition, but I couldn't test it, so I wasn't sure. Fortunately, I found a man at the auction who had worked with the saw. He informed me that the saw had hardly been used. He also said it was wired to run on single-phase current. The list price for the saw and accessories was over $1,600. My limit was $400. It was getting dark when the auctioneer finally got to the saw. Only one other person bid against me, and he dropped out at $375. I got the saw for my planned price of $400. Today that saw is still one of my main shop power tools, every bit as good as a new one. —*C.S.*

Portable Power Planes

How these versatile tools can true framing lumber and clean up trim

by Geoff Alexander

Portable power planes can solve many of the problems that come up on the construction site during framing and finish carpentry. By removing a thin layer from a piece of wood, power planes can improve the appearance of the surface by taking out saw marks, dings and other blemishes. And they're good for fitting and scribing trim. With repeated passes, power planes can straighten or taper studs, joists, rafters and beams, as shown in the photo at right. A lot of this work would be unnecessary if all framing lumber were dimensionally stable and free of twists and bows, if houses were built perfectly square, plumb and level, and if all carpenters, sheetrockers and other tradesmen did flawless work. But they don't. So my power planes get steady use.

In size, shape and function, power planes resemble hand planes, but they work like machine jointers, turned upside down and held by hand. On a hand plane, the sole is a single flat surface with a slot, or throat, through which the blade protrudes. You adjust the depth of cut by moving the blade up or down in relation to the sole of the plane. But the sole of a power plane, like the bed of a jointer, has two separate surfaces, one in front of the knives, one behind. The cutting edge of the knives is always aligned exactly with the plane of the rear shoe, and you change the depth of cut by raising or lowering the front shoe. When the front and rear shoes are in exactly the same plane, the knives will just skim the work surface and make no cut at all.

What's on the market—In my view, there are four types of power planes, with slight design variations among manufacturers. The planes all have the same basic working parts (drawing, facing page)—a motor, a rotary cutterhead that holds either fixed or adjustable knives, a two-piece shoe, one or two handles

Geoff Alexander is a carpenter and woodworker in Berkeley, Calif.

Power planes are the best tools for truing up and trimming framing members that have been nailed in place. Here a carpenter levels a crowned gluelam beam to align it with the second-story floor joists.

and a mechanism for adjusting the depth of cut and for aligning the rear shoe with the cutting arc of the knives (or knives with the shoe). Most planes have a detachable fence to help guide the tool past the work, an especially useful feature for trimming or beveling the edge of a door, window or board.

The four types differ by size and by the kind of cutterhead-drive system they have. In my business, we do everything from the rough framing of additions and new construction to finish work and architectural detailing. So I own one of each kind of power plane.

The first type is the direct-drive, or sidewinder (top photo, p. 131). The motor hangs down below the level of the surface being planed because the cutterhead is mounted directly to its rotor shaft. The direct-drive model is designed for edge planing, and its sole is only 2 in. wide. The low-slung motor helps stabilize the tool during long passes down the

edge of a door or a joist or rafter. The sidewinder I own is a Rockwell (now Porter-Cable) 126 Porta-Plane. Of all the power planes I have, this is the one that I use most.

All of the other three types of power planes have their motor mounted above the cutterhead, with a drive belt connecting the motor shaft to the cutterhead arbor. This arrangement gets the motor away from the sole, and makes the tool suitable for surface planing, even in the middle of a wide board or panel.

I group the belt-driven power planes by size because I use each type in a very different way. The smallest and lightest is the Porter-Cable 167 Power Block Plane. It's designed for one-handed use. The on/off switch is right at your fingertip, just where it ought to be. Its light weight and maneuverability let you work in situations where using a larger plane would be awkward or impossible.

The second type of belt-driven power plane is available from many manufacturers. I call it the standard size. It has a 3-in. cutting width, and a shoe length of from 11 in. to 18 in. I use the Makita 1900B, which is on the short end of this range, but others have nice features, too. Porter-Cable's 653 Versa Plane is also standard size. The great virtue of these planes is that they can perform a very wide range of tasks. They are small enough to be highly mobile, yet substantial enough to do fairly precise work; light enough to hold overhead for a short time, yet powerful enough to shave down protruding framing members. If I had to get by with only one power plane, I'd probably choose a standard size.

The fourth type of power plane simply makes possible tasks that otherwise could not be done, or that would be so prohibitively difficult or expensive to accomplish that I would not attempt them. I'm talking about the Makita 1805B. It can remove a swath of wood 6⅛ in. wide and ⅛ in. deep in a single pass. The 1805B was the first power plane of its size to be made available in the United States. Hitachi's six-inch is now being sold by many tool suppliers.

The Makita 1805B and other planes like it are made for heavy-duty surfacing work on large beams and timbers. If you work often with gluelams or heavy framing, you will want one of these big planes for sure. It also makes a dandy job-site jointer for one who does a lot of finish carpentry.

Another uniquely Japanese aspect of all the Makita planes is that the knives themselves are made of laminated steel. The cutting edge is a relatively small piece of hard, brittle high-carbon steel, while the body of the knife is a softer low-carbon steel. In theory, the harder edge can take and hold a razor sharpness, but the tough body will still be able to withstand shock and abuse.

Cutterheads—There are two basic kinds of cutterheads: those with straight knives and those with spiral (helical) knives. Most power planes have straight knives, but those on Porter-Cable planes are helical. Planes having straight knives have fixed rear shoes, so the knives themselves are adjusted up and down for proper alignment of the cut. Helical knives are permanently fixed to the cutterhead, which means the rear shoe of the plane must be adjustable.

There are advantages to each type of knife. Helical knives have a lower cutting angle than straight knives, and will cut more smoothly, more quietly, with less power consumption and less wear and tear on the machine. Helical knives stay buried in the cut for a longer part of each revolution than straight knives do, and properly sharpened, will leave behind a cleaner, less scalloped surface. Helical knives have a shear-cutting action, which means less chance of tear-out and pecking in woods with irregular grain.

While either style of knives is easy to sharpen with a grinder, razor sharpness requires honing after grinding, and honing is much easier to accomplish on the removable straight knives.

Porter-Cable's 653 Versa Plane has carbide-tipped knives. You have to send the cutterhead out for sharpening.

Principles of use—Like all other cutting tools, power planes work remarkably better when they are sharp. Power planes are designed to work properly when the cutting edges align perfectly with the rear shoe. Set the cutters so that they just touch a straightedge held against the rear shoe. You will always have to make this adjustment after you've sharpened the cutters. Locking the adjustment in place is easy on all the planes, but Makita has wisely made it more difficult to re-install the knives incorrectly than to install them the right way, though you must be sure to get the mounting screws very tight.

In most situations, you want to use a power plane to create a smooth, unbroken surface from one end of a board to the other. To do so, make each cut in one continuous pass along the full length of the board. Begin by entering the cut with firm downward pressure on the front shoe. Then as the entire sole comes to bear on the stock, shift your downward pressure to the rear shoe. Maintain the pressure on the rear shoe until the cutter has cleared the end of the work.

There is a knack to making smooth cuts, and I usually warm up on scrap stock to check the adjustments and to recapture the rhythm of a smooth stroke before starting a new job. It almost always improves the quality of the cut if you keep two hands on the plane, one in front and one in back. If you have to move your feet during the cut, make sure that your path is clear, and that the cord can't catch on anything (including your feet) during the cut. You don't want to have to stop the cut halfway through and then start up again. Many of the planers have a device for directing the cord away from the path of the cut, and these can be useful, but when I'm making a long cut, I almost always carry the cord over my right shoulder and across my back.

Tapering—In some cases you can't get the desired result by planing from one end to the other in a continuous pass. To cut a slight taper, snap a chalkline down both sides of each piece of lumber,

secure it on edge, and make the first pass 6 in. to 12 in. away from the end from which most of the stock will be removed. Back up another 12 in. or so for the second pass. Increase the length of subsequent passes until your planed surface is parallel to the line, at which point you keep removing wood until you've cut halfway through the chalkline. Snapping a line on both sides of the lumber helps guard against planing an out-of-square edge. For greater tapers, rough-cut close to the line with a circular saw and clean up with the power plane.

Truing framing members—To straighten joists, rafters or beams, it's sometimes necessary to flatten the crown of the bow, which means that the dimensions will remain true at both ends of the member. Snap a chalkline on both sides of the lumber to guide the cuts. Make the first pass about 8 in. in front of the center of the crown to produce a straight surface on the top edge parallel to the line. Increase the length of subsequent passes by 6 in. to 8 in. until you split the chalkline with an unbroken pass from one end to the other.

Concave cuts—If you are trimming to a scribed line that is straight, or nearly so, any of the planes except the big one will do a good job. For irregular or concave cuts, however, I prefer the Porta Plane. I misadjust it so that the cutters protrude slightly below the rear shoe. With this setup, you can remove a lot of wood in a hurry, so start with the depth of cut adjustment fairly shallow and experiment to find the best setting for making concave cuts. Be sure to return the rear shoe to its proper adjustment when you are through. Another method that works more slowly, but with less risk of error, is to leave the rear shoe adjusted correctly. Then, with your left hand on the depth-of-cut adjustment knob, lower the cutters as you pass over low spots in the line, and raise them to skim over the high spots. With both systems, I usually give the cut a final touch with a hand block plane held slightly askew as it runs down the wood.

Surfacing large timbers—The plans for a house I built last year called for a 6x14 exposed ridge beam, but I wanted something

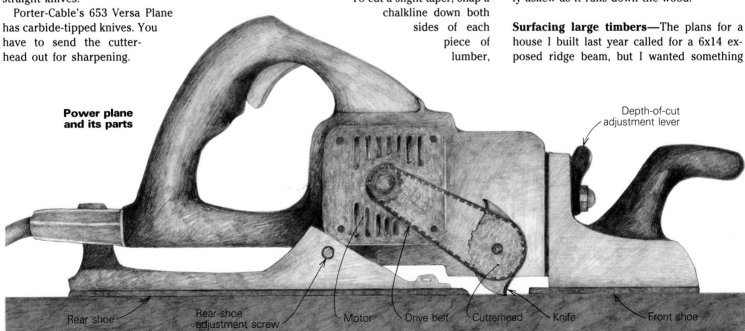

Power plane and its parts

Depth-of-cut adjustment lever

Rear shoe   Rear-shoe adjustment screw   Motor   Drive belt   Cutterhead   Knife   Front shoe

much more massive, and I found it at a used building-supply yard—23-ft. redwood 10x16s whose surfaces were so battered that the dealer sold a pair of these to me for $40. I used the Makita 1805B (photo, bottom) to take ½ in. off the bottom edge and ¼ in. off of each side to reveal unmarred wood. The result was spectacular. The beams were almost totally clear, virgin-growth redwood, so straight that I could have milled door stock from them. This one job alone justified the substantial cost of the 6⅛-in. power plane. It took only eight passes to remove ½ in. of wood 10 in. wide—about 15 minutes of work. Without the big power plane, I wouldn't have even considered the job. Not that there aren't other ways to surface a 10x16 beam. I have a friend with a huge 16-in. wide jointer who would have been happy to give it a try. But who would be crazy enough to try pushing a 600-lb. beam down a jointer?

For the rafter beams in the same house, we used 4x12s. Because the drywall was going to come up around the rafter beams, I didn't want to use green lumber, as it would inevitably shrink away from the drywall and leave gaps. New, dry 4x12s are expensive, but used ones aren't, and the Makita 1805B made quick work of surfacing the ones I found.

When you're working with used lumber, search carefully for broken nails or other debris lurking at or beneath the surface. Pore over the wood from one end to the other. And do it again. An unseen nail or staple can nick or, in extreme cases, ruin a set of knives.

Fitting doors—Here the power plane shines. Except for prehung doors, almost every door I hang has to be trimmed to fit, beveled a few degrees on the hinge side, and beveled 4° on the latch side. For trimming and beveling the edges of a door, I like to lay it flat over two sawhorses, with its best face down, then make all the trim cuts with my Porter-Cable 126 Porta-Plane (photo facing page, top). With the door flat, I can trim all four edges without having to move anything but the tool.

The Porta-Plane adjusts for bevels quickly and accurately, and when I'm edge-planing on a horizontal surface, the weight of the Porta-Plane's direct-drive motor is almost centered over the edge of the cut. In this position, the plane handles well. I keep my left hand on the fence, at the front end of the plane. And I use both hands to ensure that the shoe and the fence make snug contact with the work.

When I'm planing the top and bottom edge of a door, the cut begins and ends on end grain. Cutting the end grain is no problem, but it can chip out at the end of the cut; so I either stop the cut shy of the end and come back from the other direction, or score the far side of the cut deeply with a utility knife, and plane right through.

Correcting framing errors—Let's say you are getting ready to hang drywall, and you discover that one end of a 4x8 window header stands proud of the wall by ¼ in. The window is in place and the exterior siding is on, so you can't bash the offending member into place. The sheetrocker would keep right on hanging rock, but if you're the guy who did the framing and you're also going to hang the drywall and trim out the window, it's time to reach for the power plane. The smallest standard size you've got is best here. You need some power, but you'll be working on a vertical, overhead surface, so light weight is a big plus. Be sure to set all the nails at least ⅜ in. below the surface before you begin to make repeated passes on the face of the header, using a shallow (¹⁄₃₂-in.) setting, and entering the header from the protruding end.

Standard-size power planes are also useful for trimming studs standing proud of a wall, for evening up stair stringers, and for correcting other framing irregularities.

Exterior siding—In sidewall shingling, if you are weaving inside or outside corners, the power block plane is a natural for trimming to fit. If you're siding with plywood or any type of rabbeted horizontal siding, you may well have to custom-rabbet some of the joints. The block plane and several of the standard-size planes can cut crude rabbets easily, if you work carefully and have a steady hand. I use a table saw or router for visible joints.

Removing saw marks—In custom finish work, stock often has to be ripped to width. If the ripped edge shows, the saw marks must be removed. With a power plane, you can clean up skillsaw rips effortlessly. All of the planes are good at this job. For freehand work the power block plane is easiest to handle (photo left), but the Porta Plane makes a smoother cut.

Fitting trim—Let's say that the sheetrockers beat you to that protruding header I mentioned earlier. If you are mitering the joint between side casings and head casing, and the shoulders of the door frame are not flat in the plane of the wall, then you may well have difficulty getting the miters to fit. Patience and a sharp power plane can solve the problem. You need to shape the back face of the stock—it's an ordinary scribing problem turned 90°, and your scribing line will be on the edge of the board. Mark the edge to fit the wall and then plane the back to your line. If you have to remove a little extra material in the center of the board, drop the cutter slightly below the rear shoe. But be careful.

It may also help to back-bevel the miter a few degrees using the Power Block Plane. Work down from the mitered corners, keeping the meeting parts the same thickness. Remember that the goal of finish carpentry is to create the illusion of perfection, not perfection itself. Therefore, if you get the miters to fit tight and flat, quit fiddling. Squirt in some glue, and start nailing. Be sure to get at least

Cleaning up sawn edges with Porter-Cable's Power Block Plane, left, is easy. Because it's light and compact, this plane is well suited to working overhead, and in tight spaces.

For surfacing large timbers, the Makita 1805B, shown below, can cut a path 6⅛ in. wide in a single pass, and makes recycling used materials an attractive alternative to buying new stock.

one good nail through the miter itself. If the wood is going to be oiled or stained, sand the glued miters immediately after nailing. I use 100-grit garnet paper and sand until there is no glue residue on the surface. Sanding before the glue dries not only provides a final flattening of the joint, but also removes any excess glue and help fill any remaining gaps in the joint with a mixture of sawdust and glue.

Most of the planes have an optional adjustable fence that allows beveling and chamfering. The Makita 1900B has a groove down the center of the front shoe which makes it easy to cut a chamfer without a fence. It's fairly easy to plane skillsawn plywood edges clean enough to glue on nosing. I prefer the Porta-Plane for plywood, because it's the easiest for me to maintain a square edge, and because its high rpm and its helical knives handle the mixed grain directions with little tear-out.

Keep in mind that there's no substitute for a sharp block plane—the hand-powered type—for the final cuts on pieces of trim. You get greater control, and produce a smoother, more polished surface. Power planes are fine for most work and gross stock removal, but a hand block plane will refine your results.

A job-site jointer—Makita makes a planer-stand accessory for each of its power planes. The planer stand is designed to hold the plane securely upside-down for use as a jointer. These have some merit even for the standard-size planes, and when the big Makita is mounted on its stand, it becomes a very reasonable 6-in. jointer (photo below right) that can be carried to almost any job site comfortably by one person in one trip. This has proved so useful as an on-site jointer that I have built a support table on which the planer stand is permanently mounted, and which doubles as a carrying case for the planer stand and other accessories. To improve the plane's capability as a jointer, I have added an auxiliary wooden fence to the stock fence. The new fence contacts the shoe of the plane to prevent the cutterguard from jamming under the fence. I have also painted markings with fingernail polish on the underside of the adjustment knob to make it easier to set the depth of cut while the plane is on its stand.

Sharpening—Porter-Cable and Makita, as well as some of the other manufacturers, sell sharpening kits for their power planes. The sharpening kits that I know about use the plane motor as a power source for a small grinding wheel, and use the plane's body for mounting a jig that holds the knives. Having the sharpening kits on the job is best if you use your power plane on a regular basis.

The Porter-Cable sharpening device is

made to hold the entire cutterhead on a mandrel that is moved laterally and rotated at the same time. This compound action is necessary because the knives are helical. The Makita sharpening attachment works in a more conventional way. It's a hooded grindstone/tool-rest assembly that mounts on the rear of the plane. The knives are clamped in a bar, which slides along a track on the tool rest.

All of the sharpening systems work well, and all are fairly straightforward to set up and operate. I can disassemble, sharpen and reassemble my Power Block Plane in around 15 minutes. The big Makita takes half an hour. I try to keep two sets of cutters on hand for each power plane, and try to sharpen them both with one setup.

The Makita planes also come with a "sharpening holder"—a simple but effective jig for honing the knives on a Japanese waterstone (not included, but an inexpensive accessory). The jig, of course, would also work if you were using an oilstone, but Japanese waterstones

cut fast, stay cool, don't require oil, and are cheaper than oilstones. The jig holds the two knives in such a position that if both knives are kept in contact with the stone, they will be honed at the correct angle. If you avoid planing rough materials and lots of used lumber, and keep your knives free of nicks, you can keep them sharp by honing, something you can do several times before you have to regrind the bevels. In most of my work, I use edges right from the grinder. But if I'm doing pretty work (as opposed to surfacing used timbers), I hone my knives on the stone after grinding all of the nicks out of them.

Safety—Jointers are notorious for eating fingers, and power planes are portable jointers that can be set down on things. A rotary cutterhead can chew away flesh quickly, even when it's coasting to a stop. So be careful. Unplug it when you're fooling with its knives or adjusting the rear shoe. Watch where you put it down, and keep it out of the dirt. □

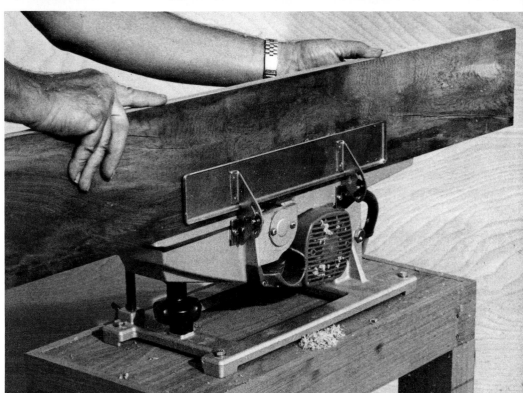

The Porter-Cable model 126 Porta-Plane, above right, is designed for trimming the edges of doors, sash and framing lumber.

The Makita 1805B, shown at right, can be inverted and mounted on a stand to become a job-site jointer that is good for truing and fitting trim. Photo: Geoff Alexander.

Acrylic Glazing

How and where to use this plastic instead of glass

by Elizabeth Holland

Commonplace now as the stuff of automobile lights, bank security windows, gas-station signs, camera and contact lenses, TV screens, and even paint, blankets and carpets, acrylic plastic has been around a long time. Although development of this highly elastic substance began back in the 19th century, it wasn't until the 1930s that chemical firms first began producing commercial quantities of acrylic, which can be manufactured as a liquid, as fibers or in sheets. And it took World War II, when the War Department started testing and using acrylic extensively in aircraft, to push the technology into the applications familiar to us today.

The larger family of plastic glazing materials has been closely scrutinized over the last decade by solar designers and builders searching for the least expensive material for collectors, greenhouses, windows, skylights and water storage. Most plastic glazings are flexible, lightweight, impact-resistant and light-diffusing. Acrylic stands out because it will not degrade or yellow in ultraviolet light. Along with high clarity and an impact resistance of 15 to 30 times greater than that of glass, acrylic offers a lifetime gauged at 20 years.

Given a burning rate of Class II in the codes, acrylic burns very rapidly, but does not smoke or produce gases more toxic than those produced by wood or paper. The ignition temperature is higher than that of most woods, but acrylic begins to soften above 160°F.

Used for exterior and interior windows, doors, skylights, clerestories and greenhouse glazing, acrylic sheets can be molded into various shapes and contours, as shown in the photos at right. Both single-skin and double-skin versions of the material are available. Single-skin acrylic is clear and comes in sheets or continuous rolls of various thicknesses. Extruded into a hollow-walled sheet material, double-skin acrylic has interior ribs, spaced ⅝ in. apart, running the length of the sheet. It is translucent, but not transparent.

The debate—Builders and designers who have worked with acrylic fall into two camps: they either hate it or love it. Any type of glazing is ultimately compared with glass, and those who like acrylic, whether single or double skin, offer these reasons:

It's versatile. Single-skin acrylic can be cut into a multiplicity of shapes, either for pure design reasons or to meet the demands of an out-of-square solar retrofit. Single-skin acrylic can

Photos: Valerie Walsh

Acrylic is light, easy to cut, and has more impact resistance than glass, though it scratches easily and has a high rate of expansion. Double-skin acrylic sheets are translucent, and the diffuse light is good for plants. Designer Valerie Walsh thermoforms these sheets into curved roof sections for custom sunspaces.

be cold-formed into curves; both single and double-skin sheets can be heat-formed.

Acrylics are easy to cut and can be site-fabricated. Lighter in weight and easier to carry, the double-skin sheet is more convenient to install than glass. The flexibility of single-skin sheets varies with their thickness; longer sheets of thinner acrylic require more people to handle them.

Acrylic has high transmissivity, and better impact strength than glass. Double-skin acrylic has an R-value competitive with insulated glass units. It is safe in overhead applications, because it will not shatter. Instead it breaks into large, dull-edged pieces.

On the other hand, builders who prefer glass offer these reasons: Acrylic has a high rate of expansion and contraction, requiring careful attention to keep an installation leak-proof. As it moves, the acrylic sheet makes a noise described as ticking or cracking. And acrylic scratches easily. The extent to which this is considered a problem, or even an annoyance, varies from builder to builder.

Costs—Acrylic used to be much cheaper than glass, but now single-skin acrylic is competitive only when purchased in bulk. Any cost advantage is likely to be lost if you attempt to double-glaze with single-skin acrylic. This is a labor-intensive process, and it's tough to eliminate condensation between the panes. The price of double-skin acrylic is close to that of insulated glass, but the cost is higher if the price of a compression fastening system is figured in. (Some builders expect the price of double-skin acrylic to drop in the future when more companies begin to manufacture it.)

If acrylic is used for a roof in a sunspace or greenhouse, however, its availability in assorted lengths can cut down on labor costs. Designer Larry Lindsey, of the Princeton (N.J.) Energy Group (PEG), points out that long pieces eliminate the need for horizontal mullion breaks, and so can be installed less expensively than several smaller ones. They're also cheaper. An uninterrupted piece of glazing can run the full length of the slope, supported by purlins underneath.

Professional use—Architect David Sellers of Sellers & Co., an architectural firm in Warren, Vt., explains his extensive use of acrylic: "Our whole plastics experiment has been an aesthetic means of expanding the type of architecture we do. With acrylic we could push the house beyond what it was already, both the inside and the outside experience of it." In the process, the firm has developed a spectrum of applications for single-skin acrylic (sidebar, below right).

Designer-builder Valerie Walsh, of Solar Horizon, Santa Fe, N. Mex., uses double-skin acrylic for a portion of the roof in the custom-designed greenhouses and sunspaces that are her firm's specialty. She first used single-skin acrylic because it was slick, clean-looking, and didn't degrade in the Southwestern sun. She began to explore unusual shapes, such as a wheel-spoke roof design. Then she turned to using double-skin acrylic. Walsh thermoforms acrylic in her own shop—curved pieces that are as wide as 5½ ft. and typically 6 ft. to 7 ft. long, although she has done 8-footers.

Safety and economics figured prominently in the Princeton Energy Group's decision to use acrylic glazing overhead in their greenhouses and sunspaces.

"The whole issue is a matter of expense," says Larry Lindsey. "In order to have glass products we feel comfortable installing overhead, we have to pay two penalties, one in transmittance and one in bucks. At present, there is no laminated low-iron glass available at a reasonable cost."

For those who have years of experience with acrylic, a willingness to experiment and to learn from mistakes has produced a valuable body of knowledge about working with the material, its design potential and its limits.

Movement—Leaking is a particular concern with acrylic glazing because it moves a lot, expanding and contracting in response to temperature changes. To avoid leaks, design principle number one is to try to eliminate horizontal joints. And wet glazing systems that may do a perfect job of sealing glass joints will not work at all with acrylic. Its movement will pull the caulk right out.

"Acrylic has a tremendous coefficient of expansion—you have to allow maybe an inch over 14 ft. for movement," cautions Chuck Katzenbach, construction manager at PEG, where they have worked with double-skin acrylic for exterior applications and single-skin for interior ones. "No silicones or sealants we know of will stretch that potential full inch of movement." Indeed, one builder tells a story about using butyl tape for bedding: The acrylic moved so much in the heat that the tapes eventually dangled from the rafters like snakes.

Room for expansion must be left on all four sides of an acrylic sheet, because the material will expand and contract in all directions. The amount of movement depends on the length of the sheet and the temperature extremes it will be subject to.

Acrylic glazing can be installed year round, but it is vital to pay attention to the temperature when it is put in place. Katzenbach explains that if it's 30° outside, then you have to remember to allow for expansion to whatever you figure your high temperature will be. If it could go from 30°F to a peak of 120°F in your greenhouse, you have to make provisions for a

Photos: David Sellers

The pliant possibilities of acrylic. For several years now, the architects at Sellers & Co. have been toying with supple single-skin acrylic to carve shapes and sculpt spaces that abandon the simple linear notion of a house. The concepts that have developed, both successful and unsuccessful, are abundant: sliding doors and windows, cylindrical shower stalls, fixed curved windows, curved windows that spring open at the bottom, bus-style fixed windows, skylights, removable windows held in with shock cords, a continuous window up the front of a house and back down its other side, and an entire roof double-glazed with ⅜-in. acrylic.

In the late 1960s, curved windows, shapes that would curl in and out from a house, began to fascinate Sellers. First he tried using a heat lamp to form a 12-in. radius curve in a ⅜-in. acrylic sheet. It worked, but the heat produced some distortions. Then he found he could cold-form the sheet into an absolutely clear curved window, just using the building's structure to hold the sheet in the desired shape (photo, top). The first attempt at curving a piece of ³⁄₁₆-in. acrylic into a shower stall, 8-ft. tall and 6-ft. in diameter, revealed that curving acrylic gave it amazing strength.

From windows, the designers turned to curving continuous sheets of acrylic, some as long as the 45-ft. strips on the sculpture studio at Goddard College, in Plainfield, Vt. The strips stretch from the peak of the roof down to a slow bend at the bottom edge. The limits of curving acrylic became apparent: in a long vertical piece with a curved bottom, the sheet is exposed to incompatible bends, horizontal on the top and then vertical on the bottom, a situation that leads to cracking where the curve begins. In addition, there is the stress of the predominantly vertical movement of such long pieces.

At the Gazley House (photo above), a new detail was tried to support the curve and prevent the cracking. The center bay of the house is glazed with three 23-ft. long strips of single-skin acrylic, fastened with a commercial compression system. Inside and right behind the glazing, a series of 6-ft. tall wooden ribs, similar in appearance to inverted wishbones, support the curve. The carved wooden ribs double as planters. For an extra measure of support, the designers also installed a horizontal metal crosspiece under the acrylic at the point where the curve begins.

The experimentation with acrylic continues: The firm has begun using a strip heater to bend the edges of single-skin acrylic. Installing adjacent strips would require simply bending up the side edges and then capping the adjacent ones after they were in place. Site-fabricated skylights with wrapped edges would be simple to finish and much less expensive than commercial units. The ideas are just beginning to suggest themselves, and designer Jim Sanford thinks the potential could revolutionize what they do. *—E.H.*

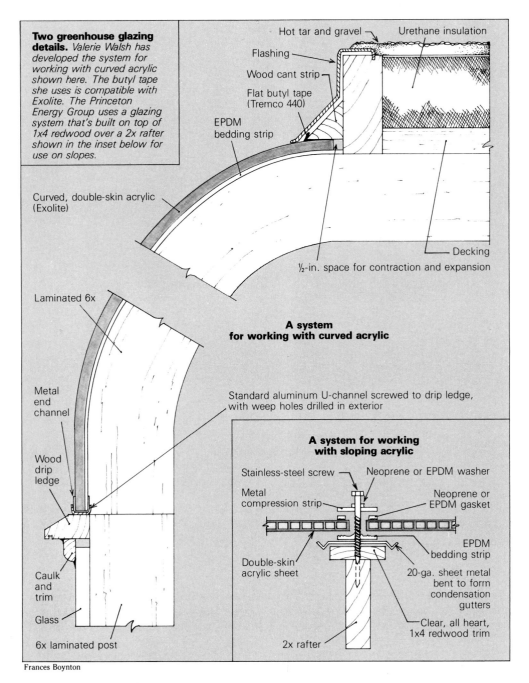

Hot tar and gravel — Urethane insulation

Flashing

Wood cant strip

Flat butyl tape (Tremco 440)

EPDM bedding strip

Curved, double-skin acrylic (Exolite)

Decking

½-in. space for contraction and expansion

A system for working with curved acrylic

Laminated 6x

Metal end channel

Wood drip ledge

Caulk and trim

Glass

6x laminated post

Standard aluminum U-channel screwed to drip ledge, with weep holes drilled in exterior

A system for working with sloping acrylic

Stainless-steel screw — Neoprene or EPDM washer

Metal compression strip

Neoprene or EPDM gasket

Double-skin acrylic sheet

EPDM bedding strip

20-ga. sheet metal bent to form condensation gutters

Clear, all heart, 1x4 redwood trim

2x rafter

Frances Boynton

Acrylic sheets can be cut on site with a sharp, fine-tooth, carbide-tipped blade in a circular saw. Work very slowly to avoid pressure-cracking, then plane, file or sand the new edges smooth. Be sure to support the sheet along both sides of the cut. Photo: Flora LaBriola.

90° change. PEG uses the formula below to calculate the extra space to allow for double-skin expansion:

$$K \times L \times \Delta F, \text{ where}$$

K = coefficient of expansion (manufacturer's specs)

L = the length of the glazing (in inches)

ΔF = the difference between the lowest and highest temperatures you expect.

Cutting—To cut a sheet of acrylic, use a fine-tooth carbide-tipped blade set for a shallow cut, and move like a snail. This is important because speed will cause little pressure cracks to appear on the bottom edge. While cutting, make sure that the sheet is firmly supported on both sides of the cut. Sharpness is vital, so use that blade only for working with acrylic. When the acrylic is cut, it heats up and the edges melt, but the wider kerf of a carbide blade will prevent the newly cut edges from melting back together again. After the cut, the edges can be planed, filed or sanded.

As the acrylic is cut, little fuzzy pieces will fly up. Some will reglue themselves to the edges and can be broken off when the cutting is completed. With double-skin acrylic, the fuzz tends to fill the ⅝-in. dia. columns between ribs. Use an air gun to blow it out.

If you drill acrylic, support the sheet fully, and use a very sharp spade bit, ground to a sharper angle than for drilling wood. The sharper angle helps prevent cracking. Drill very slowly, and slow down even more just before the drill breaks through the sheet. Be prepared to break some pieces, no matter how careful you are.

Acrylic sheets come protected with an adhesive masking, which exposure to rain or sunlight makes quite difficult to remove. Leave the protective masking on the acrylic as long as possible, and be prepared for a good zap from static electricity when you pull it off.

Fastening—For years it has been common practice to fasten single-skin acrylic by screwing it down. The designers at Sellers & Co. developed a pressure-plate fastening system to distribute the pressure evenly, and drilled the holes for the bolts or screws an extra ⅛ in. wide to allow for movement. But after ten years or more, the hole has shifted and started pushing against the screw in some installations. Cracks developed where none had existed.

Small cracks in single-skin acrylic can be stopped if the force on them isn't too great. Although his firm now uses installation details that don't involve drilling the sheets, Jim Sanford at Sellers & Co., recommends stopping cracks by drilling a ¼-in. dia. hole at the end of the crack and filling it with silicone. Designer-builder Alex Wade, of Mt. Marion, N.Y., who still uses screws, suggests drilling a tiny hole at the end of the crack, too. He then widens the crack slightly with a knife and fills it with silicone. Finally, he removes the offending screw. Wade suspects that many builders don't take into account the season of the year in which they are working when they drill the holes for screws. When installing acrylic in the extremes of summer or winter, Wade drives the screw either to the inside or the outside of an oversized hole in the acrylic, to allow for subsequent contraction or expansion when the temperatures change. It's important to space the holes evenly (about 2 ft. apart) and to tighten the screws uniformly to distribute the pressure equally. The sheet must be held down firmly, but still be able to move.

To avoid taking a chance with cracking, however, most builders have abandoned screws. Instead they use a compression system of battens that hold the plastic sheet down on a smooth bed of ethylene propylene diene monomer (EPDM). It comes in strips and is supplied by several manufacturers. The acrylic can easily slide across the EPDM as it moves.

Manufacturers recommend a 3x rafter to support the bedding in the compression glazing system. PEG installs an interior condensation gutter on greenhouse rafters that doubles as a smooth, uniform bed for the EPDM gasket in the glazing system. The 20-gauge sheet metal straddles the rafter and is bent into a ⅝-in. lip for the gutter on each side.

PEG has also developed a system for a standard 2x. A clear all-heart redwood 1x4 trim piece is screwed on top of the 2x, widening the bed and providing a smooth surface (inset drawing, facing page). Concentrated stress on the acrylic sheets is as important to avoid with a compression system as it is with screws. If one point is fastened tighter than the others, the acrylic will bow in and leak or crack.

Larry Lindsey recommends aluminum battens on south-facing roofs, because wooden ones will eventually cup upward, creating a leak. Aluminum battens can be purchased with various finishes, or they can be capped with a strip of redwood.

Double-skin acrylic needs to be supported at its base or it will bow instead of moving within the compression glazing system.

PEG lets the sheets hang over the roof's edges as a shingle would, sealing it underneath. For the bottom edge on installations with curved roofs, Valerie Walsh has developed a system with no damming problems. She slides on aluminum terminal section (ATS) from CYRO (155 Tice Boulevard, Woodcliff Lake, New Jersey 07675) on the bottom of the double-skin acrylic, then snugs the acrylic into a larger aluminum U-channel that is in turn screwed into the wood beam. Walsh then caulks the inside and drills weep holes through the outside of the U-channel (large drawing, facing page).

In one of his designs, David Sellers decided to glaze the south-facing roof area with long strips of single-skin acrylic. To avoid leaking, he encouraged the tendency of the ¼-in. sheets to

Single-skin acrylic is flexible, and some designers take advantage of this to encourage drainage. Extra blocking along the sheets encourages a sag that carries water away. Photo: David Sellers.

sag slightly. Small blocks under the edges of the sheets accentuate the dip. Melting snow or rain flows to the center of each panel and then drains off the roof. On the bottom edge, an angle keeps the acrylic from slipping, as shown in the photo above.

Glazing materials—Acrylic is fussy stuff. Chuck Katzenbach reels off a list of materials to avoid using with this plastic. Vinyl leaches into acrylic and weakens its edges. Some butyls have plasticizers that may also leach into acrylic. In these cases, either the acrylic will eventually fail or the butyl will become very hard. The plasticizer in most neoprenes is not compatible, so check with the manufacturer.

Compatible glazing materials are few: EPDM heads the list. Silicone caulk is okay, but sooner or later the acrylic's movement will pull it loose. If it's installed on a cold day, the silicone may pull out on the first really warm one. Many urethane foams can be used, but the manufacturer should be consulted about compatibility (see *FHB* #8, p. 16).

Support—The double-skin acrylic can bow in over the length of the roof, and the sheet could conceivably pull out of the glazing system under a very heavy snow load, according to Larry Lindsey. So as a cautionary measure, PEG figures on a 30-lb. snow load and installs purlins 4 ft. o.c., about ⅝ in. below the sheet.

Condensation—Acrylic transpires water vapor, so a double-skin unit typically will have some cloudy vapor inside. Double-skin acrylic should be installed with its ribs running down the slope so that any condensation inside the channels will collect at the bottom edge of the sheet. This edge needs to be vented. Double-skin sheets arrive with rubber packing material in both ends of the channels, to keep them free of debris. PEG's construction crew just leaves it there and perforates it with a scratch awl to allow air movement.

Cleaning—Never use abrasives, ammonia-base glass cleaners or paint thinner on acrylic glazing. Mild detergent, rubbing alcohol, turpentine and wax-base cleaner-polishers designed especially for plastic are safe when applied with a soft cloth.

Cleaning brings up the controversial issue of scratching. "The scratching drives me crazy, though other builders don't seem to care as much," says Valerie Walsh. Whether she is storing the sheets she has heat-formed into curves or transporting them to the building site, she keeps thin sheets of foam padding wrapped around every piece.

"With double-skin acrylic, people's tendency is not to expect to be able to see through it. They're not looking through it as they would through a window, and so they're not seeing the small imperfections in the surface itself," argues Chuck Katzenbach.

Even though single-skin acrylic is transparent, many builders who work with it say that scratching just isn't a significant problem, particularly if the glazing is kept clean. The only serious scratching problem is likely to be the work of a dog. Most scratches can be easily removed with a Simonize paste-wax buffing. And now some single-skin acrylics are available with a polysilicate coating that makes them abrasion-resistant and also improves their chemical resistance. □

Elizabeth Holland, of West Shokan, N.Y., is a contributing writer to this magazine.

Sources of supply
Acrylite (single), **Exolite** (double): CYRO, 155 Tice Boulevard, Woodcliff Lake, N.J. 07675.
Lucite (single): DuPont Co., Lucite Sheet Products Group, Wilmington, Del. 19898.
Plexiglas (single): Rohm and Haas, Independence Mall West, Philadelphia, Pa. 19105.
Acrivue A (single, with abrasion-resistant coating): Swedlow Inc., 12122 Western Ave., Garden Grove, Calif. 92645.

Concrete

Understanding the characteristics of this material can take some of the anxious moments out of your pour and ensure a finished product of high quality

by Trey Loy

Concrete is a remarkable material that can be cast into almost any shape. It will sustain and transmit tremendous loads, and once hardened, it is practically indestructible. Yet few builders feel as affectionate about concrete as did Slim Gaillard and Lee Ricks in their scat tune from the 1940s.

> Cement mixer, put-ti, put-ti,
> Cement mixer, put-ti, put-ti,
> Puddle-o-votty, puddle-o-goody,
> Puddle-o-scooty, puddle-o-vett.
> Who wants a bucket of cement?
>
> First you get some gravel,
> Pour it in the vout.
> To mix a mess of mortar
> You add cement and grout.
> See the mellow rooney come out.
>
> Slurp, slurp, slurp.
>
> Cement mixer, put-ti, put-ti,
> Cement mixer, put-ti, put-ti,
> I can never get enough
> of that wonderful stuff.

Cement mixers have been largely replaced by huge batch plants that measure out hundreds of cubic yards of ready-mix a day to waiting transit mixers. The leisurely pace of pouring concrete is also a thing of the past. The distant rumblings of an approaching concrete truck can strike fear into the heart of a carpenter still bracing the forms. At nearly two tons a cubic yard, concrete has to be poured immediately, with no time to ponder problems or locate tools. For many people, a pour is considered successful when forms don't break; and a sigh of relief can be heard when a slab is smooth and unblemished by cracks the next day.

Contrary to its reputation, concrete reacts predictably, and the builder can regulate many of the variables that affect its working properties while plastic, and its strength when hardened. Some practical knowledge of the kinds of cements and aggregates and the correct proportions of each, admixtures, slump, and the effects of weather during pouring and curing can make the difference between feeling confident about a pour, and feeling you are constantly dodging disaster.

Concrete is a mixture of water, portland cement, and fine and coarse aggregates. The active ingredients are water and cement. They combine in a chemical reaction called hydra-

The advent of ready-mix has brought a change to the quality, quantity and pace of concrete pours. The physical properties of concrete in its plastic and hardened states are well understood by researchers, engineers and batch-plant operators, but this information seldom trickles down to the builder who is actually working with the material. Knowing how mixes are designed, and the different kinds of cements, aggregates and admixtures that are used to alter the concrete can give the builder the ability to predict how it is going to react and why. For more on hand and machine-mixing concrete on site, see pp. 142-143.

tion. Although concrete that is beginning to set appears to be drying out, or dehydrating, about half the water is actually incorporated in the hydration process and becomes a permanent part of the bonding paste. This is why concrete needs to be kept moist during the first few days of curing. In fact, concrete will harden quite effectively under water.

Aggregates—The major function of aggregates is to make concrete more economical. While the cement paste binds the aggregates together in a solid mass, the aggregates keep the concrete from shrinking and cracking as hydration and evaporation take place during setting and hardening. Neat, or pure, cement is not nearly as strong as concrete correctly proportioned with aggregate.

Fine aggregate can be either sand or rock screenings. Fine-aggregate particles range in size from very fine sand to ¼ in. Coarse aggregate is either gravel or crushed stone ¼ in. to 1½ in. in diameter. The aggregate mix should be proportioned so that the smaller particles fill in the voids between the larger ones. For thick foundations and footings, gravel or rock with a diameter of 1½ inches is used. For ordinary walls, the largest pieces should be not more than one-fifth the thickness of the finished wall section. And for slabs, the maximum thickness of the rock aggregate should not be more than one-third that of the slab.

The amounts and sizes of the sand and gravel are also adjusted for the strength and workability of the mix. The plasticity of concrete in its wet state, as well as its ultimate density, depend in part on the aggregates meshing. How well the mud moves down the chute, how easily it fills the forms, and how well it finishes are measures of its workability.

Cement—An English stonemason, Joseph Aspdin, patented the process for manufacturing portland cement in 1824. He used this name because it produced a hardened concrete with a color that reminded him of the natural grey stone on the Isle of Portland. Portland cement was first produced in North America in 1872. Interestingly, the use of steel reinforcing bar was introduced a few years later, around 1880. And the first patent for prestressed concrete using steel-wire rope was issued in 1888.

Today, portland cement is produced in huge rotary kilns at temperatures of nearly

Cement Mixer, copyright 1943 by American Academy of Music, Inc. Copyright renewed. Used with permission. All rights reserved.

2,700°F. At this temperature, lime, silica, alumina and iron oxides, which are derived from limestone, oyster shells, marl, shale, iron ore and clay, undergo a kind of molecular reformation called calcination. After cooling, the resulting greenish-black clinkers are pulverized with small amounts of gypsum to control the set time of the cement.

The American Society for Testing and Materials (ASTM) recognizes five types of portland cement. Each is intended for a specific purpose, although they all achieve about the same strength after curing for three months.

Type I. This is the most common type of general-purpose cement, and is used when a specific type isn't called out. Most residential construction uses Type I.

Type II. A moderately sulfate-resistant cement, it is sometimes specified for walkways where de-icing chemicals will be heavily used. It sets more slowly than Type I, an advantage during the summer, when getting a finish on concrete can be a real race. Also, because it generates less heat during curing than Type I, it is better suited for mass pours, where heat radiating from hundreds of yards of curing concrete can cause problems.

Type III. This type is called high-early-strength cement because it achieves most of its strength within the first week of curing. This is useful if the concrete has to be put under full load within a few weeks of pouring, or if forms have to be stripped early. It is not widely stocked by concrete companies, but adding an extra bag of Type I or Type II cement per cubic yard of concrete and mixing at high speeds will produce similar results.

Type IV. A slow-curing variety that generates very little heat by hydration, it is used exclusively in mass concrete, such as dams.

Type V. This cement will withstand severe sulfate action that occurs in heavily alkaline soil or groundwater. Concrete can deteriorate because of physical and chemical reactions between sulfates and compounds formed by hydrated portland cement. Type V gains strength much more slowly than Type I.

The Canadian Standards Association (CSA) has three categories—normal, high-early-strength, and sulfate-resisting. These correspond to ASTM Types I, III, and V.

There are a number of ways that these cement types can be altered to meet special conditions. Portland cement is normally grey. White portland cement, light in color because it's made with a minimum of iron and magnesium oxides, can be tinted by adding pigments or used as is. Blast-furnace slag and pozzolan are two materials that are ground up and blended with portland cement to bring down its cost. When pozzolan is added to Type I it is designated IP; for slag, the abbreviation is IS. Another application of portland cement is

Jitterbugging a slab settles the large aggregate just below the surface to allow a smooth, troweled finish. Most problems in residential concrete are surface faults—crazing, dusting, scaling, honeycombing and shrinkage cracking— rather than strength failures.

lightweight concrete, which uses artificial aggregates and gas-forming admixtures to make it lighter. Types I, II, and III are available with air-entrainers (discussed under admixtures) interground, and designated with an A after the type number.

Portland cement is usually packaged in paper bags. Each bag weighs 94 lb. and contains 1 cu. ft. of cement. A common unit of measure in the past was the barrel, which contained the equivalent of four bags.

The cement content of a mix has a lot to do with its strength. One method of ordering ready-mix concrete is to specify how much cement should be used for each cubic yard of concrete. Producers of ready-mix prefer that you give cement content by weight (such as 470 lb. per cu. yd.), but ordering a certain number of bags, or sacks, of cement per cubic yard (such as a five-bag mix) is still very common. A four-bag mix is the minimum for most residential uses; five-bag is better. You should ask for six-bag or seven-bag if you use smaller aggregate or if you want greater strength, waterproofing and durability.

Water—For mixing concrete, it's best to use water that is fit to drink. It should not contain any oil, alkali or acid. In hydration, the water and cement in concrete combine chemically to form a paste that binds the aggregates together. It can be thought of as a glue.

The more water added to a given amount of cement, the weaker the concrete will ultimately be. This relationship of water and cement is known to concrete engineers as Abrams' law, or the water/cement ratio (W/C). It is a central factor in the design of the mix. The W/C ratio needs to be adjusted for a large number of variables, including the quantity of water that the aggregates are carrying, and the ambient temperature and humidity at the time of the pour.

The strength of concrete decreases as the W/C ratio increases, as seen in the chart above. The first column expresses the ratio in

weight (pounds of water divided by pounds of cement). The second column gives the same ratio in gallons of water per bag of cement.

W/C (weight)	Gallons per bag	Approx. 28-day strength (psi)
.45	5.0	5,000
.49	5.5	4,500
.53	6.0	4,000
.57	6.5	3,500
.62	7.0	3,000

Strength and quality—The strength of a given sample of concrete is measured by how much compression-loading a test cylinder of concrete 6 in. in diameter by 12 in. high can take before fracturing. The testing is done in a laboratory using a hydraulic piston hooked up to a meter that measures pounds per square inch (psi). Compression-strength figures for concrete refer to tests conducted after 28 days of curing unless otherwise noted. These figures indicate how well a batch of concrete will stand up to vertical and lateral loading. It is also a measure of durability and watertightness. Most engineers require that any load-bearing concrete achieve a minimum 28-day strength of 2,500 psi.

Although most concrete is batched to exact engineering standards at the plant, a lot can happen to affect its quality before, during and after the pour. How much water is added to the concrete after it's initially mixed is a good example of this.

Some drivers will ask if you want the mix stiff or sloppy, and then judge how much water to add, according to the slope (if any) involved in the pour, the angle of the chute from the truck to the forms, the depth of the forms, the amount of rebar, how long the pour is taking, and the weather (hot, dry days call for more water). A thin, watery mix is easier to handle, but much of the water added to the concrete at the job site may not have been figured in the mix design, and the resulting concrete will be less durable and much more subject to cracking.

Slump—To regulate the amount of water in ready-mix, specify slump. If you are working from a set of engineered plans, this may already be listed in the specs. Slump is a measure of the consistency of concrete—the higher the slump, the wetter the mixture. Slump is measured with a 12-in. high truncated metal cone with a base diameter of 8 in. and a top diameter of 4 in. The cone is filled with concrete right off the truck and rodded with a tamping rod. Then the cone is lifted free, inverted and placed beside the sagging pile of concrete for comparison. The distance the

Mechanical vibrators should be used with caution, and low-slump concrete. Prolonged vibrating can cause the concrete to segregate, bringing the fines—cement and water—to the top, leaving the heavier aggregates below. This can lead to surface failures on a slab.

Photo: Bob Syvanen

Concrete is one of the few residential building materials that have no forgiveness, and any pour will have its frantic moments. But good forming, careful planning, ordering the right mix, and knowing how the mud will react can keep the panic to a minimum.

concrete subsides from the top of the cone, measured to the quarter inch, is the slump.

Roadways, industrial floors and any concrete that is consolidated with mechanical vibration requires a 1-in. to 3-in. slump. Most foundations, slabs and walls consolidated by hand methods such as spading can have a slump between 4 in. and 6 in. On most residential pours, taking the time for a slump test isn't feasible. I usually just eyeball the mix. A good 3-in. to 4-in. slump mix will stand in a pile. A 5-in. to 6-in. slump sags into a blob, and a 7-in. to 8-in. slump just flattens out.

Many problems in residential concrete are the result of high slump. Unlike public projects such as bridges and highways, where compressive strength is essential, residential pours seldom suffer failures under a load. Instead, it is surface faults—honeycombing, crazing, dusting, surface cracking and scaling—that cause the problems.

One serious failure of concrete associated with high slump is segregation, which is a reseparation of concrete back into sand and gravel. Bleeding is another kind of separation that can be serious if it occurs on a large scale. Bleeding is the emergence of water on the surface of newly poured concrete. This occurs when the large aggregate settles within the mass, displacing the water in the mix. Heavy bleeding greatly dilutes the cement particles on the surface of the concrete, making it susceptible to abrasion. When this bleed water is troweled into the surface of a slab during finishing, it can result in crazing lines, shallow parallel fissures called plastic shrinkage cracking, a powdery dusting on the surface, and even scaling, which is the flaking of the finished concrete. If you keep slump low, and finish and cure the resulting concrete with care, you can avoid these problems.

The mix—Although all concrete consists of the same basic ingredients, how they are proportioned can make a huge difference in strength and workability. Because of all the variables involved, there are hundreds of possible combinations that will produce concrete with a wide variety of characteristics. As explained on p. 141, it's very important to mention the conditions and requirements of a pour when ordering concrete, so that a mix can be designed or chosen from standard designs that will give you the kind of concrete you need.

The design of a mix is complex because of the interrelationships of the materials. For example, if the size of coarse aggregate is limited in order to pour a thin slab, this will affect the amount of cement needed to reach the necessary compressive strength. The amount of cement used in turn affects the amount of water to be added, as well as the

size and proportion of fine aggregate. The mix will also have to be adjusted for slump, workability, admixtures and the weather.

Admixtures—There are four kinds of admixtures that can give concrete specific qualities. The first, called air-entraining agents, are known to most builders who work in areas with hard freezes. This admixture is a material that stabilizes bubbles formed by air incorporated in the concrete during the mixing process. The bubbles create tiny voids that act as expansion chambers or shock absorbers, which allow the concrete to withstand freeze-thaw cycles. The amount of air in the mix is a variable of mix design. It is typically 5% to 7% by volume.

Although these air bubbles make the mix slightly weaker, they also have beneficial effects. Air-entrainers increase the workability of the mud (so less water can be used for a given slump), make the mud more resistant to salts, and produce a more durable concrete.

A set-retarder may be added to ready-mix to prolong its setting time by 30% to 60%. If you need extra working time on hot days, tell the dispatcher when you order, and the plant engineer will determine the exact amount according to the weather and the mix.

Concrete companies are also likely to add a water-reducing agent, sometimes called a plasticizer, which may allow as much as a 15% reduction in water content for a given slump. Water-reducing agents can help minimize problems relating to an excess of water, such as segregation, plastic shrinkage cracking, crazing and dusting. It can also increase the concrete's strength and its bond to steel reinforcing rod.

Probably the best-known admixture is calcium chloride. This chemical is an accelerator, used to get an early set in freezing weather. Ideally, concrete should be poured when it's at least 50°F, with the mud maintained in the forms at 70°F. But pours in cold weather are often necessary, and quite common. If concrete freezes while it's setting or during the first few days of curing, it won't gain much strength and problems will develop. Pop-out, scaling and cracking occur when water in the concrete freezes and expands nearly 9% of its liquid volume.

Contrary to myth, accelerators aren't effective as antifreeze. Concrete, even with calcium chloride added, can freeze. Like any accelerator, it will only decrease the setting time. This allows the builder to pour, finish and insulate the concrete before the onset of freezing temperatures. Calcium chloride should be used sparingly and not just for convenience when better scheduling would solve the problem. It attacks aluminum conduit, lowers the resistance of the concrete to sulfates, increases shrinkage, and generally weakens the mix. If you need to use it to beat the weather, limit it to 2% by weight of cement. An effective alternative is to add an extra bag of cement to each cubic yard of concrete.

Working concrete—If you pour a lot of concrete, buy a pair of rubber boots. Leather work boots will rot off your feet after a few dunkings in concrete. The only way I've found to restore the flexibility of the leather is to remove all of the concrete and soak the boots in motor oil for a day—crude but effective.

Gloves are another must. The cement contains lime, an alkali that dries out skin and leaves it cracked and sore. Thick rubber gloves offer good protection, but they are awkward to wear. Lately I've been wearing doctor's disposable examination gloves. They cost less than $10 for fifty pairs, and fit like another layer of skin.

It's a good idea to keep a bottle of vinegar with your concrete finishing tools. Vinegar contains acetic acid, which neutralizes the lime. Wash your hands in it when you quit for the day, and don't rinse it off immediately. You'll smell like a salad bar, but your hands won't be any the worse for wear the next day.

There are several general procedures that should be followed to end up with strong concrete that looks good after the forms are stripped. Concrete should be poured in horizontal layers of 6 in. to 12 in., depending on the stiffness of the mix. Start in the corners of the forms and work toward the middle. Concrete should not be dumped into separate piles and then leveled and worked together. Pouring it near its final place will save your back and keep the mud from separating.

If you are ordering ready-mix concrete,

check the approaches that the transit mixer can make, and calculate how many chute sections you will need. Most trucks carry 10 ft. of chute. You can usually arrange with the dispatcher to have the driver bring another 10 ft. If this doesn't do it, you will have to hire a pumper, build a chute, or truck the wet mud around the site in wheelbarrows. These alternatives are listed in order of preference. Although everybody has had to use a wheelbarrow to complete a pour on occasion, it is slow, risky work.

Most concrete companies will allow between 30 minutes and an hour to empty a full truck (about 8 cu. yd.), before they charge overtime. Money isn't the only issue. Time is also a critical factor. Depending on the air temperature, the mix can agitate in the drum of the truck for up to an hour after batching. After that, more water has to be added, which will weaken the mix. Under average conditions, concrete that is left in a truck for longer than 90 minutes is considered unusable.

Pumpers should be considered for a job where ready-mix is used and some part of the pour is inaccessible by ordinary means. Hillsides, high walls, muddy ground where a fully loaded truck could become stuck, and sites with dense trees or landscaping are all good candidates. In addition to the huge pump trucks with articulating booms or snorkels used for big commercial jobs, there are smaller portable pumping machines and trucks that are suitable for residential jobs. In most areas, they can be hired with an operator for $100 to $200 for an average pour.

Using a pumper can cut down the number of people needed to make the pour, and still give better results. Small pumpers use a 3-in. or 4-in. diameter hose. This requires using pea gravel as large aggregate and adding an extra bag of cement for each yard. This mix is rich and easy to work. If you need to build a chute for the site, use at least ¾-in. plywood and lots of bracing. Pitch it at about a 5-in-12.

Once the mud is in the forms, it needs to be tamped or vibrated to eliminate voids around rebar or against the form faces. Spading the sides of the form just after the mix is placed will minimize honeycombing and sand streaking, and keep aggregates from showing on the surface of a poured wall. You can buy special spading tools—thin, flat pieces of metal mounted on long handles—but 1x4s or long pieces of rebar work fine. Another good technique to eliminate honeycombing is to rap the forms sharply with your hammer, moving up and down the forms. This brings the cement paste out to the surface of the wall to cover the aggregate.

The best tool for settling the concrete into forms is a vibrator, a portable motor with a long, flexible, waterproof shaft. You can rent one for less than $20 a day. Use a vibrator as you are pouring on each level, particularly at the face of walls, corners and around rebar. Plunge the shaft into the mud every few feet along the length of the form, and let it vibrate for 5 to 15 seconds. Prolonged vibrating is not good because it can cause segregation. In fact, a vibrator shouldn't be used on any mix that can be placed and consolidated readily by hand tools.

With a slab, it is common practice to tamp the wet concrete with a Jitterbug®—a tubular-steel frame with a mesh bottom and waist-high handles—to settle the aggregate below the surface (photo, previous page). This aids in floating and finish troweling, and brings excess water to the surface to evaporate.

Curing—Proper curing is essential in achieving high-strength concrete and durable surfaces. As long as it is kept moist and warm—above 80% relative humidity and 70°F is ideal—concrete will harden indefinitely at a diminishing rate, as shown in the chart below. If the humidity drops below 80%, the surface of the concrete begins to lose moisture more rapidly than the interior of the pour. This causes the surface to shrink, and results in a soft, dusty skin that is less resistant to abra-sion. Surface hairline fissures (crazing) are sure signs that concrete dried out too much during its initial curing. Plastic shrinkage cracks, which seldom have any structural significance but mar the appearance of a finished slab, are also caused by allowing the surface of the concrete to dry out. To get the best cure, the surfaces of the concrete should be kept moist for at least three days. After seven days it will have attained about 60% of its eventual 28-day strength.

Concrete can be kept wet by spraying it lightly with water several times a day; or, in the case of a slab, by maintaining a pond of water on the finished surface. How much moisture it will need depends upon air temperature and humidity. It's impossible to keep concrete too wet during curing. Spreading burlap or straw on the surface and soaking it with water will help hold moisture, as will covering the surface with plastic sheeting.

Membrane curing compounds, which can be sprayed on the surface, are also available. Clear compounds are preferred for surfaces that will be exposed. Black curing compounds have an asphaltic base and are used when staining isn't important. The black compounds will also hold in heat as the concrete cures. White compounds are effective in hot weather to reflect heat from the sun.

In cold weather, concrete slabs should be protected against freezing with straw covered with plastic sheeting or insulating blankets. If temperatures are in the 40's, maintain this insulation for at least 48 hours. Insulated forms used for columns and walls should remain in place for at least a week. If temperatures are lower, keep the covering on even longer. Since hydration is an exothermic reaction, the primary concern should be holding this heat in. Only if the air temperature drops below 0°F should you supply heat. □

Trey Loy is a designer and builder who lives in Little River, Calif. He spent five years pouring concrete professionally.

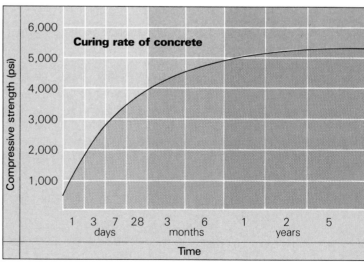

Curing. **Concrete acquires most of its strength in the first month after the pour. This graph plots the approximate compressive strength of a six-bag mix using five gallons of water per bag over time, assuming the concrete is maintained at about 70°F.**

Photo: Ross Lowell

Figuring and ordering ready-mix concrete

Gulping is the initial reflex for most builders, seasoned or not, when the dispatcher asks the inevitable question, "How many yards you want?" Like it or not, you've got to give an exact figure in cubic yards, no matter how complicated or irregular the pour is. Ordering too much means the driver of the transit mixer will need a place on your job site to dump the excess, which, like the mud inside your forms, will cost you nearly $50 a cubic yard. Ordering too little is even worse. Unless the plant can send another truck with a short load right away (this costs extra), you'll have to cap off your forms, add a keyway, and create a construction joint in a pour that was designed to be monolithic.

The way to prevent these problems is to figure your needs precisely, and then cheat the moment of truth as much as you can. If you have to reserve the concrete several days ahead, give the dispatcher a figure 15% higher than what you think you'll need, and tell him you'll call in the exact number of yards an hour or so before the pour. This way you'll be calculating trench depths and form widths as they exist. If you are figuring a big job, put off estimating the final load until after you have finished pouring from the fully loaded trucks. You won't have a lot of time in which to make that final calculation in order to send it back with the last driver or call it in to the dispatcher, but you will have cut down your margin of error by dealing with only the few yards that remain rather than with the whole job.

On the day of the pour, leave yourself enough time for careful figuring and double-checking. Last-minute concrete panic has a long tradition, but punching the keys on a pocket calculator while you are trying to get the building inspector to sign off the formwork will make for miscalculations every time. Have a helper figure independently how much concrete you'll need, and then compare notes. This should pick up careless errors in math, and the easily made mistake of forgetting to figure in some portion of the pour.

Calculating volume—The only accurate way to calculate most residential concrete jobs is to get down in the trenches and take measurements. Write down the width, depth and length of each trench or form. Having to remove a large root from a footing trench and undersupervising an overeager backhoe operator are just two reasons why the actual measurements may differ from the numbers on your blueprints enough to make a real difference.

Particularly with slabs and trenches, the depth figure can be a compromise based on measurements taken at many points and tempered by the kind of intuition that comes with experience. The same can be true of trench width, depending on the soil.

Once you've noted the measurements for each wall, footing or slab, group them into separate categories, one for each configuration or cross section. If the depth of a footing trench changes, figure this portion of the footing as a separate category, even if the width remains the same. When a configuration has more than four sides, such as a T footing, break it down into separate rectangles, trapezoids or triangles. Keep all this information on a clipboard. I number the categories, draw a small section of each one and fill in the dimensions so I don't get confused. List the different lengths for each cross section below the appropriate drawing, and note its location on a plan drawing of the pour. This really helps later in the pour when you have to calculate the last load in a hurry. You can now add the lengths of all the footings and walls in each category to get the total lineal feet for each configuration.

Concrete is ordered by the cubic yard. Unfortunately, the width and depth of footings, walls and slabs are often in inches, and length is in feet and inches. Calculate each cross section—width x depth—in inches if the increments are small, or in feet and decimal feet if the numbers get too big. Just don't mix the two. To convert your square-inch answers to square feet, divide by 144. List the total area of each cross section on your clipboard. Use a pocket calculator to grind out these numbers.

To get the total concrete needed in each category, multiply the cross-section figure, now in square feet, by the total of all the lengths in the category, which is already in feet. The product will be in cubic feet. Adding all of the totals for the categories together, and dividing by 27—the number of cubic feet in a cubic yard—will yield a total in cubic yards.

You can order concrete in fractional yards, but remember to round off high, not low, to get to the nearest large fraction. I usually order a few extra cubic feet to protect against being short, in addition to the standard 5% allowance for spillage. It's a good idea to have pier holes for decks or retaining-wall footings dug to use any excess concrete.

Placing your order—If there is more than one batch plant near the job site, ask around to see which one other builders like. Since price per yard is usually similar, their impressions will be based on phone contact with the dispatcher, and on pouring with the drivers. These opinions are useful, because the cooperation and expertise of the people in these two positions will determine how easy and successful your pour will be.

Give the dispatcher the day and time you want to see the first concrete truck, and how many yards you'll need. Then describe the mix you want, in enough detail so that the proper concrete will get sent to your job site. There are two established methods—performance and prescription.

When you order with a performance specification, you give the dispatcher a compressive strength in psi, and it's up to the batch plant to supply concrete that will test to that minimum figure in 28 days. Prescription ordering, on the other hand, puts the responsibility on you. It also lets you determine some of the variables in the mix design. You may want to duplicate a mix that worked well for you in the past, or to satisfy a restriction unique to this pour, such as a maximum aggregate size that the pumper can handle.

A minimum prescription tells the dispatcher how many pounds or bags of cement to use with every cubic yard of concrete. When you specify only the cement content, the batch plant will determine all the other variables.

If you are knowledgeable, or if there are engineering specifications on your plans you need to satisfy, carry prescription ordering a step further by specifying slump, maximum size of the coarse aggregate, or the percent of air-entrainment. If not, you should mention any special characteristics of the site or of the pour that will affect batching or delivery. If weather is a problem, ask about admixtures. If you are pouring grade beams on a steep slope, talk about slump. If you are going to use a pump, tell the dispatcher which company, confirm the time of the pour, and have the aggregate and mix adjusted.

Whatever way you order, make sure you get a batch ticket from the driver for each load you receive. This is more than just an invoice that lists the number of yards of concrete. It should also tell you the cement and water content of the mix, the size and amount of aggregates, the amount of air-entrainment, the percentage or weight of admixtures, and the slump.

Maybe most important, give the dispatcher clear directions to your job site and a telephone number where the driver or the batch-plant dispatcher can reach you or someone who can get in touch with you. It's surprisingly easy to lose a truck that weighs 27 tons, but it's more than difficult—and expensive—to deal with its load after it's sat in there a few hours.

—*Paul Spring*

Small-Job Concrete

Site-mixed mud can be batched as accurately as ready-mix, given a strong back and a few guidelines

by Bob Syvanen

Most batch plants charge extra for less than a cubic yard or two of concrete. The service you're likely to get on a small order is pretty minimal, so it often makes sense to mix on site. You can do this by hand in a trough or wheelbarrow, or you can rent a mechanical concrete mixer. You'll be using the same ingredients as the batch plant; if you measure and mix carefully, the quality of the concrete should be at least as good. For folks beyond the range of ready-mix trucks, this is the only way.

Ingredients—Clean water is a must for concrete. Sea water is okay if the concrete won't be reinforced with steel. The pour will attain high early strength, but will not be as strong in the long run. Increase the cement content and reduce the water to recover some of the strength lost to the salt.

Cement should be bought by the 94-lb. bag and kept dry. In most cases, Type I is what you need. For resistance to freeze-thaw cycles in northern climates, buy air-entrained cement, which requires using a portable power mixer, since the air-entraining agent needs vigorous mechanical agitation to be effective.

Large aggregate and sand can be purchased by the cubic yard from quarries, building-supply yards and batch plants. Large aggregate can range from ½ in. to 1½ in. It should be clean, hard, durable gravel or crushed stone such as granite or hard limestone. Most sandstone isn't usable. Crushed stone should be square, triangular or rectangular in shape. Flat, elongated pieces shouldn't be used. Sand, the fine aggregate, should be a mix of coarse and fine grains up to ¼ in.

When ordering cement and aggregates, keep in mind that the amount of concrete you get from mixing the ingredients is not nearly as much as the sum of the volume of those materials. This is because the sand in the mix fills in the voids between the gravel or stone, and the cement nestles in between the particles of sand. A rule of thumb that's sometimes used is to figure the amount of your mix

to be slightly greater than the amount of coarse aggregate you're using. The chart below will get you a little closer, so you don't end up short on the last batch of your pour.

Unless you are pouring just a few cubic feet of concrete, which can be done using dry-packaged pre-mix, have the aggregates and cement delivered. These materials are extremely heavy, and your good old pickup truck can easily get overloaded with a half-yard of wet sand. Most suppliers will deliver with a dump truck that can spot the materials almost anyplace. As near as possible to where you'll be mixing and pouring is best. Lay plas-

Cement	+	Sand	+	Gravel	=	Concrete
bag		cu. ft.		cu. ft.		cu. ft.
1	+	1.5	+	3	=	3.5
1	+	2	+	3	=	3.9
1	+	2	+	4	=	4.5
1	+	2.5	+	5	=	5.4
1	+	3	+	5	=	5.8

tic sheeting, 6 mil or heavier, on the ground for each kind of aggregate. Plywood is even better because it is a harder surface for scooping against with a shovel, but don't plan on using the plywood for anything very important afterward. You can also use old plywood to separate the sand and coarse aggregate piles vertically, so that they can be placed close together without mingling. Stack bags of cement close together, and up off the ground so they don't turn to stone. Cover them with waterproof plastic whether or not it looks like rain. Don't use cement that is so hard it won't crumble in your hand.

It's okay to scavenge aggregates for your concrete as long as you test them to be sure that they are clean and free of fine dust, loam, clay and vegetable matter. The beach is a good place to find clean sand, and old quarries and stream beds often have acceptable gravels. Aggregates taken from tidal areas contain larger quantities of salt, and should be washed with fresh water before being used in concrete. You can test both sand and gravel for dirt or loam by placing them in a glass jar filled with water. Put on the lid, shake the jar, and then wait for the water to clear. If silt covers the gravel or sand, it needs washing.

There are two tests for vegetable matter. For gravel, add a teaspoon of household lye to a cup of water in a glass jar, add the gravel and shake well. If the water turns dark brown, the gravel needs washing. This can be done with a good hosing. Sand is tested by putting it in a clear glass jar with a 3% solution of caustic soda, which can be made by dissolving 1 oz. of sodium hydroxide in a quart of water. If the solution in the jar remains colorless, the sand is in good shape. A straw color is still okay, but anything that resembles brown means finding another source for sand.

Sand shouldn't be rejected because it's holding a lot of water, but you need to know how wet it is in order to adjust the water content of your mix. Although damp sand feels a little wet, it won't leave much moisture on your hands, and won't form a ball when squeezed in your fist. It contains about ¼ gal. of water per cu. ft. Wet sand will form a ball, but still won't leave your hands very wet. Most sand falls into this category. It contains about ½ gal. of water in each cu. ft. Very wet sand is obviously dripping wet and holds about ¾ gal. of water per cu. ft.

The mix—The strength and durability of the concrete that comes out of your wheelbarrow or mixer depends on the proportion of the cement to the aggregates, and on the proportion of water to cement. Instruction manuals and construction textbooks often show concrete mixes as a ratio of cement, sand and gravel by volume, such as 1-2-4. The first number always represents the cement content, the second is the small aggregate (sand), and the third is the large aggregate (gravel or rock). The more cement used, the stronger the mix. A rich mix, 1-1½-3, is used for roadbeds and waterproof structures. The 1-2-4 mix is used for industrial floors, roofs and columns. A medium mix (1-2½-5) is used for foundations, walls and piers. A lean mix such as a 1-3-6 is used in less demanding applications.

Volume formulas like the ones above give the proportions of dry ingredients but leave the water content up to you. Start with a trial batch, and use the least water you can to get a workable mix. Add a little at a time, and keep track of how much you used.

I favor mix formulas that specify the water/cement ratio, which is called a paste. A 5-gal. paste contains five gallons of water for every bag of cement. This includes the water contained in the sand. The lower the water figure in relation to the cement, the stronger and more durable the concrete. Sidewalks, driveways and floors require a 5-gal. paste for durability. A 6-gal. paste is good for moderate wear and weathering such as foundations and walls. Where there is no wear, weather exposure or water pressure to deal with, a 7-gal. paste will do. Footings are typically poured with 7-gal. paste concrete.

Listed in the chart above are the formulas I use for 5-gal., 6-gal. and 7-gal. pastes, including volume amounts of aggregates for each mix. Each mix differs from the others not only in the ratio of water to cement, but also in the amount and size of aggregates. These adjustments are compromises between economy, strength, durability, workability and slump (stiffness of the mixture). The engineered mixes batch plants use for ready-mix concrete make the same kind of adjustments. The first formula for each mix lists the ingredients used with a single, 1-cu. ft. bag of cement. The second formula gives the correct amount of each material for mixing one cubic yard of concrete, which is useful for figuring and ordering cement and aggregates.

The amounts of water, cement and aggregates in these formulas are given by volume—gallons and cubic feet. There are other formulas that give proportions by weight, but I don't like them as well because ultimately you are trying to fill up a given space—the forms—

Five-gallon paste

1 bag cement, 4½ gal. water, 1 cu. ft. sand, 1¾ cu. ft. gravel (⅜-in. maximum);

for 1 cu. yd. of mix: 10 bags cement, 10 cu. ft. sand, 17 cu. ft. gravel.

Six-gallon paste

1 bag cement, 5 gal. water, 2¼ cu. ft. sand, 3 cu. ft. gravel (¾-in. maximum);

for 1 cu. yd. of mix: 6¼ bags cement, 14 cu. ft. sand, 19 cu. ft. gravel.

Seven-gallon paste

1 bag cement, 5½ gal. water, 2¾ cu. ft. sand, 4 cu. ft. gravel (1½-in. maximum);

for 1 cu. yd. of mix: 5 bags cement, 14 cu. ft. sand, 20 cu. ft. gravel.

---

with concrete. The formulas above also assume wet sand, with its ½ gal. of water per cu. ft. If you use damp sand, increase the amount of water you add by a quart per cu. ft. of sand. Decrease the water content by a quart for very wet sand. These proportions will yield a fairly stiff mix, depending on the size and shape of your aggregate. But you may need to make adjustments, so mix up a small trial batch first. If the concrete is too soupy, you can correct it by adding aggregates. Don't play with the cement or water content. Instead, add 2½ parts sand with 3 parts gravel in small amounts until the mud stiffens up. For the next batch, be sure to deduct the moisture carried by the extra sand from the total water to be added to the mix. If the test batch is too stiff, use slightly less sand and gravel in the next batch.

Accurate measure—The care that you take in proportioning the mix has everything to do with the quality of your concrete. If you are following a formula for mixing that is given by weight, you will need a bathroom scale for careful weighing. For measuring volume in cubic feet., make a 12-in. by 12-in. by 12-in. frame with no handles or bottom. Place it on a flat surface, fill it level, and lift. Cement is easy to deal with because it comes in 1-cu. ft. bags. You can also make a level mark on the side of your wheelbarrow to indicate the 1 or 2-cu. ft. level. In the case of a ratio mix like 1-2-4, use any convenient measure—a shovel, bucket, or box—but don't let your mind wander when you're counting. For water, mark a large bucket for half-gallons and gallons.

Mixing by hand—A lot of concrete has been mixed by hand, but it is a long, backbreaking job worth avoiding for anything more than a few cubic feet of mud. You can mix in a deep (4 or 5-cu. ft.) wheelbarrow or buy a steel or plastic mortar box (about $70) that holds 6 to 9 cu. ft., or you can make your own mixing tray. A large, shallow plywood box lined with metal so that water won't leak away works pretty well. A flat platform works even better because there are no corners for the shovel or hoe to hang up on. However, mixing must be

done carefully to avoid losing water on this flat surface.

First, load the tray with a measured amount of sand. Spread the correct amount of cement evenly over the sand and mix them together with long push and pull strokes with a hoe or shovel. Work the large aggregate into this mix with the same method. Make a depression in the center of the mix and slowly add the water. Pull the mix toward the water until the dry material is saturated, and then turn the mud over until it reaches a workable smoothness. Use this method even if you are mixing in a wheelbarrow or box. A mortar hoe is useful if your aggregate is no bigger than ¾ in. It looks like a large steel garden hoe with two holes in the blade, and costs about $15 to $20. A square-point shovel turned over so that the back of it faces away from you works too.

Mixing by machine—Machine mixing is easier than hand mixing, but it's still a lot of hard work. Electric or gasoline-powered mixers with a capacity of ½ cu. ft. to 6 cu. ft. can be rented by the day or week. Electric mixers are the least trouble. If your job site doesn't have power, then rent a gasoline-powered model. If you are going to use a mixer for more than two weeks and you do lots of small jobs involving concrete, consider buying one.

Set up the mixer right next to your sand and gravel, and run a water hose there. If you have chosen a shady spot, both you and your concrete will set less quickly. Load the drum of the mixer with all of the large aggregate and about half of the water in the formula. Start up the mixer and add the sand and cement slowly, along with the remainder of the water. Let the mixer run for about three minutes or until the concrete has become uniformly grey. When you are finished mixing for the day, add a couple of shovels of large aggregate and some water and turn the machine on one more time to scour the inside. Emptying the drum and a final rinse with a hard jet of water will leave the mixer clean.

Cold-weather concrete—Most engineers do not want you to pour concrete at air temperatures lower than 40°F. A lot of good loads have been poured when it's colder than this, but it's a bit of extra work. Both the aggregate and the water can be heated to keep the concrete warm while it's being mixed and poured, but don't heat the cement. If you heat just the water, bring it to a boil in a 55-gal. drum or other container and pour it on the aggregates to warm them. If you are heating the aggregates also, keep the temperature of the water below 175°F, or the cement will flash-set when mixed, and you won't be able to get a finish on the concrete. Aggregates can be heated on a tray of heavy sheet metal. Build a makeshift firebox out of large stones or concrete block underneath the tray, and heat the aggregates separately. Take them off the fire when they are hot to the touch. □

Bob Syvanen is consulting editor to Fine Homebuilding. *Photo by the author.*

Wood Foundations

Pressure-treated studs and plywood make an economical system for owner-builders

by Irwin L. and Diane L. Post

We first heard about All Weather Wood Foundations (AWWF) in 1978. We later learned more about this pressure preservative-treated system, and when we began to build our own house we chose it over the more common poured concrete and concrete-block foundations. There were several reasons for our choice.

First, the rugged winters here in the mountains of Vermont made the insulation of the house one of our primary concerns. Since we planned our basement to be living space, we wanted it to be as well insulated as the rest of the house (fiberglass to R-26 in the walls and R-38 in the top floor ceiling). AWWF walls are built of studs, and they can be insulated with fiberglass as easily as can any stud wall. With concrete foundations, a thick layer of insulation significantly reduces the usable space inside the basement, and it's difficult to attach insulation and the interior finished wall to the concrete. The stud walls of the AWWF allowed us to hide the wiring and plumbing, too.

The AWWF also suited our construction schedule. We were able to start on it as soon as the excavator finished the cellar hole. For a poured foundation, we would have had to hire a contractor and wait for him to work our job into his schedule. We would then have had to wait for the forms to be erected, and the concrete to be poured and to cure. For a concrete-block foundation, we would have had to pour a concrete footing and then begin the time-consuming task of laying the blocks.

The AWWF facilitated the installation of windows and doors. We simply nailed and screwed them in, as in ordinary frame construction. An error in pouring a concrete foundation can be disastrous when it comes to installing doors and windows; just a little reframing corrects an error with the AWWF. We decided to relocate a door slightly—it took us only two hours.

We wanted to use spruce clapboards for our exterior siding, and we would be able to nail the clapboards directly to the exposed portion of the wood foundation. We're not sure what we would have done with the above-grade portions of a concrete foundation.

The clincher for us was the money we saved. We were the labor force for everything but the excavation, standing-seam steel roof and drywall. Like most people who build their own homes, we did not include our labor as part of the cost. While we didn't do a detailed analysis for all the options, we estimated a savings of more than $1,000 over the cost of poured concrete. Materials for our 24-ft. by 32-ft. foundation cost us about $1,850 in August 1980.

Many people ask us if we're worried about the foundation rotting out. We aren't. The required preservative salt retention in the pressure-treated wood is 0.60 pounds per cubic foot, which is 50% higher than building codes require for general ground-contact applications. The USDA Forest Service's *Wood Handbook: Wood as an Engineering Material* indicates that test stakes in Mississippi have lasted more than 20 years at lower preservative salt retentions.

The AWWF is a stud wall sheathed with plywood on the outside. The walls stand on footing plates, which lie on a pad of gravel. A concrete slab poured inside the walls prevents the backfill from pushing in the bottoms of the walls. The tops of the foundation walls are securely fastened to the first-floor structure before the backfilling begins.

The size of the footings and studs and the thickness of the plywood depend on the size of the building, the grade of lumber, the depth of the backfill and the spacing of the studs. An industry booklet, "The All-Weather Wood Foundation: Why, What and How" ($1 from the American Plywood Association, Box 11700, Tacoma, Wash. 98411), supplied enough information for us to design our AWWF with confidence. We used 2x10 footing plates, except at the back wall, where we used 2x8s because of the smaller load on the back wall. Our design called for ½-in. plywood and 2x8 bottom plates, studs and top plates. The top plates and the plywood that was more than 1 ft. above grade were not preservative treated. We choose to use 2x8 studs on 16-in. centers so we could fit two layers of R-13 insulation inside the exterior walls. Our house design required very low-grade studs (F_b of 975 psi minimum) for adequate strength with our depth of backfill (up to 5 ft.). The fasteners were 10d stainless steel nails to connect the bottom plates to the studs and footing plates, 8d stainless steel nails to connect the treated plywood sheathing to the studs below grade, 8d hot-dipped galvanized nails to connect the untreated plywood to the studs above grade, and

Once the gravel pad is compacted and carefully leveled, left, the footing plates are set in place around its perimeter. The layout of the excavation and the drainage pipes is shown in the drawing on the facing page. Right, pressure-treated foundation framing is built 8 ft. at a time and tilted up. The interior bearing wall is framed with less costly untreated lumber because it won't be in contact with water or wet earth. Photos: Irwin and Diane Post.

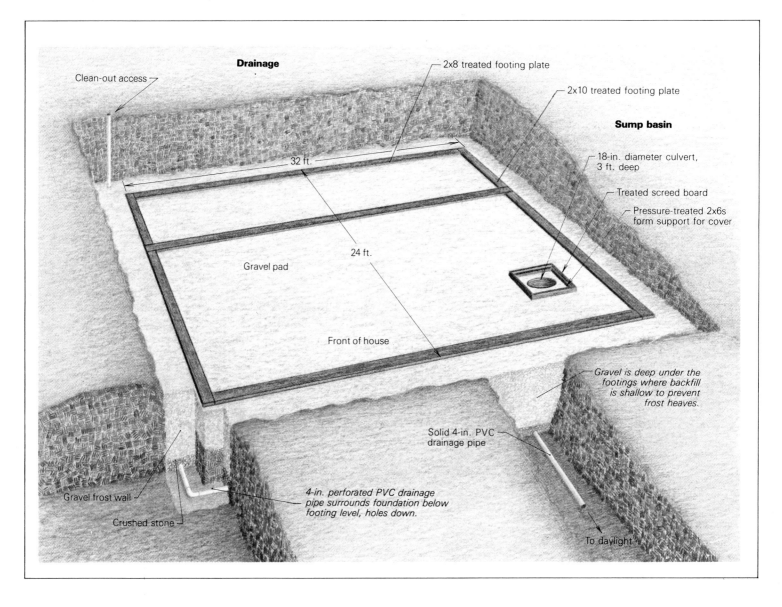

Drainage

Clean-out access

Sump basin

2x8 treated footing plate

2x10 treated footing plate

18-in. diameter culvert, 3 ft. deep

Treated screed board

Pressure-treated 2x6s form support for cover

32 ft.

24 ft.

Gravel pad

Front of house

Gravel is deep under the footings where backfill is shallow to prevent frost heaves.

Solid 4-in. PVC drainage pipe

Gravel frost wall

Crushed stone

4-in. perforated PVC drainage pipe surrounds foundation below footing level, holes down.

To daylight

16d hot-dipped galvanized nails to connect the top plate to the studs.

We found two lumberyards that were willing to bid on our AWWF materials. The better bid was for far higher-quality studs than we needed and for the ⅝-in. plywood the supplier had in stock rather than the ½-in. plywood we'd specified. We could have saved if we had been willing to wait for material closer to our specifications. The price for the stainless steel nails seemed unbelievable at $6 per pound. This works out to about 10½¢ for each 10d nail!

In ordering our AWWF materials, we specified that the wood had to be stamped with the American Wood Preservers Bureau (AWPB) foundation grademark, which ensures that the wood is properly treated for use in foundations. We were pleased to find many plugged holes in our material where samples of wood had been removed after treatment to check for retention of the preservative.

Excavation and drainage—We had two excavators bid on our job. Neither one had ever worked on a cellar hole for an AWWF. After reviewing our engineering drawings and instructions, one excavator seemed reluctant. The other showed interest, so we chose him.

The main objectives of the excavation were to lay the gravel pad on which the footings sit and

to provide good drainage around the foundation. Good drainage is an absolute necessity for any foundation. After having suffered with wet basements, we were not about to take any shortcuts with our new house. The design we settled on is diagrammed above.

In the front and sides of the cellar hole, where the footings were going to be less than 4 ft. below finished grade, we wanted a frost wall built to prevent frost heaving under the footings. We had a ditch 2 ft. wide dug to about 5 ft. below finished grade, and set drainage pipes surrounded by crushed stone in the bottom. The ditch was then backfilled with gravel.

Along with the frost wall, we had a sump basin excavated inside the foundation walls. This basin is simply a hole in the basement floor connected by pipe to the drains around the house. Groundwater normally flows to the drain-pipe outlet. If the outlet becomes plugged, the water backs up into the sump basin so we can pump it outside, keeping our basement dry. We used a 3-ft. section of corrugated aluminum culvert 1½ ft. in diameter to form the sump basin. A treated wood cover over its top is flush with the concrete slab.

Next, we worked on the gravel pad. It varied in depth from 6 in. (the required minimum) to more than a foot. We used crusher-run gravel, which does not contain large cobbles. Pea-stone,

coarse sand or crushed rock could have been used instead. The excavator drove his bulldozer back and forth over the gravel to compact it. We made sure the pad was large enough to lay out our footings at 24 ft. by 32 ft., and we marked the locations for the footing plates.

After the gravel was compacted, we leveled the pad by driving 2x2 stakes of treated wood into the gravel at 6-ft. intervals along the critical lines where the footing plates were to lie. Using a surveyor's level for accuracy, we drove the top of each stake to exactly the same elevation, which was close to the average elevation of the gravel. Then, using shovels and garden rakes, we leveled the entire pad to match the tops of the stakes. To level the footing lanes exactly, we scraped a straight 2x4 stud over the gravel between the stakes. Leveling took half a day and some patience and care.

Next we installed the ABS wastewater pipes that were to run under the basement floor. We laid out their positions precisely on the gravel surface, then we dug the ditches, put the pipes in place (gluing the joints well), covered them, and releveled the disturbed areas. We were then ready to set the footings.

Framing—The footing plates were simply laid flat on the ground (photo, previous page, left). We made careful diagonal measurements with a

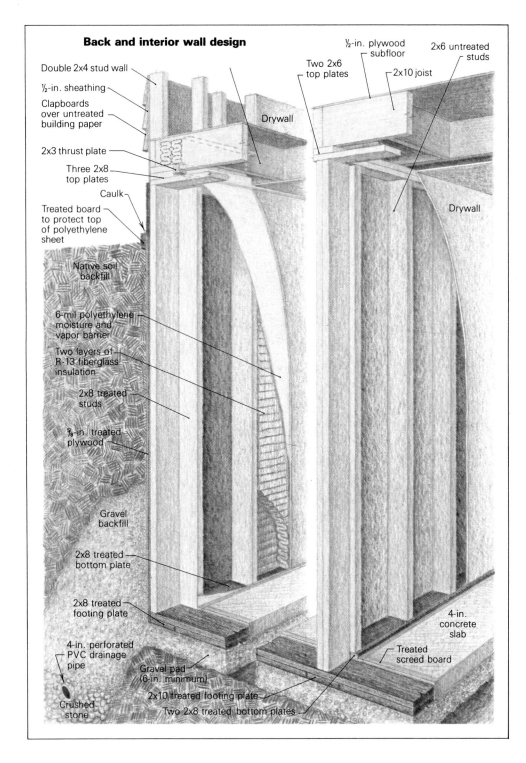

Back and interior wall design

Double 2x4 stud wall

½-in. sheathing

Clapboards over untreated building paper

2x3 thrust plate

Three 2x8 top plates

Caulk

Treated board to protect top of polyethylene sheet

Native soil backfill

6-mil polyethylene moisture and vapor barrier

Two layers of R-13 fiberglass insulation

2x8 treated studs

⅝-in. treated plywood

Gravel backfill

2x8 treated bottom plate

2x8 treated footing plate

4-in. perforated PVC drainage pipe

Gravel pad (6-in. minimum)

Crushed stone

2x10 treated footing plate

Two 2x8 treated bottom plates

½-in. plywood subfloor

Two 2x6 top plates

2x6 untreated studs

2x10 joist

Drywall

Drywall

4-in. concrete slab

Treated screed board

fiberglass tape to ensure that the footings were positioned squarely. Sections were cut out of the plates to accommodate the wastewater pipes located in the bearing walls.

There were just the two of us, so we were not able to handle long, heavy wall sections. We framed one 8-ft. section at a time, stood it in place, nailed it to the footing plates, and nailed on the top plates so as to connect adjacent wall sections. As the wall took shape, we frequently checked for plumb with a 4-ft. carpenter's level and a plumb bob.

The drawing above contrasts the design differences between the back wall and the interior bearing wall in our foundation. We did not use preservative-treated wood for the interior foundation bearing wall of 2x6 studs. By adding an extra bottom plate of treated wood and trimming off the appropriate length from each stud,

we raised the bottom of the studs above the level of the concrete slab. Using untreated studs in this wall saved us a lot of money.

When the foundation walls were all in position, we nailed sheathing onto their lower halves and fully sheathed some of the corners to stiffen the structure. Then we brushed the cut ends of the foundation wood with a generous coat of preservative, and applied a bead of silicone caulking between every sheet of plywood on the foundation.

We attached the first floor joists to the top of the foundation so the floor structure would resist the force of the backfill against the walls of the foundation. To make an especially strong connection at the back of our house, where the fill is deepest, we nailed an extra top plate and a 2x3 thrust plate onto the back foundation wall, and notched the joists to fit. To stiffen the end walls,

where the fill is deep, we added blocks between the two outer joists at 4-ft. intervals. For additional strength we glued (as well as nailed) the floor decks onto the joists.

It took us six days to erect the foundation walls, sheath their lower halves, attach the first floor joists and deck the first floor. We have read that experienced crews working with a small crane and prefabricated wall sections can completely erect AWWF walls in a few hours.

Finishing up—The concrete slab was poured after the first floor deck was completed. We prepared the floor by re-leveling the gravel in the foundation. This did not require as much accuracy as leveling for the footings—half an inch tolerance was acceptable. We shoveled excess gravel outside and laid a 6-mil polyethylene sheet on the gravel and a few inches up the walls as a moisture barrier. We also nailed screed boards of 1x3 treated wood around the sump basin and along the long sides of the two floor sections (one covering the front two-thirds of the foundation, and the other over the back third). Besides providing guidance in spreading the wet concrete, the screed boards helped hold the plastic in position.

The concrete truck pulled up to the front of our house, and the chute was put through the large window openings. We used a homemade chute extension, built from plywood and 2x10s, to reach the back third of the house. Aside from the person who delivered the concrete, we had one other to help us with the pour. The resulting slab was roughly 4 in. thick, with its surface about 1 in. above the bottom of the exterior studs. The openings between the studs provided plenty of air circulation so we used a gasoline-powered trowel for surface finishing.

After the slab was in, we finished the foundation sheathing. The excavator completed the drainage-pipe loop around the back of the house. The pipe was set lower than the footing plates all the way around. As in the bottom of the frost wall, it was surrounded with crushed stone. We had a cleanout installed in the highest section of the loop in case we ever need to flush the drainage system.

Next, we draped 6-mil polyethylene sheeting around the foundation from finished grade to just below the bottom of the footing plate, and protected its top with a 1x4 strip of treated wood caulked along its top edge. Gravel was used as backfill close to the foundation, and the finished grading included sloping all the surfaces away from the house to direct surface runoff away from the foundation.

The basement in the house we built has turned out to be very warm and dry—a very comfortable living space. The ease and speed of building the AWWF was outstanding, and the cost savings over the other types of foundations was significant. Our experience makes us wonder how long it will be until All Weather Wood Foundations displace concrete foundations, just as concrete foundations have displaced those of fieldstone. □

Irwin and Diane Post are forest engineers. They live in Barnard, Vt.

Sliding tool chest

I was able to solve the problem of keeping my tools locked up and yet accessible in my pickup camper shell with this sliding-shelf tool locker.

I mounted a piece of angle iron along each side of the bed, level with the top of the tailgate. Then I cut a ¾-in. plywood shelf about 4 ft. long and as wide as the bed, and bolted it to the angle. I hinged another piece of plywood at the back of this where it would fold down just behind the wheel wells. I added a latching hasp and now had a lockable compartment for my tools as well as a storage space above them. For inside this compartment I cut ¾-in. plywood narrow enough to slide between the wheel wells. I edged the plywood on three sides to keep toolboxes

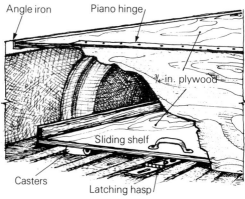

Angle iron Piano hinge

¾-in. plywood

Sliding shelf

Casters

Latching hasp

from sliding off, put a handle on the front, and attached casters to the bottom. This gave me a drawer that could be locked in place when the lid was closed. With the tailgate open, this sliding shelf will reach far enough to give access to everything on it.

Long material can be carried out the back of the camper shell supported by the top of this locker and the tailgate, which are at the same height. The entire system can be removed in minutes, if the full use of the bed is needed. Also, I have discovered that the inside of the lid is a good place to store such awkward tools as my 4-ft. level.

—*Kevin Ireton, Dayton, Ohio*

Cleaning paintbrushes

I've been a builder for many years, involved in all phases of construction, and at some point in nearly every job I've had to get out the paintbrushes. I admit I'm no artist, but I have learned a few techniques along the way.

When I purchase a new brush, I get a pure-bristle one with an unpainted handle—there's no store-bought finish to peel off the raw wood, and the oil from my hands preserves the handle. I usually put a small nail above the ferrule and bend it down toward the bristles at a 90° angle. This allows the brush to hang in a can of paint without touching the bottom, so the bristles won't warp.

When it comes time to paint, I take a 16d nail and punch a ring of holes in the deep part of the groove at the top of the opened paint can. These holes allow the paint that accumulates when you wipe the excess from your brush to drip back into the can, rather than overflow and run down the outside. When the lid is replaced, the holes are sealed inside.

Whenever I use a brush with oil-base paint, I clean it with paint thinner and liquid laundry soap. First, I pour about ½ cup of thinner into a pan, and I work the brush back and forth in it to remove most of the paint. Next, I pour this first batch of thinner into a storage container (a 3-lb. coffee can is fine). Then I rinse the brush in ¼ cup of thinner. I work the bristles with my fingers to get out as much paint as possible, and return the thinner to storage. In the container, the paint solids will settle to the bottom, allowing the clear liquid on top to be poured off and used again.

At this stage in the brush cleaning, I pour about a tablespoon of laundry soap into a pail, wipe it up with the brush and work it into the bristles with my fingers. Next, I fill the pail with water and rinse out both brush and bucket. I repeat this soap sequence one more time, and then I shake out the brush thoroughly and return it to its jacket to make sure that its bristles remain straight. Using this cleaning method, I've been able to use the same brush to apply both paint and lacquer, with no paint residue spoiling the clear finish.

—*Chris Thyrring, Halcyon, Calif.*

Fixed protractor

I use a saw protractor for many of my cut-off needs. It saves the step of drawing the cutting line on the material, and is highly accurate. I find it particularly useful for cutting bevels such as scarf-jointing 2x fascia boards.

Because most of my cuts are square, I have altered my protractor for long periods of use at 90°. To do this, I set the protractor with a reliable framing square, tighten the wing-nut clamp, and drill a pilot hole through the calibrated arc into the arm. A self-tapping sheet-metal screw holds it at a perfect 90°, ready for all square cut-off needs. The screw is easily removable for angle cutting.

—*Ralph Dornick, Altoona, Pa.*

Knock-down horses

I like the convenience of knock-down sawhorses. The commercial ones are generally too flimsy, and their small legs sink into the mud after you've got them loaded. After a lot of experimentation, I built a strong, compact, stackable, knock-down sawhorse that doubles as a toolbox (drawing, below).

I use a 2x8 or 2x10 for the top. This gives me a bench surface that I can work on or set my miter box on. I cut a hand hole in the top for moving the horse around when it's set up.

The end caps, central dividers and interior leg supports are 2x6s. The dividers hold a dowel that serves as a handle when the horse is disassembled and inverted. The compart-

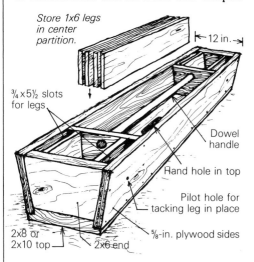

Store 1x6 legs in center partition.

12 in.

¾ x 5½ slots for legs

Dowel handle

Hand hole in top

Pilot hole for tacking leg in place

2x8 or 2x10 top

2x6 end

⅝-in. plywood sides

ment that is formed by these dividers should be the length of the legs so that they can be stored there, along with a few tools.

The slots for the legs are ¾ in. by 5½ in., a snug fit for the 1x6 legs. I drill a pilot hole through the legs and the ⅝-in. plywood sides so that I can use a 16d nail to hold the legs to the body of the sawhorse when it's in use.

I cut different-length legs for different tasks, and I build different-width horses as well so that they stack easily. Aside from being slightly heavy, there's only one drawback to these sawhorses—you can't leave them on the job.

—*George M. Payne, Olney, Md.*

A better plumb bob

I keep my plumb-bob string wound around a spool exactly 2 in. long. I drive a screw into one end of the spool, cut its head off and grind the shank to a long, diamond-shaped

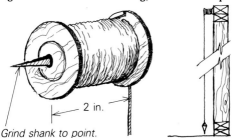

2 in.

Grind shank to point.

point. Since I usually work by myself, I can drive the point into the top plate of the wall and dangle the plumb bob off the end of the spool. When the bob hangs 2 in. out from the bottom plate, I know the wall is plumb.

—*Earl Roberts, Washington, D.C.*

Log Construction and Timber Framing

An Introduction to Timber Framing

Learning this traditional method begins with the mortise-and-tenon joint

by Tedd Benson

The standards of work in timber-frame structures aren't new; we have inherited them from a 2,000-year-old tradition of craftsmanship. The evolution of this building method (which throughout much of history was just about all there was to carpentry) resulted from pursuing a very simple goal: to put together better and stronger buildings. The proof of the success of this development can be found in the barns, houses, churches, temples and cathedrals that have become architectural treasures in all parts of the world. It wasn't until the advent of nails, joinery hardware and dimensioned lumber that true timber framing went into decline.

In reviving the craft today, we pay careful attention to the lessons and the standards evinced by these old buildings. Indeed, the thrill of practicing timber framing today lies in knowing that there is much left to learn, and in believing that each improvement brings us closer to the day when we can feel we are no longer journeymen. At the same time, we are working toward continued refinement and development. Timber-frame buildings are finding renewed acceptance for many reasons, but the integrity of the structure and the rewards to the craftsman and home owner are preserved only when high standards are maintained.

In our own shop we have learned this obvious truth the hard way—good timber frames happen only as a result of good joinery. Precise work is as important in timber framing as it is in cabinetry, so joint-making is the first thing an aspiring timber-framer needs to learn.

Mortise and tenon—This joint is practically the definition of a timber frame. Most of the joints that go into a frame are variations on the basic mortise-and-tenon joint. Once you've mastered the skills for making this joint, you should be able to execute just about any other joint in the frame. And since there are several hundred joints in the average timber frame, speed and precision are equally important.

The joint we will be working on is the shouldered mortise-and-tenon (above). It's a good example of a slightly modified mortise-and-tenon. We use this type of joint where major girts or connector beams join a post. The full width of the horizontal timber is held by the bottom shoulder on the post, and this extra bearing surface makes the joint far stronger than a straight mortise-and-tenon.

Squaring up—Before you can lay out the joint, the timber must be square at the joint area. Each face of the joint must be square if you're going to achieve a precise fit. One way to square and flatten timbers is to run them through a large thickness planer. With pine,

What makes a good framing chisel?

The framing chisel is to the timber framer what the racket is to the tennis player: You can't play the game without it. Since every phase of timber framing requires work with the chisel, you will want to own the best possible tool. A good chisel will not make you a good timber framer, but like a good racket, it will immediately improve your game and make you much happier as you learn. Those of us who work with timbers a lot look for the perfect chisel the way that King Arthur's knights used to search for the Holy Grail. It's just that elusive.

Let me describe the perfect framing chisel, sometimes called a firmer chisel or a beamer's chisel. The blade should be stout and strong, but not too long because it's difficult to control the cutting edge if your hand is a foot away. The blade should be 6 in. to 8 in. long, 3/8 in. thick at the shoulder, and 3/16 in. thick at the bevel. The steel must have a Rockwell hardness in the range of 60 (C) if it's going to hold a good edge, especially in oak or other hardwoods. The back of the blade must be honed perfectly flat if it's to cut true. To reduce friction and to enable the cutting edge to get into tight places, it's better if the side edges are bevel ground. The handle itself should be hickory or ash, and it should be fitted with a steel ring just below its striking surface. To prevent the handle from splitting, it should fit into a socket rather than over a tang. You'd order the tool as a bevel-edged socket framing chisel. Old or new, I have yet to see an unmodified tool that fits this description. The chisel you will get will probably be a compromise.

If you are lucky enough to find an old framing chisel that is still in good condition, buy it. Most of these are socket types (the old-timers were very practical) and you might well find one of the better-known brands like Buck, Witherby or White. Be careful, though. Many of these old chisels have lost their temper or have been used as a pry bar once too often.

If you're buying a new chisel, you'll probably have to choose between a Marples (heavy-duty tang), Sorby (heavy-duty tang), or Greenlee, and several brands of Japanese temple carpenter's chisels (see tool retailers list below). Almost everyone in our shop uses either Marples or Sorby chisels. They have excellent blades and are nicely balanced. Most of us have modified them by having the edges beveled at a local machine shop. All of us are frustrated by the frequency with which we have to replace split handles.

The most disappointing chisel in the group is the Greenlee. Though they do have sockets for the handles, my experience with them shows that their blades are poorly ground, and their backs are anything but flat. We've cut at least 1 in. off the tip and completely reground the back of several Greenlees to make them useful. On some, the tang is slightly off center, which makes the tool feel very unbalanced.

The newest tool on the market is the Sorby framing chisel, which is made specially for Lee Valley Tools. It's a well-made tool and is rugged enough to take demanding use. Its problems are that it is still a tang chisel, and with a length of more than 19 in., it's just too long for fine work. With such a long tool, the timber being worked must be very low so you can strike the chisel at a comfortable level. Still, it is pleasing to see a tool that is so well made, and if you don't work your beams up on sawhorses like we do, it might be just right.

We are just becoming familiar with Japanese chisels. These tools fulfill most of the requirements I mentioned earlier, and with their laminated steel blades, they have a harder cutting edge than any Western chisels. The problems are that the blade is quite short (about 3½ in.) and the metal at the socket seems to be made from softer steel, which can bend too easily. These chisels work very well as long as they're not struck, so we use them for paring.

Most chisels are bought with a factory bevel of between 25° and 30°. This angle is fine for mortising softwoods such as pine, spruce, fir or hemlock. For work in hardwood, it's better to change the angle to between 35° and 40°. Too blunt a bevel angle will crush the wood fibers in softwoods, while too shallow an angle can cause the chisel to chip if you're working in hardwood.

In either case, the honing angle should be about 5° greater than the angle of the bevel. This makes the cutting edge stronger by eliminating the feathering at the tip. It also allows you to touch up the edge quickly since there is so little surface area on the honed edge.

For honing, we use two stones: a soft and a hard Arkansas. Though we've started to experiment with Japanese waterstones in the shop, the Arkansas stones are the ones we take to the site. There are a few jigs available that clamp to the chisel, guiding it across the stone to maintain a consistent honing angle, but we haven't found much use for them. If you spend enough time using your chisel, you should be able to hone by eye, and by feel. Be sure to back off (hone the back of the blade) only on the finer stone. After a chisel has been honed four or five times, we regrind the basic bevel.

The slick—This tool is just a chisel with a wide blade and a long handle that is not meant to be struck with a mallet; it's designed to be pushed by hand. The slick is used for paring large surfaces, and it's especially good for slicing across the grain to finish the sides of mortises and tenons. Of course a chisel can be used for final paring as well, but most professional timber framers prefer a slick because its wide blade makes for quick, accurate work on broad surfaces, and its long handle provides extra leverage and control.

Since this tool is used only for paring, its bevel should be ground to a shallow angle—about 20° to 25°. There aren't many brands of slicks. Most people in our shop use the slick sold by Woodcraft; it's well made and moderately priced. However, for the quality of steel, light weight and balance, I would have to say that the best are made by the Japanese. If I can save enough money, I'll buy one.
—T.B.

Where to get them

The following companies sell framing chisels:

Frog Tool Co.
700 W. Jackson Blvd., Chicago, Ill. 60606.

Garrett Wade Co.
161 Ave. of Americas, New York, N.Y. 10013.

Lee Valley Tools Ltd.
Box 6295, Ottawa, Ontario, Canada K2A 1T4.

The Tool Works
76 9th Ave., New York, N.Y. 10011.

Woodcraft Supply
Box 4000, 41 Atlantic Ave., Woburn, Mass. 01888.

Woodline—The Japan Woodworker
1731 Clement Ave., Alameda, Calif. 94501.

Illustrations: Frances Ashforth

Squaring up. Using a flat outside face of the timber as a reference, check the other three faces around the joint area for square. The surest way to do this is to use a framing square and a combination square together, as shown at left. A small plane is fine for trimming high spots, and a tolerance of $\frac{1}{32}$ in. is acceptable.

to make all layout lines with a sharp pencil and light touch, but be sure to scribe the joint before cutting it, using a sharp awl along the grain and a utility knife across the grain. The advantage of the scribed or scored line is that it cuts the surface fibers of the wood and gives you a notch into which you can set your chisel.

Measure to the top of the mortise (the height of the tenon) and scribe line CD. Now, from the outside face of the timber, measure to the outside edge of the mortise and make two marks that can be connected by using the framing square as a straightedge. Scribing with the grain is difficult in a coarse-grained wood, so take care that lines don't wander. This gives you line EF. Measure from E and F the width of the mortise and make line GH. Then make a line down the exact center of the mortise to guide the drill bit. Mark this centerline at both ends of the mortise, measuring in half the diameter of the drill bit plus $\frac{1}{16}$ in. from AB and CD. This extra $\frac{1}{16}$ in. gives room for error and lets you to work up to the line with your chisel.

The mortise is now ready for drilling, but you should lay out the shoulder first. With the blade of the square held on the mortise side of the timber, scribe a line exactly 1 in. down from A and B to S_1 and S_2. Connect S_1 to C and S_2 to D.

The goal of drilling is to remove wood quickly, roughly excavating the mortise to its full depth and staying away from the scribed edges of the joint. We use power tools for this part of the job. I think that as we work toward the revival of the craft of timber framing in the 20th century, we should use the tools, techniques and knowledge that are available in our age. Throughout the history of timber framing, as the tools improved and as the knowledge about wood became more complete, the joints became more sophisticated and stronger; buildings became better.

C.A. Hewett, the English building historian, documents tremendous changes that took place in the evolution of timber framing as the tools became more highly evolved and easier to use. When it was a great struggle to drill one hole to remove the wood for a mortise, only one hole was drilled and the mortise was therefore approximately square and contained no housing or shouldering that might have improved its strength. It was simply too difficult to remove the wood. In the 11th century, there was a marked improvement in the quality of the mortises, which seems to have been tied to the ability to drill holes more easily. As a result, the joints became more elongated, using more of the surface area of the pieces being joined.

It is in gross wood removal that we have the greatest advantage over our predecessors. With power tools we can remove wood extremely quickly and therefore make strong, beautiful joints more efficiently than ever.

Use a drill bit that is at least $\frac{1}{8}$ in. smaller than

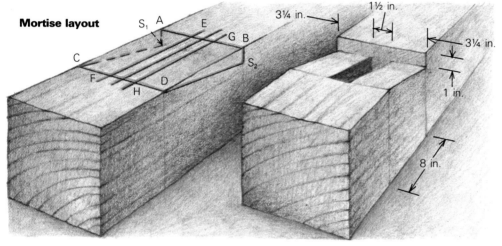

Mortise layout

3¼ in.   1½ in.   3¼ in.   1 in.   8 in.

hemlock, fir and other softwoods, this works well. But with oak, we've had no luck. There just doesn't seem to be a planer that has enough power to get the job done without driving the operator nuts. Therefore, we take a big power hand planer to the timber for rough squaring (we've had good results with the 6⅛-in. wide Makita 1805B), and then finish the job around the joint area with a hand plane.

Use a straight, flat outside face of the timber as a reference to square up the other three faces. The outside face is least likely to be worked or otherwise altered when you cut the joint. If there is no outside face (as on an interior post), then square up from a designated side to keep all the adjacent faces perpendicular, and opposite faces parallel. Interior beams such as floor joists and summer beams should always be squared from the top face. Use a framing square to check the joint area as you trim each face with a plane. By resting a combination square on the blade of your framing square, as shown in the photo above, you can check three faces of the timber at once.

The mortise—With the timber square to a tolerance of $\frac{1}{32}$ in., you're ready to lay out the joint. Whether you measure from the side of

the timber or from its center depends on the relationship between the timbers that are being joined. For example, both timbers might need to be flush with an outside face, as is the case with a post and girt. Or you may have an interior timber that is smaller in section and needs to be centered on a post or beam.

There are a number of ways to do layout. Except for very repetitious details, I favor methods that require the worker to keep thinking about how timbers relate to each other in the frame. Too many templates or marking shortcuts can lead to the belief that layout is an automatic process. When you fall into this pit, strange things begin to happen—the wrong template is used; the marking gauge doesn't get reset; you forget that this is the one layout that is different because of a sizing adjustment (a timber slightly larger or smaller than its blueprinted dimensions)—and you can't just throw timbers away when you make errors like these. For most of our work, we use the framing square. It's a simple tool that can be used to lay out any joint as long as you think while you use it.

With the blade of the square held against a side of the timber, scribe or draw the shoulder line at the bottom of the mortise (line AB in the drawing above). At this point you may choose

the width of the mortise, so you can work up to the finished surface with your chisel. If you're using a portable electric drill and you don't have a positive stop, file a mark into the drill bit to gauge your depth. In our frames, major tenons are usually 3¾ in. long, so we drill to a depth of 4 in. Bore the holes at either end of the mortise first, as shown in the photo at right. Then, using the centerline as a guide, bore a series of overlapping holes.

Now you're ready to work with the chisel. Use a good framing chisel (see p. 151) with a blade slightly narrower than the width of the mortise. Just as boring the mortise was part of the rough work, your first work with the chisel also involves removing lots of waste. So don't spend a lot of time being fussy when it isn't necessary. Get the waste wood out quickly so you'll have the time and concentration to be precise as you work toward the line of the finished mortise.

Roughing-out should take you to between ⅛ in. and 1/16 in. of the line. Strike your chisel with heavy blows from the mallet and hog off reasonably large slices of wood (photo below left), but be careful in this rough stage not to attempt too large a bite. Make sure that you are in control, not the grain of the wood. Attempting to take out too much wood usually results in back-chiseling (going beyond true) or in a stuck chisel. Make your cuts across the grain before turning the chisel with the grain so you don't run the risk of splitting the wood. This will also break pieces of wood loose more easily. A corner chisel can be used at this stage, but don't take the whole corner out at once; let the chisel drift toward the center of the mortise.

Tap the chisel more lightly as you get closer to the line. Finally, you can pare the remaining slices off by pushing the chisel or a slick with your hand (photo below right). As you work, keep checking the inside of the mortise with a combination square to make sure that the walls are straight and square.

When the mortise is complete, it's time to cut the shoulder. Set the blade of your circular saw 1/16 in. less than the depth of the finished shoulder—15/16 in. in this case. We use worm-drive saws because their extra power makes a difference in oak. When you cut across the shoulder (line AB), stay about 1/16 in. shy of the line. Then set the saw at full depth and turn the timber to each side for the two side cuts into the shoulder (S_1C and S_2D). Again, stay away from the line by about 1/16 in. You can break out the two waste pieces on either side of the completed mortise quite easily by driving a chisel into the kerf at line CD.

Staying away from the lines on these cuts means that the joint is actually finished with a chisel and a slick. The theory is the same as for the mortise—do the rough wood removal rapidly with power tools and then work up to the line with more control.

Making the tenon—Begin once again by squaring the timber at the joint area, working from a designated face as you did when you laid out the mortise timber. If necessary, square off the end of the timber so that the end of the

Drilling the mortise. The bit diameter should be about ⅛ in. less than the width of the mortise. Set the bit point on the mortise centerline, drill the two outside holes first (above) and make a series of overlapping holes to open the slot.

Cleanout. Next come wood removal, left, and paring to the line, below. Strong blows with the mallet help the framing chisel to slice out large pieces of waste to within about 1/16 in. of the line. Then more careful paring removes this last narrow section of waste.

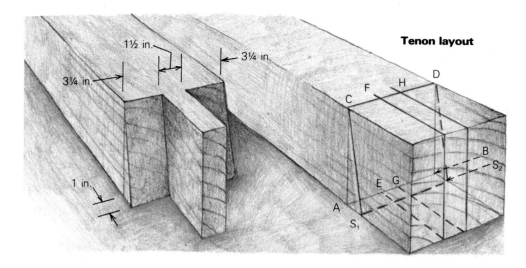

Tenon layout

1½ in.

3¼ in.

3¼ in.

1 in.

C F H D

E G

A S₁

B S₂

tenon will be square. Then measure 3¾ in. (or ¼ in. less than the depth of your mortise) back from this edge and square a line all the way around the timber.

Now you are ready to lay out the beveled shoulder of the tenon to match the angled cheek of the mortise. Mark its 1-in. depth from the squared line (AB in the drawing at left) at the bottom of the timber. This will give you line S₁S₂. Connect S₁ and S₂ with points at C and D, which correspond to the measurements used to mark the cheeks of the mortise. Remember that although the timber dimension may vary, you have to keep the measurements constant. Complete the layout by marking both sides and the end of the tenon, as shown in the drawing. To ensure that the tenon will be perpendicular to its shoulders and properly aligned with its mortise, use a single outside face of the timber as a measuring edge.

Cut along the beveled shoulder lines (CS₁ and DS₂) first. Set the saw depth to leave ¹⁄₁₆ in. of waste outside the tenon line, and stay about the same distance from the shoulder line when you make the cut. Make the next cuts—four in all—from the end of the timber, sawing in on the top and bottom of the tenon with the blade at full depth. Again, stay shy of the line. Then use these saw kerfs and the twin vertical lines on the end of the timber to guide the blade as you make drop cuts on either side of the tenon (photo far left). Now you can remove the waste wood by driving the chisel into the kerfs at the end of the timber. This completes the rough wood removal.

Working with the tenon on its side, chisel to the beveled shoulder line first. Start by establishing the line with a series of shallow chisel cuts and then work the rest of the surface to this edge (photo left). It's always a good idea to cut across the grain like this before cutting with the grain to remove wood. It doesn't hurt to back-chisel about ¹⁄₃₂ in. from the edge to ensure a tight edge joint and to compensate for possible shrinkage near the surface of the wood. Check the accuracy of your work with a combination square.

Pare the surface of the tenon to the line with a slick (photo bottom left) or a rabbeting plane. You can use calipers to check the tenon thickness, or sight beneath the blade of your combination square by eye. To finish, bevel the end of the tenon slightly so that it will start easily in the mortise.

The last step is assembly and pegging. When the timbers meet, the tenon will fit tightly into its mortise, the beam will rest squarely on the full surface of the shoulder, all face edges will meet precisely, and both timbers will be in proper alignment with the rest of the timber frame. We usually use a come-along and rope to pull joints tight and square; then we hold them that way until the hardwood pegs are driven. For this shouldered mortise and tenon, we'll drill out a pair of 1-in. dia. holes 2 in. from the top and bottom of the mortise and 1¼ in. from the beveled face on the post. □

Roughing out the tenon. **Shoulder and side cuts have already been made to within ¹⁄₁₆ in. of the joint line. The drop cuts at the end of the tenon, shown above, are the last ones to be made, using the kerfs from the side cuts and the penciled joint lines as guides.**

Paring to the line. **Set the chisel edge in the razored scribe line and make shallow finish cuts all the way across the face of the shoulder (right). Below, the easiest way to complete the tenon is with a slick. Work the blade across the grain until the surface is flat and true.**

Tedd Benson's timber-framing company is based in Alstead, N.H.

Tools for Timber Framing

A housewright's implements for measuring, marking and cutting, and how he keeps them sharp and true

by Edward M. Levin

The recent interest in timber framing has led to the revival of woodworking techniques long in decline or disuse. These methods call for implements not found in most carpenters' tool kits since before the turn of the century, along with new uses of more familiar tools. Here is an introduction to the use and care of some of the fundamental tools of the housewright's trade.

Measuring—The framing square is the essential tool. Apart from its chores as a straightedge, and in squaring up assembled framework, we use it to measure and mark across timber, as in

the photo above, to test flatness and, with screw-on fixtures, to repeat roof and stair angles. Two framing squares, one set at either end of a beam, make excellent winding sticks to test for and quantify twist (drawing, below).

A sliding bevel complements the framing

square and fixtures for laying out joints, testing angles and setting up machinery. Use a bevel with a slotted screw-tightening mechanism. If your sliding bevel has thumbscrews or wing nuts, it's best to replace them with slotted screws or hex nuts, and tighten with a screwdriver or a wrench to fix the angle securely. A bevel that slips is worthless.

Combination squares have largely replaced the traditional try square for checking mortises

(as in the photo above), notched and lapped joints, the shoulders of tenons, and for squaring across end grain. The 12-in. size is standard, but the 6-in. size is handy to square drill bits to a surface and to work in tight spots. The blades of combination squares can serve as straightedges to test surfaces for flatness. To check inside different-sized housings and gains, we often supplement them with a variety of special straight-

edges made of ⅛-in. by 1-in. brass or aluminum with edges filed flat.

We use templates to lay out frequently repeated joints like notches, lap joints and dovetails, as well as decorative chamfer stops. The ideal template material is thin (about ⅛ in.) but stiff and can be cut with a knife, but resists damage by knife or scribe while in use—I haven't found it yet. Brass, aluminum or steel work well for permanent templates, but they must be cut out tediously with hacksaw and file. Wood and hardboard must be sawn out and erode away in use. For the most part we use throwaway tem-

plates of dense grey cardboard (newsboard or chipboard), like the one in the photo above.

For measuring lengths in timber work, I have found a 1-in. by 25-ft. power return tape to be best. The 1-in. width makes it easier for a person working alone to extend the tape without its

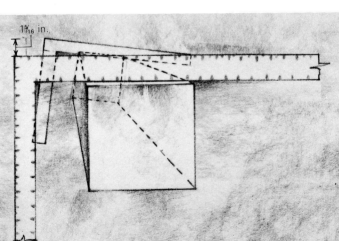

1⁵⁄₁₆ in.

1. Set squares at both ends of a timber (say, an 8-in. wide beam). Their blades will sit at different angles if there is any twist.

2. To find the amount of twist, position yourself to site down one edge of the timber, aligning the top end point of the far square with the top edge of the closer one.

3. Moving your eyes but not your head, site across the closer square and take a reading at the point its top edge seems to intersect with the vertical scale of the far square (1⁵⁄₁₆ in. in this example). This is the difference over the full 24-in. length of the rear square. But you want to know the deviation (twist) over the 8-in. width of the beam itself.

4. Divide the length of the square blade by the width of the beam to find the ratio of one to the other: 24/8 = 3.

5. Since this beam is ⅓ as wide as the square blade is long, its twist will be ⅓ the deviation you noted on the square, or 1⁵⁄₁₆ in./3 = ⁷⁄₁₆ in.

folding over. Try to use the same tape throughout a frame, since graduation may vary significantly from tape to tape.

We use an unchalked carpenter's line to check beams and assemblies for straightness and alignment and to measure the crown in bowed timbers. A chalked line comes in handy when we have to lay out a beam from its centerline because its edges are irregular.

Marking—Gauges are best for laying out the sides of mortises and tenons, the backs and bottoms of notches, the depths of housings and to mark any straight line parallel to the edge of a timber. Standard marking gauges have a useful reach of about 4 in. Beyond this point, switch to a panel gauge.

Gauge stems carry three spurs to scribe the centerline and both sides of mortises, as in the

photo. I make these spurs from hardened sheet-metal or drywall screws chucked in an electric drill and pointed on a grinding wheel while both wheel and drill are spinning. The screw thread allows you to adjust the heights of the spurs to keep all three marking evenly.

Once the centerline is determined, use an awl or an automatic center punch to mark drill-centers. The dimple will engage the worm of the auger used to bore out waste.

A knife is best for marking across the grain, as in the first photo on p. 155, and a scribe for marking with it. Pencils won't stay sharp on rough lumber, and they leave fat, ambiguous lines that make for sloppy fits. (They are, however, useful for marking out decorative chamfers and stops, and in other applications where a residual scribed line is undesirable.) Cut and scribed lines can be split cleanly by setting the edge of your chisel right in the line when paring or chopping a joint. There are other advantages: Cut lines aren't erased by light surfacing and can be recovered or "darkened" without benefit of straightedge since the knife or scribe will track in the path of the existing line. An incorrect mark can still be changed—the sideways pressure of the tool cutting a new line will close up the grain in the adjacent old one. Finally, grain will not chip up beside a cut line when you're crosscutting with a circular saw. This not only makes for neater work, but also signals when you are cutting off the line, since in that case the saw will begin to raise flecks of wood.

Marking knives, awls and striking tools incorporating both knife blade and scribe are available from mail-order tool houses. For cut lines, I use an ordinary utility knife, and for scribing, a homemade awl made from a hardened nail pointed like the gauge spurs and hafted in an adjustable file handle or a pin vise. Both knife blades and awl points are easily reground or replaced when they wear away or break.

To identify timbers and joints for later assembly there are several options: The traditional tool is the timber scribe or race knife, which consists of a handled blade, often made out of an old file, bent back on itself in a hook or V shape. It is pulled toward the user, and scoops out a groove (the race) in the wood. You can use a chisel for the same purpose, or turn to number and letter stamps or lumber crayons.

Knives and planes—If your marking knife can't do double duty, a pocket or utility knife is as essential for miscellaneous tasks in timber framing as in all carpentry. A drawshave (also called a drawknife) will remove bark and, along with a spokeshave, will clean up wane (rounded or bark corners), and cut decorative chamfers and stops on beams.

The carriage or coachmaker's rabbet plane (photo below, right) is the unsung hero among timber-framing tools. We use this versatile plane (Stanley #10, Record #010) to knock off bumps around knots and checks, to level crowned timber surfaces before marking out, to dress housings or gains, finish cheeks and sometimes shoulders of tenons, and to surface scarf joints.

For hand-planing timber, a short, wide smooth plane like the Stanley #4½ or Record #04½ (photo below, left) is best, but standard smooth

or jack planes are adequate. Grind the iron in a shallow curve so the corners won't dig in.

The photo above shows other useful planes: A fillister or duplex rabbet plane (Stanley #78, Record #778) with 45° wooden fences attached (right) makes an excellent chamfer plane. And (left) a hand router (Stanley #71, Record #071) will clean up stopped or blind housings that the bench rabbet can't get into. An electric router will also do this, but its voracious appetite for wood is not always satisfied with waste. If the gain is wider than the tool, attach a wooden base or leave intermediate ridges in the work for the router to ride on. You can chisel these off later.

Among tools for power planing, I find the Makita model #1805B (photo above) unsurpassed. It cuts a 6⅛-in. swath and, when sharp, leaves a glassy surface on which tool marks are barely discernible.

Chisels—Several different patterns of chisel are handy in timber framing. The standard tool is the framing chisel, a category that ranges from tanged firmer chisels like the one in the photo on the facing page, bottom right—a tang is a spike-shaped extension of the blade, which extends into the handle—through heavier socketed varieties, all the way to stout mortise chisels and millwright's chisels. This last is often a two-man tool (one to hold, one to strike) capable of chopping a mortise straight out of the solid wood with little or no pre-drilling. It has pretty much vanished from the scene. Advances in drill-bit and drill technology make it more convenient and more efficient for the carpenter to bore out most of the waste in a mortise.

Tanged chisels used for chopping the ends of mortises should be somewhat narrower than the mortise width, lest the tool become wedged in the joint. Thicker socketed framing chisels can take the prying force necessary to free them, and so can be used full width. The sharper the corners of the top of the chisel (the beveled side), the better it will cut the sides of the mortise when levered forward.

One caution: Many carpenters take a straight-backed chisel for granted. Since most old and many new framing chisels are not flat on the back (unbeveled) side, check carefully when purchasing a tool. The English chisels available through mail-order tool catalogs are usually reliably flat. And you can repair most others—more about this later.

A large (1-in. to 1½-in. wide) bevel-edged chisel is useful for cleaning up inside acute angles (as in dovetails) along with a small one (⅜ in. to ½ in. wide) for even more restricted areas. The corner chisel, as its name implies, cuts two adjacent perpendicular surfaces at once. Using this chisel requires an unusual combination of concentration and dexterity. Since corner chisels are a nuisance to sharpen, I don't recommend them. A sharp framing chisel is perfectly adequate.

The slick—3 in. to 4 in. wide and 30 in. long—is the king of chisels. With its handle offset to clear the work, the slick is used to surface large areas, often for roughing out the area in advance of a plane. Slicks are paring chisels—they are always pushed, never struck.

Saws—Contractor's circular saws get a real workout in timber framing. The ideal power saw for this kind of work will cut at least halfway through the narrow dimension of your largest stock, without being unduly heavy. Four candidates are the Rockwell 510 Speedmatic 10¼-in. saw (which cuts 3⅝ in. deep at 90°, but weighs 35 lb.), the Makita 5201N 10¼-in. (3¾ in., almost 20 lb.), the Makita 5103N 13-in. (5 in., 22 lb.) and the Skil 860 10¼-in. (3⅞ in., 17½ lb.).

An 8¼-in. worm-drive saw makes up in tenacity what it lacks in depth of cut. And don't despair of your 7¼-in. saw. There are many operations in timber framing where a small, light saw is best (roughing out housings, cutting tenon shoulders and so forth). However, you will have to finish cutoffs and deep shoulder cuts by hand.

For sawblades, I recommend the chisel-tooth combination pattern, filed as a ripsaw. If your saws are in heavy and continuous use, you might consider carbide. After many years of using steel blades, we switched over several years ago when I realized that saw sharpening had become an almost daily chore. Our saws range from an 8¼-in., 18-tooth, 14-gauge blade which makes a .125-in. kerf, to a 12-in. 24-tooth, 12-gauge blade with a .159-in. kerf. The tooth pattern on all our blades is left/right/raker, which is free cutting on both rip and crosscuts, yet doesn't tear or leave a ragged cut.

A basic handsaw kit includes a 7 or 8-point crosscut, a 4 to 6-point crosscut and a 3½ or 4-point ripsaw. Finding coarse-toothed saws can be a problem. A pruning saw makes an acceptable heavy-duty crosscut. Docking saws can be filed to rip or to crosscut. Most of my saws with fewer than 7 teeth per inch started life with a finer tooth pattern and were retoothed and refiled by the local saw doctor. Cutting edges should be straight or slightly crowned in order to work down to a line when sawing joints.

A well-sharpened two-man crosscut saw makes fast and surprisingly accurate work of cutoffs and tenon shoulders.

Drills—These days, most of the work of cutting peg holes and boring out waste in mortises and notches is done with hand-held electric drills. Use a powerful (6 amp or more), slow-spinning

(300 to 600 rpm) reversible, ½-in. or larger drill with long handles. Guard against binding, lest the drill spin instead of the bit. You can guide the drill by eye, with a small square on the surface or a batten attached to the side of the work, or by attaching a circular bubble level to the tool. For the first and last of these methods, it's important to have leveled up the timber beforehand.

The 19th-century boring machine (photo above) is a sensible hand-powered way to drill timber. Out of production for many years, such machines are scarce, but they are a joy to use. The coffee-grinder handles give the workman tremendous leverage. When the desired depth is reached (indicated by a scale on the side of the machine or by a built-in adjustable depth stop), a set of gears is engaged to retract the bit and most of the chips. A 2-in. hole can be drilled 4 in. into oak in less than a minute.

A drill press is useful for repeated operations on smaller pieces such as diagonal bracing and light floor joists or purlins.

The Scotch-pattern auger is the appropriate bit for most timber-framing applications. It is identical in most respects to the more familiar Jennings pattern—a double-twisted body, a nose with flat cutting edges and a screw lead—but instead of downward pointing spurs, the Scotch auger has upward-facing side wings at the outside of the cutting edges (the difference is apparent in the drawing at left). This leaves a rougher hole, but makes for a freer-cutting bit with less drag. A Jennings auger is an acceptable substitute if the Scotch is unavailable.

Other bit patterns include the ship's auger (single twist, single cutter and side wing with worm) for peg holes, a long barefoot auger to finish deep holes, brad-point twist drills for use in a drill press, and the Irwin bit (single twist, solid center) for general work. Drills without screw leads (multispur, Forstner, power bore, spade, and so forth) have trouble clearing chips in all but the shallowest holes and are not recommended for timber framing.

Mallets—Wooden mallets are kinder to trunnels, tool handles and timber surfaces than steel hammers, but don't stand up well to heavy use with steel-hooped chisels. We like to use mallets with rawhide-faced iron heads (photo below). Carver's mallets have a cylindrical or conical

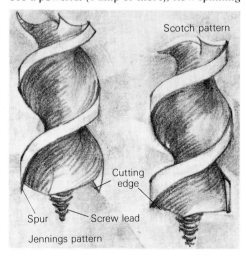

Scotch pattern

Cutting edge

Spur   Screw lead

Jennings pattern

head that is parallel to the handle. Carpenter's mallets have a cylindrical or rectangular head that is perpendicular to the handle. The carpenter's mallet is more common in timber work, but personal preference rules. Some carpenters prefer to use dead-blow mallets of shot-filled plastic for driving pins or knocking light framework together, and in other situations calling for a heavy blow without bounce. The commander is the grandfather of wooden mallets. It weighs

Photo: Richard Starr

20 to 30 lb. and is used to drive beams or wall sections together and shift them into place, as shown in the photo above.

Tool rehabilitation—There is no guarantee these days that new tools from the manufacturer will arrive in good working order. They are often poorly tempered and carelessly ground (or belt sanded instead). They may have been machined while their castings were still green. There are also a lot of old tools out there in attics, antique shops, junk stores and flea markets. They may be rusty or out of adjustment, but many of them, along with lots of less than perfect new tools, can be made perfectly serviceable with a little tinkering.

Metal straightedges can be trued up with a long flat or mill file. A small wooden guide block

clamped or held to the file (photo above) will help maintain square. Alter your filing pattern for even wear on the file, and clean the file regularly with a file card and a brush. This is especially important when you are filing brass or aluminum, which clog up ordinary files very

rapidly. (If you do a great deal of soft-metal filing, get special-purpose files.)

To test tools for flatness and square, keep a master straightedge and square. You can also test squares by marking across a straight board and reversing the body of the square, as in the

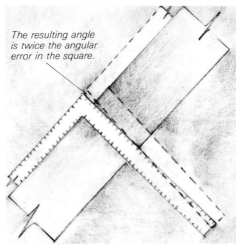

The resulting angle is twice the angular error in the square.

drawing. If the blade of the tool and the line still coincide, the square is true. If they don't, the angle they form is twice the error in the square.

Out-of-square combination squares can be repaired with a small, thin file (a 4-in. flat or mill file will do). File the base of the slot in the body of

the square (photo above). The drawing below shows how to true up framing squares. Scratch a line between the corners at the juncture of the body and the tongue. Place the square on an anvil or the other solid support and, using a hammer, strike on the scribed line.

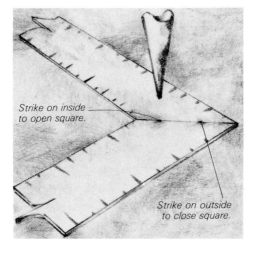

Strike on inside to open square.

Strike on outside to close square.

mer and center punch, strike on the scribed line. A blow toward the inside corner of the square will open the angle, and one near the outside will close it.

Small deviations from flat in the backs of chisels or bottoms of planes can be eliminated in several ways. Joint the tool on sandpaper or an abrasive belt attached with spring clamps or with rubber cement to a flat surface like the bed

of a jointer (photo above). Or you can lap it with abrasive compound (tripoli, carborundum powder or valve-grinding compound) on a flat piece of plate glass. Polish the tool with fine-grit paper on the jointer top, polishing compound on the glass, or on a large sharpening stone.

More pronounced curvature in chisels and planes can be taken out on the flat side of a grinding wheel, but this is a touchy business, and it's probably best to consign these tools to the surface grinder of a competent machinist. The same applies to rabbet planes whose sides are not square to their bottoms.

Here is a procedure to follow for resuscitating an iron plane:

1. Disassemble the plane. Examine and treat the parts separately (drawing below).

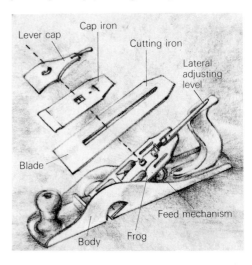

Lever cap
Cap iron
Cutting iron
Lateral adjusting level
Blade
Body
Frog
Feed mechanism

2. Remove the cap iron. Clean up rust and flatten the back of the cutting iron on an India stone, removing scratches with an Arkansas stone. Reassemble and check for daylight between cap and iron. There should be no play between the two when the screw is tightened. Twist or insufficient spring tension in the cap can be cured by placing it in a vise and bending it slightly. If the

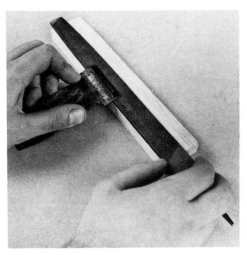

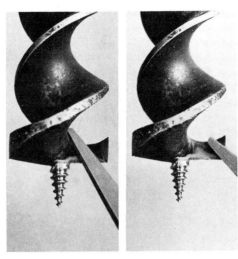

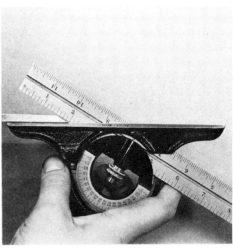

Grinding and sharpening—To sharpen drill bits, file the top side of the flats (photo below left) and the inside of the spurs or side wings (photo below right) with a bit file. Be sure to remove material evenly from both sides.

cap still does not touch the blade along its full width, then it must be jointed. Block a file or stone up off the bench, as in the photo above, and flatten the contact surface of the cap with the heel resting on the bench. This should ensure that cap and iron meet at the leading edge of the cap, preventing shavings from becoming wedged between the two. Grind and hone the cutter (see grinding and sharpening, at right).

3. Clean up the frog. If the feed mechanism doesn't work freely, spin the knob off and examine it and the stud it rides on. Damaged male threads can often be repaired with a needle file or a thread-repairing tool or file. The knob or stud can also be replaced, either by cannibalizing a spare-parts plane or by ordering from the manufacturer (Stanley maintains a comprehensive repair and spare-parts catalog). Recutting the threads is difficult as they are often odd-sized or left-handed. The lateral adjusting lever can usually be freed up by lubricating and working it vigorously back and forth.

4. Remove rust from the body using penetrating oil and a stainless-steel sponge. For more serious cases, use a light application of the jointing techniques described above. Ease sharp corners with a file or emery cloth. Replace the frog (set at the widest opening) and install the iron. You may need to file the forward edge of the throat if the opening is too narrow for timber work. (The same may be necessary with the apertures for chip clearance in the sides of a rabbet plane.)

To remove rust or pitting on drill bits, carefully stone or file the bottoms of flats and lightly polish the outside of the spurs or side wings with a stone or buffing wheel. This is the only time these surfaces should be touched. If a bit drags or won't feed in test holes, try the following:

1. Check chip thickness. If one side of the bit is taking a thicker chip, file back the leading cutter until both cut to an even depth. Match the height of the spurs on Jennings augers. On a Scotch auger, the leading edges of side wings and edges should coincide so that both cut simultaneously. Remember to file only the tops of cutters and the insides of wings or spurs.

2. Check the screw lead. Damaged or shallow threads can be filed with a needle file.

3. Check the cutter angle, and file it down if it appears too steep. For timber work, 30° to 35° is about right.

4. File down the spurs on Jennings augers to reduce drag.

The big danger in grinding edge tools is burning the tool and drawing its temper. To avoid this, use a sharp wheel. Dress it with a carborundum, a star wheel or a diamond dressing tool when it becomes uneven, glazed or clogged. White aluminum oxide wheels wear quickly, but stay sharper and are an improvement over the older-style grey wheel. Avoid excessive pressure, and check the tool regularly to see that it doesn't overheat. One way to do this is to dip the plane iron or chisel into water and grind only until the droplets just begin to steam off the edge before dipping again.

If you do burn a tool (you can tell by the discoloration at its edge) or have one that is too soft (the edge rolls over and nicks easily) or too hard (the edge chips or corners break off), you can usually have it retempered by a competent blacksmith or metalworker.

We grind large tools on a belt grinder with an 11½-in. contact wheel (photo below). Plane irons

and smaller tools are ground on a 6-in. hand grinder since the large wheel doesn't leave a sufficient hollow in the narrower bevels. A 7-in. or 8-in. grinder is a reasonable compromise between the two.

First check and correct the edge for square. If the sides of the tool are not parallel, split the difference. We grind most timber-framing tools at

an angle of 30°. Check the angle with a sliding bevel or bevel protractor as in the photo above. Another test for a 30° angle: The length of the bevel is just twice the thickness of the tool. Experience may dictate a larger or smaller angle on certain tools. Experiment. Grind until you raise a wire edge, checking for straightness and square, and then hone at the same angle, with the bevel laid flat on the stone, until a shiny narrow line appears at the edge. Whet the tool only until all the scratches from grinding are clear of the edge. This should leave enough steel for several additional honings before the bevel is flat and has to be reground.

Hone plane irons on the standard 6-in. or 8-in. stones. With careful grinding you can go directly to a soft Arkansas stone (photo below). If this

works too slowly, take a couple of strokes on an India stone first. For a razor edge, finish with a hard Arkansas stone. Whet the back of the tool last to remove the burr. Light mineral oil makes an excellent and inexpensive honing oil.

Since it can be very difficult to hand-hold framing chisels and other large edge tools at a steady angle while moving them over a sharpening stone, try moving the stone rather than the tool. Lay the chisel on its back and hone with small India and Arkansas files. This method works for all edge tools and is especially handy when retouching edges away from the workshop, since you don't have to tote around several large and fragile stones. □

Ed Levin, of Canaan, N.H., designs and builds timber-framed houses.

Sizing Roughsawn Joists and Beams
Methods and formulas for engineering your own timber frame

by Ed Levin

Structural engineering in timber design figures the allowable loads in building members, with an eye to staying safely within the elastic range of the material. The greatest challenge in designing and building a timber frame is to accommodate timber placement to the floor plan of the structure. In this process, the carpenter should always be at the service of the architect, developing a framing skeleton so that no diagonal braces have to be ducked, and no posts obtrude into the living space. Structural engineering lets you work to the designer's specifications, choosing scantlings (timber sizes) and deciding upon joist and rafter spacings.

Intuition and experience—Our medieval ancestors didn't have refined engineering tools, yet they created monumental wooden structures unmatched in modern times. An experienced carpenter seldom needs tables, and with some conventional building experience under your belt you should be able to extrapolate to dimension timber in other framing situations. For instance, in a floor where 2x8s 16 in. on center are adequate, then 3x8s 24 in. o.c., 4x8s 32 in. o.c. or 6x8s 48 in. o.c. will do just as well, with heavier flooring as spacing increases.

If you wonder whether a certain beam will be springy, block the ends up off the ground or floor and jump on the middle. With some blocking, beams and planks, you can quickly mock up a floor and try it out. Vary the timber spacing and see what happens. A little empirical knowledge goes a long way in timber design, and leaves you better prepared to deal with the complexities of the more rigorous engineering approach.

Using structural formulas—When you meet up with a problem you can't solve empirically, or you want to check on a particular frame element, it's nice to have the resources of modern structural engineering to fall back on. Joist and rafter tables in the standard carpentry texts aren't much help, because they are keyed to dimensioned lumber. There are a few books geared to the needs of the timber framer (see the bibliography at the end of the article), but even the best reference has limitations. For example, if you are checking deflection in 16-ft. long, 7-in. by 9-in. hornbeam timbers on 37-in. centers with a 95 pound per square foot (psf) live load, chances are all the tables in the world won't help

Ed Levin, of Canaan, N.H., designs and builds timber-framed houses.

you. You'll need to go directly to the formulas upon which the tables are based. These are the fundamental engineering tools in timber design.

Some pointers: allowable stress values (modulus of elasticity, extreme fiber stress, compression parallel to the grain, horizontal shear stress and so forth) for particular species and grades of wood can be found in tables in several of the books in the bibliography on p. 164. If working stress values for a particular species of wood are unavailable in the standard references, clear wood strength values can be obtained from the USDA *Wood Handbook*.

Most timber frames are built of green wood, which is both heavier and weaker than dry. The increased dead load due to the initial high moisture content, combined with decreased stiffness (increases of 20% to 30% are typical after green wood dries) can cause a joist or girder to dry with a pronounced sag. Take care that your floor will be strong enough when new. You may want to add temporary midspan support to beams whose green strength is questionable.

In beam design, you should check all members for shear and bending moment. An additional calculation for deflection is necessary for floor joists and in other situations as called for by code. Shear is the limiting factor in short, heavily loaded spans or if you're using narrow, deep members. Deflection governs in long, lightly loaded beams, as well as in broad, shallow ones. For intermediate situations, bending moment is usually the limiting factor.

For sizing floor joists, some codes specify live load deflection only, while others call for deflection due to both live and dead loads. Typical allowable deflections in floor joists and beams are $\frac{1}{360}$ of the length of the unsupported span if you're figuring live load only, or $\frac{1}{240}$ of the length of the span if you're figuring combined live and dead loads. The minimum standards for uniformly distributed live loads are 40 psf for first floors, 30 psf for upper floors and inhabitable attics and 20 psf for uninhabitable attics.

When you plug values into the formulas, use actual and not nominal sizes of members. Finally, lengths in formulas are usually given in inches, not feet, so keep your units straight.

The formulas here will let you size a simply supported rectangular beam that bears a uniformly distributed load or a concentrated load at the middle of a span, the two most common loading conditions on floors and roof. The formulas cover 90% of the bending and deflection problems you'll encounter in floor framing.

Deflection
Live load deflection—The dimensions of floor joists and beams are usually governed by deflection. Let's use the most common standard for a floor: a minimum first-story live load of 40 psf with a live load deflection of $\frac{1}{360}$ of the length of the unsupported span (this is the only computation in which you don't need to use combined live and dead loads). We'll run through an example, sizing red oak floor joists, 36 in. on center over a 12½-ft. span, to be covered with nominal 2-in. spruce decking.

The formula for finding the deflection resulting from a uniformly distributed load is:

$$\Delta = \frac{5wL^4}{384EI}, \text{ where}$$

Δ = The maximum deflection (in inches) at midspan;

w = the weight each lineal inch of each joist must support (in lb./in.);

L = the length (in inches) of the unsupported span;

E = the modulus of elasticity (psi);

b = the width of the joist in inches;

d = the depth of the joist in inches; and

I = the moment of inertia (in in.⁴), $bd^3/12$ (see glossary).

Begin by substituting

$\frac{L}{360}$, the maximum allowable deflection,

for Δ, and juggling the formula algebraically to solve for I, which will let us determine acceptable joist dimensions. So:

$$I = \frac{75wL^3}{16E}.$$

We already know some values:

L = 150 in. (12½ ft.), and
E = 1,500,000 psi for red oak.

To calculate w, we first have to figure out the load each joist must carry. Each joist supports half of the uniformly distributed load on each side of it. for interior joists, this is the same as the distance between two joists (see the drawing on p. 163). In our example, this is 36 in.

$$w = \frac{40 \text{ psf (uniform load)} \times 36 \text{ in.}}{144 \text{ in.}^2/\text{ft.}^2 \text{ (so } w \text{ is in lb./in.)}},$$

Glossary of Structural Terms

Tension

Compression

When a simple beam is bent downward under load, its top is in compression and its bottom is in tension. Tension is the state of stress in which particles of material tend to be pulled apart. Ropes and cables are purely tensile elements and will not assume any compressive or bending stress. The opposite of tension is compression, the state of stress in which particles of material tend to be pushed together. Concrete resists compression but doesn't cope well in tension. That's why it is reinforced with steel rods in areas of tensile strain. It's easy to see that knots are a greater

disadvantage in tension. This explains why they were traditionally placed in the top surface of joists and girders.

Bending stresses are greatest at the top and bottom surfaces. They decrease toward the center. Fibers in the middle of the beam are neither tensed nor compressed, and are said to lie along the neutral axis (see deflection).

Neutral axis

Extreme fiber stress

Bending strength is limited by the maximum safe extreme fiber stress. This is the capacity of the wood to resist powerful compressive and tensile forces at the upper and lower faces of a loaded timber. For a given load, bending strength varies directly with the breadth and the square of the depth of the timber, and inversely with the length of the span

$$\left(\text{strength} \propto \frac{bd^2}{L}\right).$$

Section modulus

Bending moment

In bending calculations, timber size is usually expressed as the section modulus *(S)*. For rectangular beams, $S = bd^2/6$. Bending force is quantified as the bending moment. For our simple beam, the moment is greatest at midspan and drops to zero over the supports.

Elastic

Plastic

Wood is an elastic material, one which deforms under load, but whose deformation vanishes when the load is removed. Materials that remain deformed after loads are removed are described as plastic, and as with any elastic material there are limits to the loads wood can bear without behaving plastically.

Stress

Strain

A compressive or tensile force acting on an elastic material sets up stress (force per unit area, usually expressed in psi). This causes the material to be slightly shortened or stretched. This is strain, the lengthwise deformation per unit of a material under load (expressed in inches of deformation per linear inch of material, or in./in.). The set of stress values that wood responds to elastically is called its elastic range. The point at which it begins to exhibit plasticity is its yield load. In most cases, persistent plastic behavior immediately precedes the failure of the timber.

Elastic range
Yield load

Within its elastic range, wood deformation under load is directly proportional to that load. This proportion of load to deformation, written as the ratio of stress to strain for a particular species of wood, is called its modulus of elasticity *(E)*, the constant used to calculate stiffness in beams. It expresses the linear relation between a given stress (load) and the resulting strain (deformation) in the material:

Modulus of elasticity

$$E \text{ (psi)} = \frac{\text{stress (psi)}}{\text{strain (in./in.)}}$$

Stiffness
Deflection

While a timber's bending strength is its ability to carry a load without breaking, stiffness is its ability to remain rigid in use. Stiffness is measured by beam deflection, the amount a loaded beam will bend below the horizontal. For a fixed load, stiffness

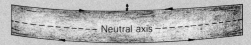

varies directly with the breadth and the cube of the depth of the timber, and inversely with the cube of the length of the span (stiffness $\propto bd^3/L^3$).

Moment of inertia

In deflection calculations, timber size is usually expressed as moment of inertia *(I)*. For rectangular beams, $I = bd^3/12$.

Vertical shear
Horizontal shear

Shear is the state of stress in which particles of material tend to slide relative to each other. Horizontal timbers under load tend to break or shear off at the edge of their supports. This vertical shear is always accompanied by horizontal shear. You can understand this phenomenon if you take a half-dozen or so

Carrying beam   Beam in span

Horizontal shear along neutral axis

pieces of wood about ⅛ in. by 2 in. and lay them flat one on top of the other. Support the ends and depress the middle. You'll notice that as the center bends downward the individual strips of wood tend to slide along one another. This horizontal shear force operates the same way in a solid timber, except there adhesion between the wood fibers keeps them from sliding, which causes shear stress in the timber. The shear stress acts to split the timber along the grain, the direction in which wood is weakest. This points up another property of wood—it is anisotropic. Unlike steel, concrete, aluminum or plastic, the structural properties of timber are not identical in every direction.

Anisotropy

Dead load

Live load

Dead load is the weight of the structure itself and all loads permanently on it. For our purposes this generally means the weight of the timber frame plus flooring or roofing and insulation. Live loads are all loads other than dead loads—usually the weight of people and their furniture as well as wind and snow loads. You need dead load values to design a structure, but you can't determine them until after the structure is designed. So even engineers have to start with an educated guess.

Roof rafters must be strong to take snow loads, but stiffness is not a requirement unless you plan to finish the underside of the roof. So scantlings are often determined by bending strength. But floor joists and the timbers that carry them must not only be strong, but stiff as well, because you don't want a springy floor, so sizes are limited by deflection. A certain amount of deflection is inevitable in rafters, joists and beams, but you can take advantage

Camber

of bowed timbers to introduce an upward camber in the floor or roof. Then as the member settles under load it will not assume a negative or downward curvature. —*E.L.*

d / b	3	3½	4	5	5½	6	7	7¼	8	9	9¼	10	11¼	12
1½	3 / 2	5 / 3			21 / 8			48 / 13			99 / 21		178 / 32	
2			11 / 5		36 / 12			85 / 21			167 / 33			288 / 48
3	7 / 5	11 / 6	16 / 8	31 / 13	42 / 15	54 / 18	86 / 25	95 / 26	128 / 32	182 / 41	198 / 43	250 / 50	356 / 63	432 / 72
3½	13 / 7				49 / 18			111 / 31			231 / 50		415 / 74	
4	9 / 6		21 / 11	42 / 17		72 / 24	114 / 33		171 / 43	243 / 54		333 / 67		576 / 96
4½		16 / 9			62 / 23			143 / 39			297 / 64		534 / 95	
5			27 / 13	52 / 21		90 / 30	143 / 41		213 / 53	304 / 68		417 / 83		720 / 120
5½		20 / 11			76 / 28			175 / 48			363 / 78		653 / 116	
6			32 / 16	63 / 25		108 / 36	172 / 49		256 / 64	365 / 81		500 / 100		864 / 144
7				73 / 29		126 / 42	200 / 57		299 / 75	425 / 95		583 / 117		1008 / 168
7¼					101 / 37			230 / 64			478 / 103		860 / 153	
7½					104 / 38			238 / 66			495 / 107		890 / 158	
8						144 / 48	229 / 65		341 / 85	486 / 108		667 / 133		1152 / 192
9							257 / 74		384 / 96	547 / 122		750 / 150		1296 / 216
9¼								294 / 81			610 / 132		1098 / 195	
10						180 / 60	286 / 82		427 / 107	608 / 135		833 / 167		1440 / 240
10½							333 / 92			693 / 150			1246 / 221	
11¼											742 / 160		1335 / 237	
12									512 / 128	729 / 162		1000 / 200		1728 / 288

Moments of Inertia (I) and Section Moduli (S)

$$I = \frac{bd^3}{12}, \quad S = \frac{bd^2}{6}$$

b = breadth of timber (in.)
d = depth of timber (in.)

Moments of inertia, at the upper left of each small box, are expressed in in.⁴ Once you have figured what value is required for maximum allowable deflection, check to see that I for the timber dimensions you are considering is higher. Section moduli, at the lower right of each box, are expressed in in.³. After you've figured the value for maximum allowable bending, check to see that S for your timber dimensions is higher.

so w = 10 lb./in. (each inch of each joist must carry the load of 10 lb.).

Now we plug these figures into our formula:

$$I = \frac{75wL^3}{16E} = \frac{(75)(10 \text{ lb./in.})(150 \text{ in.})^3}{(16)(1,500,000 \text{ psi})}$$

$$= 105.47 \text{ in.}^4$$

By checking the table at left for moments of inertia or by substituting for b and d in the equation $I = bd^3/12$, we see several joist dimensions that would work:

3x8: $I = \frac{3 \times 8^3}{12} = 128 \text{ in.}^4$

4x7: $I = \frac{4 \times 7^3}{12} = 114.33 \text{ in.}^4$

6x6: $I = \frac{6 \times 6^3}{12} = 108 \text{ in.}^4$

Combined load deflection—The other common code standard for floors is combined live (40 psf) and dead loads with deflection of $\frac{1}{240}$ of the length of the unsupported span. To find the dead load, we have to know the weight of the decking (about 4 psf) and of the joists themselves (48 lb./ft.³ for red oak at 25% moisture content). A 6x6 joist, the heaviest of the three listed above, contains 432 in.³ of wood per linear ft. (6x6x12), or .25 ft.³ of wood. Multiplying this by the value of 48 lb./ft.³ for red oak gives us a weight of 12 lb. per linear ft. Since the joists fall on 3-ft. centers, they contribute

$$\frac{12 \text{ lb./ft.}}{3 \text{ ft.}} \text{ or 4 lb./ft.}^2 \text{ to the dead load:}$$

$$
\begin{aligned}
& 4 \text{ lb./ft.}^2 \text{ (Joist weight)} \\
&+\ 4 \text{ lb./ft.}^2 \text{ (Decking weight)} \\
&=\ 8 \text{ lb./ft.}^2 \text{ (Total dead load)} \\
&+\ 40 \text{ lb./ft.}^2 \text{ (Live load)} \\
&=\ 48 \text{ lb./ft.}^2 \text{ (Combined load).}
\end{aligned}
$$

Finding w as we did in the first case:

$$w = \frac{(48 \text{ psf}) (36 \text{ in.})}{144 \text{ in.}^2/\text{ft.}^2} = 12 \text{ lb./in.}$$

And for $\Delta = L/240$, $I = 25wL^3/8E$. So

$$I = \frac{(25)(12 \text{ lb./in.})(150 \text{ in.})^3}{8 (1,500,000 \text{ psi})} = 84.38 \text{ in.}^4$$

all four previously selected joist sizes pass the combined load, $L/240$ test. So do a couple of new choices:

3x7: $I = \frac{3 \times 7^3}{12} = 86 \text{ in.}^4$

5x6: $I = \frac{5 \times 6^3}{12} = 90 \text{ in.}^4$.

To check deflection for any concentrated mid-span loads, use the formula $\Delta = PL^3/48EI$, where P = the weight of the concentrated load in lb. The inverted form of this equation is:

$$I = \frac{15PL^2}{2E} \text{ when } \Delta = L/360, \text{ or}$$

$$I = \frac{5PL^2}{E} \text{ when } \Delta = L/240.$$

Maximum deflection in both concentrated midspan and uniformly distributed loads occurs at midspan. To find deflection resulting from a combination of the two, we can do the separate calculations for each and simply add the results.

In a floor with three or more members on centers of 24 in. or less, allowable stress may be increased by 15%. See values for repetitive member uses in the supplement of the *National Design Specifications for Wood Construction*.

Deflection figures are based on so-called normal loading, which assumes that a structural member will carry its full design load for an accumulated total of ten years during the life of the building. Where you expect that a beam will be fully loaded, continuously or cumulatively, for longer than that, you should reduce the allowable stress values by 10%. Timbers permanently stressed at full design load can deflect as much as 1½ times (2 times for unseasoned wood) the amount calculated for normal loading, because of long term inelastic deformation, or *creep*.

Bending

Having satisfied the deflection criteria, we should verify that the joists are strong enough in bending as well. In bending:

$$f = \frac{M}{S},$$

where

$$M = \frac{wL^2}{8} \text{ for uniform loads}$$

and

$$M = \frac{PL}{4} \text{ for concentrated loads,}$$

where

- f = maximum extreme fiber stress in psi (see glossary);
- M = maximum bending moment in inch-pounds;
- P = weight of the concentrated load in lb., and
- S = section modulus $\frac{bd^2}{6}$ (in.³).

Solving for uniform loads:

$$S = \frac{M}{f} = \frac{wL^2}{8f} .$$

Given w = 12 lb./in.,
  L = 150 in., and
  f = 1,500 psi for our red oak joists:

$$S = \frac{wL^2}{8f} = \frac{(12 \text{ lb./ft.})(150 \text{ in.})^2}{(8)(1,500 \text{ psi})} = 22.50 \text{ in.}^3.$$

This time we check the table on the facing page for section moduli:

3x7: S = 25 in.³
3x8: S = 32 in.³
4x7: S = 33 in.³
5x6: S = 30 in.³
6x6: S = 36 in.³

All six choices are adequate in bending.

As with deflection, you can find bending stress

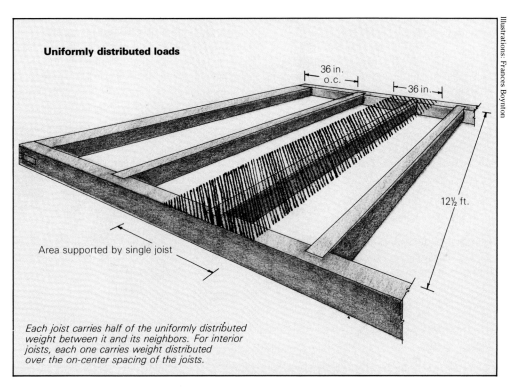

Uniformly distributed loads

36 in. o.c.
36 in.
12½ ft.

Area supported by single joist

Each joist carries half of the uniformly distributed weight between it and its neighbors. For interior joists, each one carries weight distributed over the on-center spacing of the joists.

Summary of most useful equations	
Uniformly distributed load	Concentrated midspan load
Deflection: $\Delta = \dfrac{5wL^4}{384EI}$	$\Delta = \dfrac{PL^3}{48EI}$
for $\Delta = L/360$, $I = \dfrac{75wL^3}{16E}$	for $\Delta = L/360$, $I = \dfrac{15PL^2}{2E}$
for $\Delta = L/240$, $I = \dfrac{25wL^3}{8E}$	for $\Delta = L/240$, $I = \dfrac{5PL^2}{E}$
Bending: $f = \dfrac{M}{S}$ $\;$ $M = \dfrac{wL^2}{8}$ $\;$ $S = \dfrac{M}{f} = \dfrac{wL^2}{8f}$	$f = \dfrac{M}{S}$ $\;$ $M = \dfrac{PL}{4}$ $\;$ $S = \dfrac{M}{f} = \dfrac{PL}{4f}$
Shear: $V = \dfrac{wL}{2}$ $\;$ $H = \dfrac{3V}{2}$ $\;$ $h = \dfrac{3V}{2A} = \dfrac{3V}{2bd}$	$V = \dfrac{P}{2}$ $\;$ $H = \dfrac{3V}{2} = \dfrac{3P}{4}$ $\;$ $h = \dfrac{3V}{2A} = \dfrac{3P}{4bd}$
For notched beams: $H = \dfrac{3Vd}{2b'(d')^2}$ or $d' = \sqrt{\dfrac{3Vd}{2b'h}}$	$h = \dfrac{3Pd}{4b'(d')^2}$ or $d' = \sqrt{\dfrac{3Pd}{4b'h}}$

Δ = Maximum deflection (at midspan) in inches	M = Bending moment (lb.)
w = Weight per unit length (lb./in.)	f = Extreme fiber stress (psi)
L = Unsupported span (in.)	S = Section modulus (in.³)
E = Modulus of elasticity (psi)	b' = Net width (in.), as in dovetail waist
b = Beam width (in.)	d' = Net depth (above notch) in inches
d = Beam depth (in.)	v = Vertical shear (lb.)
I = Moment of inertia ($bd^3/12$)	H = Horizontal shear force (lb.)
P = Concentrated load (lb.)	h = Horizontal shear stress (psi)
	A = Area of beam cross section (in.²)

due to combined uniform and centrally placed concentrated loads by addition.

Shear

Once you've chosen timber dimensions, you should check their resistance to horizontal shear stress. For the loading discussed here, shear stress is greatest directly over the supports and negligible at midspan. With loads placed symmetrically, the vertical shear force (V) is equal at both ends. The formulas for finding V are:

$$V = \frac{wL}{2} \text{ in continuous loading, and}$$

$$V = \frac{P}{2} \text{ in concentrated loading.}$$

For rectangular beams, horizontal shear force (H) is $1\frac{1}{2}$ times as great as the vertical shear force ($H = 3V/2$). Horizontal shear stress (h) is equal to the horizontal shear force divided by the area of the beam cross section (A), taken over the supports ($h = H/A$ where $A = bd$). So:

$$h = \frac{3V}{2bd} .$$

Our oak joist with the smallest cross section was the rough 3x7.

Checking it for shear stress, we find:

$$V = \frac{wL}{2} = \frac{(12 \text{ lb./in.})(150 \text{ in.})}{2} = 900 \text{ lb., and}$$

$$h = \frac{3V}{2bd} = \frac{(3)(900 \text{ lb.})}{(2)(3 \text{ in.})(7 \text{ in.})} = 64.29 \text{ psi.}$$

An allowable h value for red oak of 100 psi puts this scantling well within safety limits. Safe h values for most species used in framing range from 65 psi to 150 psi.

Shear with notching—When a notch is taken from the bottom of a joist or beam, the cross-sectional area is diminished while the shear force remains constant, resulting in increased horizontal shear stress. The concentration of stress at the inside corner of the notch (which often lies close to the neutral axis where shear is greatest) makes splitting and eventual failure at this point a danger.

To allow for this weakness, the shear formula ($h = 3V/2bd$) is modified: The net effective depth of the timber above the notch (d') is substituted for the total depth (d). The value of the resultant shear stress is then further increased by multiplying it by a safety factor equal to the ratio of the full depth to the net depth (d/d'). The revised formula looks like this:

$$h = \left(\frac{3V}{2\,bd'}\right)\left(\frac{d}{d'}\right)$$

Going back to our floor-framing example, suppose that we use a 4x7 joist and half-lap it into its carrying timber (drawing, below). Solving the formula for an acceptable d' and plugging in the acceptable h value of 100 psi, we have:

$$d' = \sqrt{\frac{3Vd}{2bh}} = \sqrt{\frac{(3)(900 \text{ lb.})(7 \text{ in.})}{(2)(4\text{in.})(100 \text{ psi})}} = 4\tfrac{7}{8} \text{ in.}$$

So when using a 4x7 in this application, a net depth of at least $4\frac{7}{8}$ in. must be left above the notch. Checking the other timbers which were possible solutions to the joist problem, we find that for a 3x8, $d' = 6$ in.; for a dressed 4x8, $d' = 5\frac{5}{16}$ in.; for a 6x6, $d' = 3\frac{11}{16}$ in.; for a 3x7, $d' = 5\frac{5}{8}$ in.; and for a 5x6, $d' = 4\frac{1}{16}$ in.

It is apparent that the wider the timber, the deeper the notch that may be taken out. This sheds some light on the apparently illogical use of square-section joists and beams in traditional structures where it appears that narrow and deep pieces would have served better structurally. Not only were square timbers easier to hew from the tree, but they were also better suited to the notching necessary to fit them into the frame. It is often the case in timber framing that when a particular member seems poorly designed in structural terms, it makes perfect sense when the demands of joinery are considered.

One way of relieving the stress concentration in soffit notches, when there is no aesthetic objection, is to cut away the material below the inside corner of the notch in a gradual curve (drawing, below). This brings the strength of the joist in shear back up close to the value indicated by the net depth, unaffected by notching.

Notching the side of beams, as in dovetail lap joints, for example, also reduces the area of the cross section and increases shear stress. For these situations, substitute b' (the net effective width of the timber) for b (the total width of the timber). For a dovetail lap joint, b' is the thickness of the waist of the dovetail, and

$$h = \frac{3Vd}{2b'(d')^2}$$

For a dovetail lap joint with a 4-in. waist in a 6x6 joist:

$$d' = \sqrt{\frac{3Vd}{2b'h}} = \sqrt{\frac{(3)(900 \text{ lb.})(6 \text{ in.})}{(2)(4 \text{ in.})(100 \text{ psi})}} = 4\tfrac{1}{2} \text{ in.}$$

The net depth above the notch must be at least $4\frac{1}{2}$ in. ☐

For reference and further reading

Structure in Architecture by Mario Salvadori and Robert Heller ($21.95 from Prentice Hall, Englewood Cliffs, N.J. 07632).

Structures, or Why Things Don't Fall Down by J.E. Gordon ($17.95 from Plenum Press, 233 Spring St., New York, N.Y. 10013).

Both books are excellent introductions to the basic structural concepts without mathematical overload.

Wood Structural Design Data ($12.00 from the National Forest Products Association, 1619 Massachusetts Ave., N.W. 20036).

The best structural manual for timber framers, with tables, formulas and background information.

National Design Specifications for Wood Construction ($6.50 from the National Forest Products Association).

Supplement includes comprehensive tables of design values for graded lumber.

Timber Construction Manual by the American Institute of Timber Construction ($22.95 from John Wiley & Sons).

A good basic reference, with information on working stresses and loading situations.

Wood Handbook ($10.00 from the U.S. Government Printing Office, Washington, D.C. 20402).

Another good basic reference, with information on clear wood values.

Simplified Engineering for Architects and Builders by H. F. Parker ($27.50 from John Wiley & Sons).

An introductory structural text.

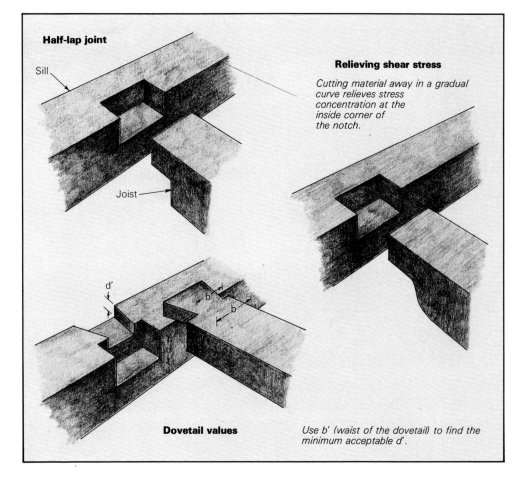

Half-lap joint

Sill

Joist

Dovetail values

d'

b'

b

Relieving shear stress

Cutting material away in a gradual curve relieves stress concentration at the inside corner of the notch.

Use b' (waist of the dovetail) to find the minimum acceptable d'.

Raising Heavy Timber

Tools and tips for maneuvering big beams

by Trey Loy

When there were many gigantic redwood and fir trees in the Pacific Northwest, huge logs were milled into massive timbers to build sawmills, bridges, wharves, warehouses and buildings for heavy industry. Lumber 12 in. square was common, though larger beams were also sawn. (The largest piece I've seen is 18 in. square and 42 ft. long, but the old-timers say they milled bigger ones than that.) The joinery of these structures was simple, relying on steel pins, bolts and plates for strength.

Today, many of the big-timber buildings are dilapidated beyond repair. Often the owner just wants to get rid of the old wreck, so salvage rights can be obtained before the wrecking crane is called. Salvaging any material is sound economy, and in recycled lumber there are some terrific finds like clear, tight-grained redwood, and well-seasoned fir that is suitable even for fine cabinetry. Used lumber, cleaned of paint and grime by rough-planing, sandblasting, and wire-brushing, reveals a new and rugged complexion that's quite pleasing to the eye, with nail holes and blemishes adding character.

We recently built a house using timbers purchased before it was designed. The timber had framed a navy warehouse in Eugene, Ore.; we bought 2,400 linear feet of Douglas fir 12x12s in 10-ft., 20-ft. and 30-ft. lengths, and 9x18s 32 ft. long. Many pieces had several coats of paint, and others were covered with dirt, grime and grease. The lumber was roughsawn and boxcut; its width sometimes varied more than an inch from one end to the other, and many beams were twisted along their entire lengths. Wide checks had further distorted dimensions. Broken nails and the torched ends of pins protruded from the surface—nasty stuff to work with. We pulled most of them with a nail puller and a crowbar. After the house was framed, we cleaned the exposed surfaces with a portable sandblaster, keeping the nozzle moving to avoid gouging grooves in the earlywood.

Moving and raising timber—Maneuvering heavy posts and beams is no great task if you've got a crane or boom truck. But the four of us on this job didn't have access to any such large equipment. So we used a few old-fashioned but effective tools: a peavey, a sweet william, a pulley, a ramp (inclined plane) and a gin pole.

A peavey is a stout hardwood pole, usually of ash or maple, about 6 ft. long and hollowed at one end to receive a pointed steel pin (photo, right). A tapered steel collar keeps the pin from

splitting the end of the handle. The upper part of the collar is fitted with two eyes through which a bolt passes to secure a large steel hook, shaped like a fishhook, which swings parallel to the pole. If you want to move a beam laterally, swing the peavey so the hook digs into the side of the timber and place the pointed end on top. Lifting and pulling on the handle pivots the timber. It is easy to flop the timber over and over until you get it where you want it. If you hook the peavey into the end of a timber, you can make a dead lift.

A sweet william, sometimes called a timber packer, is similar to ice tongs, except that the hooks are suspended so they swing and swivel from a steel collar fastened to the center of a 6-ft. wooden handle. The tongs grab opposite sides of a timber, and the scissoring action holds the timber firm. Two workers can lift the end of a beam for carrying or help drag a load up a ramp. These tools left some deep gouges

in our timbers, but the new wounds were hardly noticeable among the old scars.

Ramps are great back-savers for moving logs or timber to a higher level. To load a truck use two stout planks at least 3 in. thick and wide enough to walk on and place one at each end of the truck bed. Roll the timber over and over with peaveys, walking the beam up the ramp. If the luck of the day left you with only one peavey, tie a rope to the other side of the truck bed, run it around the center of the timber two or more turns, then back to a person standing on the bed. The turns of rope act like a continuous lever. As one person pushes with the peavey, the other pulls on the rope (drawing, top of next page).

Rollers under a timber make light work of moving a beam end first. Firewood-size logs work fine on rough ground; on the smooth surface of a ramp or subfloor we use lengths of 2-in. pipe. You alter the direction of travel by

Using a peavey in the end grain, you can either push or lift a heavy timber. For lateral movement the hook digs into the side of the beam, and leverage is applied through the hardwood handle.

Using a rope as a lever

When you've got only one peavey, a few turns of rope around the center of a timber will create the leverage you need to work it up an inclined plane.

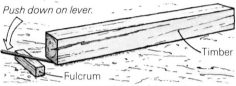

At least two turns

Lever and fulcrum

You can maneuver huge timbers with the proper application of leverage.

Push down on lever.

Timber

Fulcrum

First pry up one end, with the fulcrum between you and the weight.

Block

A small block inserted close to the timber's center becomes the next fulcrum.

Push down on timber's end.

Larger block

Press down on the end of the timber itself, insert a larger block, and so on.

Timber pivots on blocking.

Blocking in center

Once the timber is high enough you can rotate it on a fulcrum block set under its center.

The gin pole has been raised, and the trucker's hitch on one guy line is being tied off. The tackle that will raise the timber is dropping straight down from the yardarm. A ⁵⁄₁₆-in. braided steel cable supports the pole from the rear.

tapping the roller askew; the timber follows the rollers' path.

The posts for this house were light enough to be carried by three men using a peavey and a sweet william. Once on the subfloor, we man-handled each post into an upright position, plumbed it as well as possible, and braced it with 2x4s nailed temporarily to the stem wall and to stakes driven into the ground at right angles to the building's face. We moved heavier pieces onto the subfloor by rolling them up a strongly braced 2x8 ramp, using a wedge of wood behind each roller as a brake to prevent the timber from rolling back down.

Though we tried to put the lumber we'd need first on the top of the pile, invariably the timber we needed was at the bottom. At those times a lever and fulcrum came in handy. Using a lever, it's best to have the load on the other side of the fulcrum so you are pushing down with your weight to raise the load. With the load between you and the fulcrum, you have to lift up, and that's the kind of lift that can bust something loose inside. For levers we used steel bars, the peavey, and lengths of lumber. By prying up the end of a beam so it is slightly raised and slipping a block of wood underneath as far toward the middle as possible, you turn the beam itself into a lever. I'm always amazed at the small effort needed to seesaw a half-ton of wood back and forth. By alternately placing fulcrums of increasing height on either side of the balance point, you can raise the beam higher and higher, as shown in the drawing at left. With the fulcrum at the balance point, the timber can be swiveled in a new direction.

The gin pole—To raise the top plates, ridge beams and rafters into place, we used a gin pole. This is an upright pole with three guy lines for support; a block and tackle hung from the top does the lifting. A gin pole works only for vertical lifts and must be repositioned for every piece, but it can be moved around the site easily by two workers, though four are required to raise loads. You can set up the gin pole anywhere there is a solid place for its butt.

Our pole was a fir sapling 22 ft. long, straight, true and measuring 4½ in. in diameter at the butt and 3½ in. at the top. We passed a ⅝-in. steel pin 18 in. long through a hole drilled 1 ft. from the narrow end to serve as a yardarm for the rigging; a loop of chain hung over the top of the pole and resting on the yardarm supported the upper block of the block and tackle. When we moved or set up the pole, we ran out the lower block and tied it to the bottom of the pole. The guy lines are spliced with eyes that slide over the pole and rest on the yardarm. We used ⁵⁄₁₆-in. braided steel cable 120 ft. long for the main guy line directly supporting the pole (cable won't stretch) and two lengths of ½-in. rope 100 ft. long for the side guy lines that position and brace the pole.

To set up a gin pole, first find the balance point of the timber to be raised, and then determine where this point will be after the piece is in place. Directly under this imaginary point, mark an X on the floor or ground. Four or five feet back from the X, make a chock to hold the

butt of the pole by nailing two pieces of lumber on the floor in a V-shape, or by digging a shallow hole in the ground. Then lay the gin pole over the X, with its butt end resting in the chock. Raise it to about 75°, with the tackle plumb above the X. This will keep the load from rubbing against the pole during the lift.

The main guy line should run in a straight line behind the pole. Stretch the side guy lines out to either side of the pole, slightly behind the chock. The farther from the pole you anchor the lines, the smaller the angle of pull, which means less force is needed to raise and secure the pole. We usually fasten the main line at least 100 ft. from the pole and the support lines at least 50 ft. away. Loop the lines around something stable, like a tree or solid framing, or drive stakes in the ground at an angle away from the pole. We use 2-in. steel pipe 4 ft. long, driven 2 ft. into the ground with a maul.

The main guy line is run to its stake and given a couple of turns around the pipe or anchor. The side guy lines are secured with a knot capable of retrieving slack called a trucker's hitch (drawing, facing page, top). The pipe and the loop of the hitch act like two pulleys, and though there is some friction, this method makes it easier to haul the pole up. In fact, the friction works to keep the load from slipping back when the lines are held together.

On our job, one of us manned each of the three lines and a fourth worked the pole, lifting the end of it over his head and walking toward the butt. After the pole reached about 45°, we raised it the rest of the way with the lines alone, as the fourth man made sure it didn't slip out of the chock.

When the gin pole is nearly in position, untie the lower block and holding the falls (the rope you pull on) firmly with one hand, attach a weight to the lower block hook (we used a chunk of timber) for a plumb bob to center the block and tackle over the X. Once the tackle is plumb, draw the guy lines taut, secure them, and double-check everything. Tie off the lower block again to keep tension on the pole. Now roll the timber in place, positioning it with its mid-section over the X.

Lifting—It took quite a bit of time to prepare for each lift, and because of the size of the timbers, some days we got only one rafter into place. (The gin pole can work well and quickly to raise lighter weights like standard ridgepoles). We tried to maintain an even pace, with two of us preparing the rigging and two working on the next piece. Dealing with these tremendous weights requires teamwork. First we chained the timber to the lower block hook. Those of us who had to lift used our legs, and a short countdown was called out so we could heave in unison. As two of us hauled in on the falls, the other two guided the timber into place with tail lines tied to the ends of the timber. It's easy to swivel the beam in an arc and rock it like a seesaw to maneuver it around stuff that is already in place. □

Trey Loy is a carpenter. He lives and works in Little River, Calif.

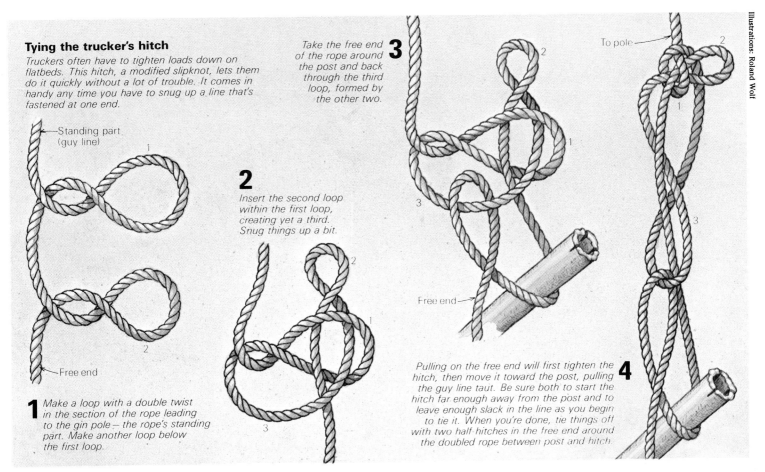

Tying the trucker's hitch

Truckers often have to tighten loads down on flatbeds. This hitch, a modified slipknot, lets them do it quickly without a lot of trouble. It comes in handy any time you have to snug up a line that's fastened at one end.

Standing part (guy line)

Free end

1 Make a loop with a double twist in the section of the rope leading to the gin pole — the rope's standing part. Make another loop below the first loop.

2 *Insert the second loop within the first loop, creating yet a third. Snug things up a bit.*

3 *Take the free end of the rope around the post and back through the third loop, formed by the other two.*

Free end

To pole

4 *Pulling on the free end will first tighten the hitch, then move it toward the post, pulling the guy line taut. Be sure both to start the hitch far enough away from the post and to leave enough slack in the line as you begin to tie it. When you're done, tie things off with two half-hitches in the free end around the doubled rope between post and hitch.*

Illustrations: Roland Wolf

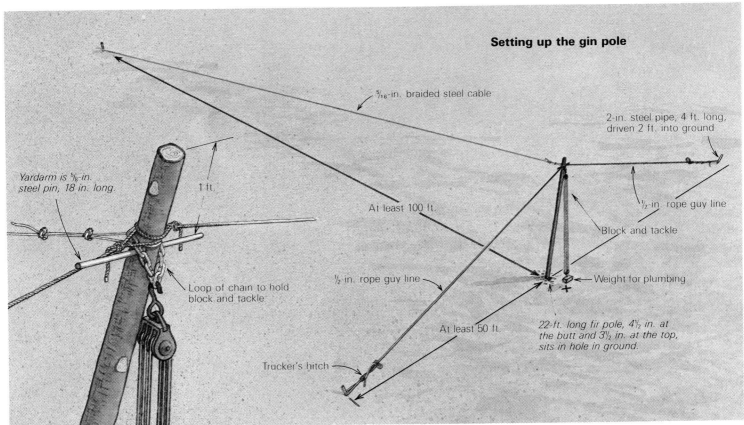

Setting up the gin pole

Yardarm is ⁵⁄₈-in. steel pin, 18 in. long.

1 ft.

⁵⁄₁₆-in. braided steel cable

2-in. steel pipe, 4 ft. long, driven 2 ft. into ground

At least 100 ft.

½-in. rope guy line

Block and tackle

Loop of chain to hold block and tackle

½-in. rope guy line

Weight for plumbing

At least 50 ft.

22-ft. long fir pole, 4½ in. at the butt and 3½ in. at the top, sits in hole in ground.

Trucker's hitch

How strong? A civil engineer calculated for us some of the forces working on the gin pole and rigging. He considered a 3½-in. diameter pole 20 ft. long leaning at an angle of 77°. The load on the block and tackle is 1,000 lb. The main line, which is 100 ft. long, keeps the pole from bending toward the load and has to resist a force of 288 lb. The resultant force of the cable and the load on the pole is calculated to be 1,088 lb. The force trying to kick the pole out of its chock is about 270 lb. The force on the side guy lines is negligible, but exists. Thus the gin pole carries most of the burden and will continue to do so unless force surpassing the buckling strength of the wood is applied. The buckling point for our clear fir pole is 2,145 lb., and theoretically it could be used to lift timbers as heavy as 1,800 lb. It might work, but when we raised the 32-ft. 9x18 ridge beam weighing around 1,400 lb.—I hauled on the block and tackle with my '51 Plymouth—the gin pole twanged like a freshly plucked guitar string. To lift heavier loads a stouter pole and stronger rigging are required. A surfaced 6x6 of clear-grained fir 20 ft. long, for example, has a buckling strength of 22,000 lb. —*T.L.*

Timbers and Templates

An unconventional router bit and particleboard templates make shallow mortising fast and accurate

by Mark Songey

Last summer I worked on a house that required extensive, accurate and repetitious mortising for joints in heavy timbers. Designed by architect and engineer Lawrence Karp to be his residence and studio, the house combines a concrete foundation and first floor with cantilevered steel I-beams supporting a timber-framed second story. The timbers were to be mortised together and left exposed, so it was important to join them precisely. The need to mortise accurately for hundreds of joints of different sizes and for steel fastening straps led me to use in a new way a router bit designed for flush-cut trimming.

The bit, called a TA 170, is made by the Oakland Carbide Engineering Co. (1232 51st Ave., Oakland, Calif. 94601). It consists of a ¼-in. shaft threaded at one end to receive a ½-in. dia. bearing and a ½-in. dia., two-flute cutter (drawing, facing page). Unlike most trim bits with pilots attached below the cutter, the TA 170 has a ball-bearing pilot mounted above the cutter. This allows the bit to be used without a guide bushing (template guide) when routing with a template. The pilot is the same diameter as the bit's cutting arc, which means that the resulting mortise is precisely the same size as the template you make. The bit cuts exactly the same path traced by the bearing. Using the bit successfully for cutting mortises requires making an accurate template the same size as a cross section of the member to be let in.

I like to use ½-in. particleboard for template material; it is easy to cut, has no voids and it's cheap—no small advantage, since I went through 12 4x8 sheets on this job. The photos on the facing page show what's involved in making and using a template. It's important that the template be set square on the work; it must also stay firmly in position while you're routing. I allow at least a 5-in. wide border around the template edges to give the router base ample bearing surface, and I draw perpendicular guidelines on the blank before laying out the template. To hold the template on the work, I use 1-in. blue lath nails. They go in like needles and come out fairly easily.

The flush-cutting trim bit cuts accurately, but not very quickly. For this reason, I use a 1½-hp router with a ½-in. two-flute straightface bit to remove most of the waste in the mortise. To prevent this bit from accidently gouging the template, I attach a ⅝-in. dia. guide bushing to the router base (drawing, facing page). Once the mortise has been roughed out, the flush-cutter can finish the job. I make each mortise ½ in. deep, cutting in increments of ³⁄₁₆ to avoid burning the wood and overheating the cutter. Once the cut is ½ in. deep, the mortise acts as its own template because the pilot rides against the already established mortise walls.

Inevitably, the template will get a ding or two that must be patched to restore the smooth profile. This can be done with fast-setting autobody putty. Overfill the ding and then trim flush with a file or with router and piloted bit. As a precaution, as soon as I make one template and test it for fit, I make an exact copy. This clone saves a lot of trouble. □

Mark Songey is a carpenter and cabinetmaker. He lives in Walnut Creek, Calif.

Lawrence Karp

Songey positions a router template for cutting a sill mortise.

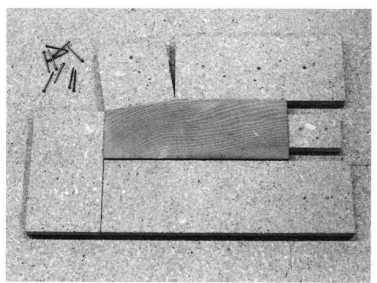

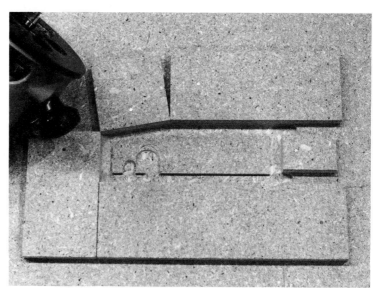

Making the template. Step 1—Cut out a cross section of the piece to be mortised, and lay it on a sheet of ½-in. particleboard. Then surround it tightly with guide strips nailed to the board with 1-in. blue lath nails.

Step 2—Remove the section and cut into the blank using the router and flush-cutting bit. Each clockwise pass should remove about ³⁄₁₆ in. of material. Continue lowering the bit until the center section falls out.

Cutting the mortise. Remove the guide strips and nail the template over the member where the mortise will be cut. The ½-in. wide gaps on the left edge allow the bit to exit and enter, and still leave a straight-walled mortise rather than a radiused cut. A particleboard frame keeps the bit from digging into the template on the first pass. The rough cut can also be made with a guide bushing as shown at right.

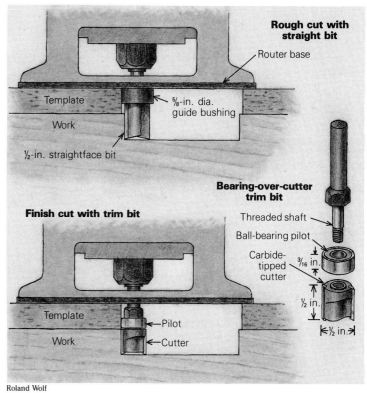

Roland Wolf

The finished mortise. Most of the wood was removed from the inside of the mortise with a straight-face bit and guide bushing. Next, the flush-cutting bit trimmed the mortise to its final dimension (drawing, above right). These cuts were made in depth increments of about ³⁄₁₆ in. to avoid splintering and tear-out, and straining the equipment.

The assembled sill-to-post joint (far right) is tight, thanks to an accurate template and the special flush-cutting bit.

Appalachian Axman's Art

From two decades of hewing and notching experience, some innovative techniques for log-home builders

by Drew Langsner

Peter Gott is both my friend and teacher. He came to western North Carolina 21 years ago, forsaking a college degree in math and intent on learning the folkways of his neighbors. He remained, building a homestead of remarkable log structures. Isolated in the mountains, he worked to master the craft of log building, and in the process, developed methods of hewing and notching that have made traditional Appalachian log building faster and more accurate. Gott's techniques combine two basic innovations: using a spirit level to define precise horizontal and vertical planes on the log, and measuring lines for hewing and notching from chalked centerlines, instead of laying out from the outer dimensions of the

timber. These techniques, combined with a standardized way of calculating allowances for chinking width, notch slope and log diameter, enable a log builder to notch hewn, tapered logs accurately and efficiently before they are raised on the walls.

Notching has traditionally been done by eyeballing or in-place scribing. Both of these methods have the same drawback. They require test-fitting the logs while they are perched on the wall—a dangerous, time-consuming and awkward procedure. Most of Gott's work is done on the ground, and with very little guesswork. Although many of these techniques have been discovered by other log builders working in isolation, Gott's mastery

of them has resulted in log structures of unusual grace (photos, facing page).

Unlike the round-log styles of the north, southern Appalachian cabins are usually made of logs that are hewn flat on the two vertical sides. The spaces between them are chinked with clay. Gott got his first lessons in this kind of log building from Daniel O'Hagan in a Mennonite section of Pennsylvania. When Gott and his wife Polly moved to the hills of North Carolina, he picked up more tips from his neighbors, who were accomplished in logging and axwork.

At first, Gott used hand tools to hew and notch, but the constraint of time has changed his approach. He now uses a chainsaw in most

Using a barking spud, Peter Gott removes the bark from a short poplar log. Gott, who moved to the mountains of North Carolina 21 years ago to study the local folkways, has mastered Appalachian hewing and notching, adding his own layout techniques for speed and accuracy.

Color photos: Drew Langsner
Illustrations: Christopher Clapp

The Gotts' Appalachian Home

The Gott homestead began in 1965 as one small log structure on 40 acres of steep mountainside, miles from the nearest paved road. The basic plan consists of a 14-ft. by 18-ft. log "pen," with a shed addition along each eave wall. The stone fireplace, centered along one gable wall, is a direct descendent of the Scotch-Irish heritage of southern Appalachia.

Also in traditional Appalachian style, Gott has built other small log structures near the original cabin. These include a combination barn and studio, a workshop and storage building, and a poultry house made from logs originally used in a neighbor's corn crib.

The roofs are massive, influenced by the farmhouses of Germany's Black Forest, where Peter and Polly Gott studied folk culture for a year after dropping out of college in the late 1950s. The oak shingles that cover the roofs are split by hand with a froe. Metal eave gutters keep the exterior walls dry.

The only level ground at the Gotts' homestead is the small area where the cabin sits; the other buildings are tucked into various slopes and hollows. The intimate scale of the structures, their relationship to each other and the thick forest around them give approaching visitors the impression that they are coming upon a tiny village. —D.L.

of his work, not just for scoring the logs before hewing, but also for cutting and even dressing the notches.

Like most log builders, Gott cuts trees that grow relatively straight, with few knots and not much taper—pine, fir, spruce, poplar and oak—and works them green. Once the logs are felled, he checks them for straightness, length and girth to determine where in the house they will be used. At this point, logs can be matched up for each course, or round. The logs should measure the distance between the notches on the inside of the house—plus another foot on each end to account for notches and notch overhangs. Minimum tip diameters also have to be figured to make sure that

there is enough wood on the tapered end to cut the full notch. For a half-dovetail, a notch used frequently with hewn logs, the minimum tip diameter for a log hewn to 5 in. thick is 8 in. Nine inches is about right for a log hewn to 6 in. thick.

Barking—Peeling the logs, the next step, makes the chalked layout lines easier to see, and keeps axes and chainsaws from being dulled by bark that has picked up sand and dirt from felling and skidding. Barking begins by chopping a 3-in. to 5-in. wide strip of bark the full length of the log with an ax. Then the log is peeled with a barking spud, a steel bar resembling a large cooking spoon that's been

flattened, curved slightly downward and attached to a handle. It's inserted between the sapwood and the bark and used to pry (photo facing page). If the log was cut in the spring, the bark often comes off in big slabs. If it doesn't, Gott removes a second strip from the other side of the log, or uses a drawknife.

Snapping hewing lines—Accurate hewing lines are the first effort to standardize the log by giving it an exact width and flat planes. If the width varies and the hewn sides aren't parallel, then the notches will be difficult to fit, and the logs in the finished walls may not be in the same plane.

Once the log has been barked, its ends

With the log set up on cribs, and its crown down, Gott snaps chalk guidelines for hewing. These lines ensure an even hewn thickness despite irregularities in the log. He uses a level held plumb to transfer the hewing lines from the bottom of the log across the ends to the top.

should be supported on cribs—short cross-logs with saddle notches chopped at their midpoints. The log should be placed crown down (photo top left). This way, the two sides to be hewed are positioned on either side, where they can be worked on. Although the log will eventually sit on the wall with its crown up, twists and crooks that need to be anticipated to find the center of the log are more visible with the crown down.

Locating the practical center of the log requires the kind of intuition that comes from experience. Gott sights along the edge of a 2-ft. level held plumb against the end of the log to help determine the centerline down the length of the log. He scribes this vertical centerline against the level on each log end.

The hewing lines will be parallel to this centerline. To lay them out, Gott measures half the distance of the hewn width to each side of the center line, on each end of the log. With a jackknife, he cuts small notches where these vertical lines meet the edge of the log. Using these notches as guides, he drives a scratch awl into the end grain of the log to hold the string in the right place, then snaps a chalkline along the length for the two top hewing lines. On long logs, Gott holds the level plumb on the top of the log in the middle of its length to guide the pull-stroke when he snaps the chalkline. This way the two lines will be true and parallel. To complete the hewing lines, he rotates the log 180°, and repeats the procedure on the crowned side.

Scoring and juggling—Before getting down to hewing to the line, the great bulk of wood from the outside of the log has to be removed. This is called juggling. It is made easier by rotating the log 90° and scoring it across the grain to the depth indicated by the pair of hewing lines on each side of the log. Gott uses a chainsaw to score the log (photo center, far left); it is faster and usually more accurate than the traditional double-bit ax. The scoring cuts should be 5 in. to 10 in. apart, except at knots and irregular grain, where they should be spaced more closely.

Juggling is done with a single-bit ax. Gott rotates the log another 30°, and standing behind it, swings freely along the hewing plane removing a chunk of wood with every few strokes (photo center left). The first juggle on each end should be split off with a froe or a single-bit ax struck with a maul. This offers more control than swinging away when there isn't any established surface to use as a guide. These end areas have to be particularly flat and even for laying out the notches.

Once the juggles have been removed, Gott rescores the log lightly with an ax. This makes the finish hewing easier because it cuts the wood fibers, and it gives the hewed log its

Scoring. With a chainsaw, Gott scores a log for hewing. In the past, log-builders made these cuts with an ax. Gott will rescore the log lightly with an ax after juggling to give it the traditional appearance, and to make hewing easier.

Juggling. Big chunks of poplar fly as Gott uses a single-bit ax to prepare the log for hewing. The chunks of wood on the ends of the log were split with a froe for accuracy, since this is where the notches will be laid out.

Hewing. A broadax is used for final hewing. The handles of these tools are bent, so that knuckles won't get torn and bruised by the log. It takes a good axman only a few minutes to hew both sides of a log that has been juggled to a flat plane.

traditional textured appearance. The cuts should be shallow, using even strokes 4 in. apart, and perpendicular to the log's length.

Hewing—The log should be shifted from cribs to trestles—sturdy log sawhorses—for the hewing, which is done at waist level with short controlled strokes (photo facing page, bottom). Hewing is akin to planing, although it is done with a broadax. It pares away the last layer of wood for accuracy and relative smoothness. The results of good broadax work are thin shavings that remove the hewing lines and leave a flat surface. A broadax blade is beveled on one side only—the side away from the log—so that it cuts at a very narrow angle to the log. Like all log-builder's tools, it needs to be very sharp. The inside face of a broadax curves away from the work so the corners of the blade don't catch the log as the ax is swung down along the log's length. The handle of the broadax, which is shorter than that of a single-bit ax, also bends sharply away from the work, just below the ax head, to prevent scraped knuckles.

The stance for hewing feels a bit odd at first. The thumb of your dominant hand is placed along the top edge of the handle up near the head of the ax. Your other hand goes down the handle where it's most comfortable. With your knees slightly bent, and your trailing arm resting on your thigh, you make controlled chopping motions. Your back should be in line with your center of gravity. Hewing is very tiring at first—a broadax head weighs 6 lb. to 12 lb.—but comfort and stamina come with experience.

Notches—Conventional log building is based on the simple idea of stacking horizontal logs to form a rectangular enclosure. To stabilize the structure, the corners are usually inter-locked. This also lowers the logs, so that the need for chinking—the horizontal infill between log courses—is reduced.

The two basic kinds of notches, simple and compound, are shown in the drawing below. In a simple notch, as much as half of the log diameter is removed from the bottom of each log. These notches are generally used in round-log construction. They are laid out by placing one log across another, and tracing with a scribing tool the upper contour of the lower log onto the lower half of the upper log (for this technique, see pp. 176-179).

With compound notches, typically used with hewn logs, waste is taken from the top and bottom of each log. The upper and lower segments each account for one-fourth of the log diameter. This kind of notching has several advantages. Notch cutouts are comparatively shallow—half the depth of simple notches—so there is less fragile short grain at the extension beyond the corner. Also, compound notches last longer because notching removes sapwood, leaving mostly heartwood, which is more decay-resistant.

The half-dovetail notch—Along with the V-notch, the half-dovetail is one of the most common ways of notching corners in Appalachia. Like the V-notch, it locks the corner together with each cross-log. It also allows for some adjustment. When the inevitable log shrinkage takes place, you can force the joints tighter with a sledge hammer, something you can't do with a V-notch, which has opposing 90° faces. Gott used the V-notch for most of his homestead because he likes its appearance and the stringent accuracy it demands, but he uses the traditional half-dovetail in teaching and building for others.

The upper cutout on a half-dovetail notch slopes down from interior to exterior. This helps shed rainwater. The lower cutout slopes upward along the length of the log to accommodate the upper notch in the cross-log below. The degree of slope for each cutout is somewhat arbitrary. Gott likes to use a slope of 1:3 for a half-dovetail notch. This makes a good-looking joint that has enough slope to shed water, and leaves as little sapwood in the joint as possible.

Calculating notch depth—How deeply a log is cut out on the top and bottom for the notch is referred to as notch depth. However, because Gott measures from the centerline of the log rather than from the outside, this term can be confusing. Notch depth, when it is used for laying out a log, refers to the amount of wood that remains above and below the centerline on the log's exterior face.

Three variables affect the depth of a half-dovetail notch—chinking height, rise height and log height, as shown in the drawing, below right. Once the builder determines these, then allowances for them can be made in the depth of the notch. Reductions for chinking, rise and log height affect how far apart the logs will sit when stacked, and each one affects the notches differently, depending on the number of times the allowance must be made. Understanding why these reductions are made will help you in cutting notches for any kind of log building.

Chinking height is the actual space between logs after they are stacked on the wall. It can range from ½ in. to 3 in. or more. The exterior of the chinking itself will be much wider than this, because the chinking gains height as it conforms to the radius of the logs. The choice of chinking height depends on the uniformity of the logs in the round you are working, and on personal style. Increasing the chinking height means decreasing the depth of both

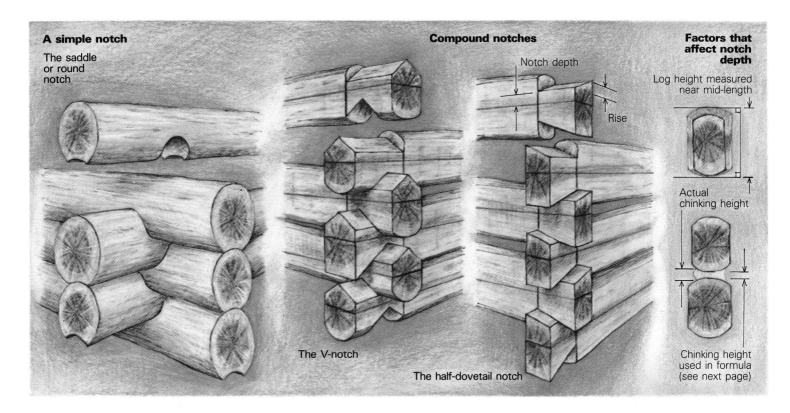

A simple notch

The saddle or round notch

Compound notches

The V-notch

The half-dovetail notch

Notch depth

Rise

Factors that affect notch depth

Log height measured near mid-length

Actual chinking height

Chinking height used in formula (see next page)

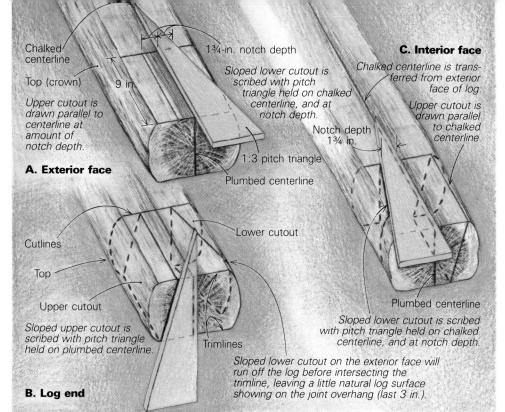

A. Exterior face

Chalked centerline

Top (crown)   9 in.

1¾-in. notch depth

Upper cutout is drawn parallel to centerline at amount of notch depth.

Sloped lower cutout is scribed with pitch triangle held on chalked centerline, and at notch depth.

C. Interior face

Chalked centerline is transferred from exterior face of log.

Upper cutout is drawn parallel to chalked centerline.

Notch depth 1¾ in.

1:3 pitch triangle

Plumbed centerline

B. Log end

Cutlines

Top

Upper cutout

Lower cutout

Sloped upper cutout is scribed with pitch triangle held on plumbed centerline.

Trimlines

Plumbed centerline

Sloped lower cutout is scribed with pitch triangle held on chalked centerline, and at notch depth.

Sloped lower cutout on the exterior face will run off the log before intersecting the trimline, leaving a little natural log surface showing on the joint overhang (last 3 in.).

Laying out a half-dovetail notch. **On the exterior face of the log, measure and draw the cutlines perpendicular to the centerline. Then mark the notch depth on either side of the centerline to establish the beginning of the cutouts. Use a 1:3 pitch triangle to scribe the rise (a). Lay the log ends out next (b), and then carry the layout onto the interior face (c).**

notches on a log end. Since a given chinking space is affected by the top and bottom notches on two consecutively stacked logs, the reduction for chinking height is divided four ways (drawing, below).

The height of the sloping shoulder on a compound notch is called the rise. Gott's 1:3 slope will produce different rises for logs hewn to different thicknesses. For instance, the rise for a 6-in. thick log cut at a 1:3 slope

is 2 in. Since this happens only on the top of each log, the accommodation for rise is shared only by the interlocking notch on the bottom of the log above. This means that the rise is divided only two ways.

The last consideration is log height. Since most timber tapers from butt to tip, the diameter of the ends is seldom the same. Although the logs are stacked on the walls by alternating butt and tip at the corners, each round must remain level. Gott measures the height of the logs in the round, and comes up with an average height. Because he lays out from centerlines rather than from the outside of the log, this averaged figure works well. The height of the notch from the centerline remains constant, but the waste removed from each end is usually very different.

These three factors—chinking height (C), rise (R) and log height (H)—can be combined into a formula to find notch depth (D):

$$D = ¼(H + C) − ½(R)$$

Once you choose the chinking height and rise, you can calculate the notch depth on logs of any size. If you make up a chart that lists the notch depth for a range of log heights, at a given chinking height and rise, you can lay out notches without having to make a calculation each time the height of the logs changes. The chart below is for logs hewn to 6 in., and a notch slope of 1:3. This results in a rise of 2 in. The chinking height is figured for 1 in.

Understanding notching

How notching relates to chinking height and log height is best understood by simplifying a joint, reducing it to essentials. Unfold the corners so that the logs form lap joints, and change the sloped dovetail rise to a flat plane.

A. No notch cutout leaves a space between the logs that's full height.

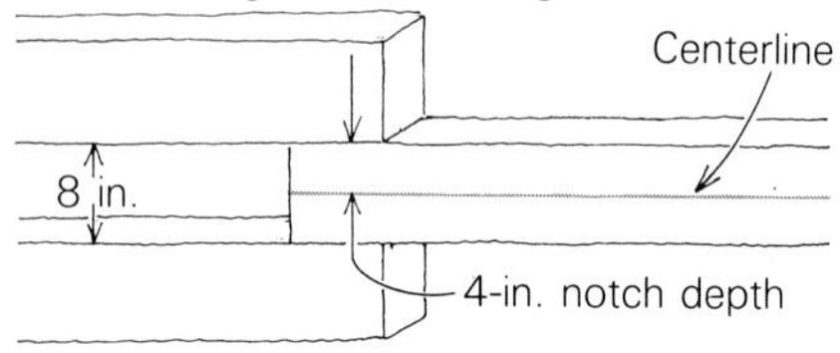

Centerline

8 in.

4-in. notch depth

B. A notch cutout of ⅛ the log height reduces the space between logs by half.

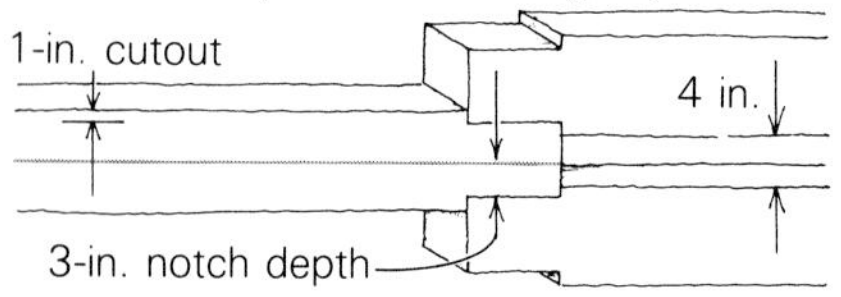

1-in. cutout

4 in.

3-in. notch depth

C. A notch cutout of ¼ the log height yields no space at all between logs.

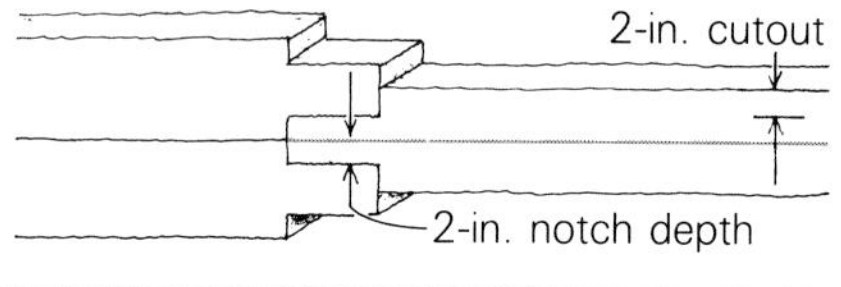

2-in. cutout

2-in. notch depth

Log height (in.)	8	9	10	11	12	13	14	15	16
Notch depth (in.)	1¼	1½	1¾	2	2¼	2½	2¾	3	3¼

Laying out the notch—If you are relying on these notch depths, you must lay out the exterior face of the log first, and then transfer the upper cutout line across the end of the log to establish its beginning layout point on the interior face, as shown in the drawing at left. These figures won't work if you use them to lay out the interior side of the log first, because the upper cutout slopes up and away from the centerline, making it wider than at the same point on the exterior side.

Gott sometimes relies on his own practiced eye to scale out notches without using the formula. When he does, he lays out the interior face of the log first to minimize adjustments to the inside corners. V-notches are also begun on the interior face of the log, since they can't be moved to tighten the corners.

The first step in laying out the half-dovetail notch is to check again for the crown, and rotate the log until its exterior face is up. Next, use a framing square to find the center of the log face about 1 ft. from each end. With the body of the square sitting on the hewn surface of the log, hold the tongue down over the top edge so that it gauges the widest point of that side of the log. Make a pencil mark that corresponds to any even increment on the blade of the square near the center of the log. Gauge the bottom edge of the log by flopping the square and using it in the same way. The result will be two marks close together on the face of the log. Eyeball or measure a point in between this set of marks, and you have the very center of the log along a horizontal axis. Do this on both ends and then connect these two centerpoints with a chalkline to get the accurate centerline.

The trimline is the line along which you will cut the log to its final length. The cutlines are the lines that form the shoulders of the notch. To determine the cutlines and trimlines, first tape the total inside measurement of the log, starting about 1 ft. in from one end. The beginning and end marks of this measurement are the cutlines. Add a standard increment, such as 9 in., to each of these marks to get the trimlines, farther out toward the ends of the log. If you are using a log hewn to 6 in., the 9-in. allowance will provide for the thickness of the cross log plus a 3-in. overhang. Using the framing square again, draw the trimlines perpendicular to the centerline, and crosscut the log at those points with a chainsaw or handsaw. Use a framing square to make sure these end cuts are perpendicular to the hewn surface, so that the interior face of the log can be laid out from these trimlines.

Next, level the log on the trestles perpendicular to the chalked centerline. Draw a plumb centerline on the ends of the log that intersects with the lines that you chalked on the face of the log, as shown in the drawing. Using the formula for a log 10 in. in height, with 1-in. chinking and a rise of 2 in., measure the distance listed on the chart (1¾ in.) in both directions along the cutlines from the chalked centerline. These points determine the depth of the notch.

On the crown or top side of the centerline,

draw a line parallel to the centerline from the 1¾-in. mark out to the end of the log. This is the upper cutout line on this face of the log. On the other side of the centerline, draw a sloped line for the lower cutout. For this, Gott uses a simple pitch board made from scrap. It is a right triangle with legs of 6 in. and 18 in. This gives him the 1:3 slope he wants. Use the triangle by holding the long leg along the centerline and sliding the triangle until the hypotenuse intersects the depth-of-notch point, and scribe a line. Using a 1:3 slope, this lower cutout line will run off the exterior face of the log before it intersects the trimline, leaving a little of the natural log surface showing on the overhang of the finished notch. This won't affect the notch itself.

On the end of the log, only the sloping upper cutout line needs to be laid out. Hold the pitch triangle with its long leg against the end centerline, and its hypotenuse intersecting the upper cutout line where it reaches the end of the log. Scribe along the hypotenuse.

When the exterior faces and ends are laid out, turn the log over. Chalk a centerline along the length by connecting the centerlines on the ends of the log. Then complete the notch layout as shown in the drawing, facing page, top. To finish drawing the cutlines around the convex top and bottom of the log, lay a square across the log and draw a line while eyeballing the points on the two hewn sides. A flexible steel ruler works even better.

Cutting the notches—Whether you choose a handsaw or a chainsaw, make the crosscuts at the cutlines first. Then make the sloped cuts. If you use a chisel for removing waste, make sure that it's sharp. A 2-in. framing chisel is the best tool for this job (see page 151). Make several cross-grain cuts with the handsaw to the depth of the layout lines on the notch cutouts, just as you did in the hewing process. This will make chiseling easier, and the results more accurate.

Gott's chainsaw technique on the sloped cuts is to begin with the heel of the blade to establish a line of cut. Then he buries the tip of the bar into the work by pulling back, sawing just halfway through (photo top right). He makes the rest of the cut from the other side of the log in the same way. This method prevents overcutting the joint, but requires some care, and locking your elbow, to deal with the danger of kickback.

Gott uses either an adze or a chainsaw to dress the bottom cutout after the waste is removed, and he cuts this surface slightly concave to ensure a tight fit on the edges with the crosslog above. The top cutout, however, is dressed with a slick (photo center right), which is a big chisel on a long handle. It removes any evidence of saw marks, and produces a flat, smooth surface. □

Drew Langsner, a homesteader in Marshall, N.C., is the author of A Logbuilder's Handbook *(Rodale Press, Emmaus, Pa. 18049). He operates a crafts school, Country Workshops, out of a log home that he built himself.*

The heel of the blade is lowered to establish the line of cut on the lower half of the dovetail notch. Plunging the tip of the bar into the work along this line will complete the top half of the cut.

Gott uses a slick to clean up the surface of the upper cutout of a half-dovetail. The cutout on the crosslog above will be sawn slightly concave in the center to ensure a tight fit. The result, below, is a joint that wouldn't accept a slip of paper when put together for the first time. The chalked center lines that indexed this accurate cutting can be seen along the hewn sides of the logs.

Round-Log Construction

Precision fitting with the chainsaw

by Alasdair G.B. Wallace

Wallace's log house features precision joinery.

One Sunday afternoon I took penknife in hand and set about prying at the dovetailed corners of a century-old hewn-log house. I was astonished to find that everything I'd read about the precise fit of the joints was true: The logs fit together so perfectly that I couldn't insert a blade between them. When it came time to build our own wilderness log home, my wife and I decided to match the precision of the old-time builders. However, the logs on our "wilderness" were too small to hew square, so we decided to build with peeled, round logs. It was a wise decision: Not only would round logs provide more insulation in Ontario's forty-below weather, they would be far more in keeping with the graceful pines and firs that studded our property.

Using a chainsaw, level and log scriber, I shaped rounded lateral grooves and corner notches in each log. The grooves and notches fit precisely over the log immediately below; each round of logs overlaps the preceding round, interlocking at the corners. This technique of fitting round logs yields a weathertight joint that improves with age. It does not weaken the log by promoting checking, and it requires no chinking, as shown in the photo below.

Many log builders advocate a straight-sided lateral V groove. It can be completed in less than half the time required for the tighter round groove, but the arguments against it are several. Unless the V groove is very shallow (in which case it tends to hang up on the inside of the shoulders of the cut), it weakens the log by putting stress at the apex of the groove.

The chainsaw— After buying cant hooks, blocks, adz, broadax, slicks, log dogs and the attendant paraphernalia of the log builder, I surprised myself by deciding to work the logs with a chainsaw. I had used chainsaws in building a 50-ft. bridge, milling wood and cutting cordwood; their performance was impressive.

The type and size of chainsaw you select can determine the success or failure of the log building project. The saw must start easily in all weather, run continuously for hours in hot weather without overheating or suffering from vapor locks, and it must be finely balanced.

Your chainsaw should be as versatile as possible. While an 8 cu. in. machine cuts with ease and rapidity, it requires superhuman stamina to wield it for hours on end. The 1.5 cu. in. displace-

ment will certainly be far more manageable as you work left handed at the top of a ladder, but it will cut agonizingly slowly. Compromise. My chainsaw, a Swedish-made Jonsereds, has a 3.1 cu. in. displacement. It handily outcuts my larger saw. You could also purchase two bars and chains for instant versatility. A 19-in. bar for felling and limbing and a 16-in. bar for notching and grooving make a good combination.

Keep the blade razor sharp. Two or three careful strokes with a file at frequent intervals are much easier on the saw and sawyer than one major session each day. Watching a professional hone a chain is more helpful than trying to follow the manufacturer's often confusing diagrams and notes on angles.

Positioning the logs—Since the first round of logs lies directly on the foundation, these logs must either be flattened on the bottom or set into a 1-in. layer of concrete. The only other fitting needed in the first round is for the corner notches. For the second round and all succeeding rounds, every log must be plumb (as in the drawing below) and approximately horizontal to the log directly beneath it. Since very few logs

The logs below are joined with round grooves and corner notches that were cut and shaped with a chainsaw. The lower surface of each log was hollowed out to fit the contour of the log beneath it. This method of fitting logs together yields a tight, weatherproof seal. As the logs shrink with the passage of time, the joint tightens even more—often creating a virtually "chinkless" surface. The logs provide enough insulation for comfortable living in Ontario's cold climate.

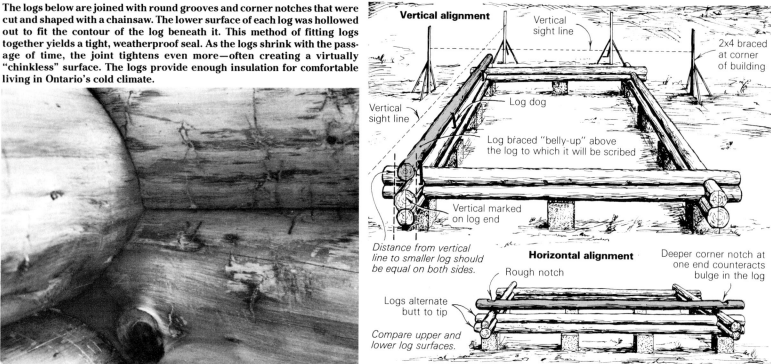

Photos: Alasdair Wallace and Wendy Wallace

are straight, vertical alignment is always a compromise. It is, however, critical that the corner notches be plumb; they will bear three-quarters of the weight of the finished wall.

Position the log to be scribed belly-up above the log to which it will be fitted, and secure it in place with log dogs. There are several methods of aligning logs vertically. I generally use a large carpenter's level to mark a vertical line on the end of each log before I scribe it. Since the log will probably bend in one or several directions, these vertical lines will not always pass through the center of the log as you look at the end surface. They should, however, form an unbroken line. Another method requires locating vertical sight lines at each corner, against which the wall can be vertically aligned by eye. Straight, securely-braced 2x4s at each corner make good vertical lines.

Once you've aligned the log vertically above the log to which it will be scribed, horizontal alignment begins. Logs will, of course, alternate butt to tip as the wall rises. Each round should approximate the horizontal, as shown below, with any discrepancies being noted and corrected in the next round. A deeper corner notch may be required at one end to counteract a bulge in that particular area; however, incorporating a thinner log in the next round would also achieve the same effect. Log builders should, in fact, always plan one round ahead.

At this stage rough corner notches are cut with the chainsaw. When the log is turned over for scribing, the rough notches will position the log in the required horizontal plane relative to the log beneath; they will also drop the upper log to within two or three inches of its partner, to accommodate the adjustable scribing tool. Turn the log over and test-fit the notches. If both horizontal and vertical planes are aligned, mark the verticals on the ends of the log with a copying pencil. Then lock the log firmly in position with log dogs, so it won't rock during scribing.

Scribing the groove and notches—While each part of the process required to achieve the precision of a weatherproof fit is exacting, the scribing process is the most crucial. No matter how closely the chainsaw approaches the scribed line, the nicety of the final fit depends on how accurately the scribed line followed the contours of the log below.

Scribers are available in many designs. They range from simple metal dividers to elaborate instruments with brass adjusting screws and fancy metal leveling devices. Since the the two points of the scriber must always remain plumb, the type with a double-bubble level is most likely to yield satisfactory results. (You can buy them from Woodcraft Supply Corp., Box 4000, 41 Atlantic Ave., Woburn, MA 01888.)

When used correctly, a log scriber faithfully transfers the contour of one surface (the top of the lower log) onto another surface (the lower half of the upper log). Set the scribers to a distance equal to the maximum gap between the two logs plus approximately ½ in. If the scriber were set only to the maximum gap, the logs at that point would only touch; there would be no cupping of the upper log over the lower. The amount of cupping needed depends on the diameter of each log (as shown in the drawing below), the minimum wall thickness required along any part of the log and the balance of the whole. A large log may require a much deeper lateral groove, resulting in a thicker wall section, to achieve balance with smaller logs.

With the log positioned and the scribers set, you're ready to scribe the lateral groove and the corner notches. Practice angling the scriber toward your body and pull, rather than push, as shown in the photo below. This prevents the points from catching or digging into the log. Initially, scribing a log may require two hours; with practice, you'll probably speed up to thirty minutes per log.

As you pull the scriber, keep the pencil tip and the scriber point vertically aligned. Even a small fluctuation from the vertical will show up later in a log that fails to seat effectively on the log beneath. It is much easier to spend extra time scribing correctly than to hurry the task and, later on, find that you must fit by trial and error or rescribe the entire log.

Realign the points of the scribers every time you reset them for a new log or sharpen the pencil. To make a simple aligning device, mark a vertical line down the center of a board. Affix the board to a static surface so the line is plumb, and adjust the scriber to the line.

The scribed line travels the length of the log for the lateral groove, around the corner notches and back down the other side of the log, as shown below. When you've completed scribing, transfer the location of major irregularities on the lower log to the scribed log. I use appropriately-located strips of masking tape coded C (center) or E (edge) as reminders to cut more deeply at the center or edge of the groove.

Next, roll the log toward the center of the building, with scribed lines up, and dog it firmly into position. Incise the scribed line with a sharp knife. (See photo A on the following page.) The cut should split the line at approximately the angle of the finished groove. This will ensure an accurate line after the final shaping.

Cutting the groove and notches—Cutting a smoothly-rounded lateral groove with a chainsaw—a process I could never understand until I

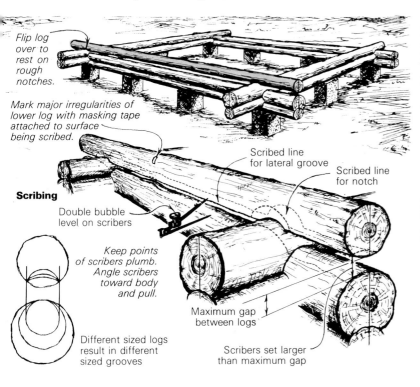

Flip log over to rest on rough notches.

Mark major irregularities of lower log with masking tape attached to surface being scribed.

Scribed line for lateral groove

Scribed line for notch

Scribing

Double bubble level on scribers

Keep points of scribers plumb. Angle scribers toward body and pull.

Different sized logs result in different sized grooves

Maximum gap between logs

Scribers set larger than maximum gap

Making rounded grooves and notches is exacting work, but the most critical part of the process is scribing the log. A log scriber accurately transfers the contours of the lower log onto the surface of the upper log, as shown above. The scriber point (on the lower log) and the pencil tip (on the upper log) must remain vertically aligned at all times. The scribers have a built-in level, which makes it easier to maintain a true vertical line. Angle the scribers toward you and pull, so they won't catch on the log. The scribed line travels the full length of the log, as shown in the drawing at left.

Cutting and shaping the lateral groove

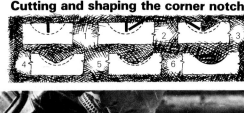

A—Incise the scribed line with a sharp knife. Split the line at the angle of the finished groove.

B—Rip a kerf down the center of the two scribed lines for the full length of the log.

C—Angle the chainsaw 60° from vertical, and cut ¼ in. inside the scribed line to the bottom of the center kerf. Gauge the depth of the cut by watching sawdust gather in the center kerf.

Cutting and shaping the corner notch

H—Cut out a wedge inside the scribing on the other side of the notch, forming a deep V shape.

G—After cutting a kerf to the deepest part of the notch, angle the chainsaw and cut out a wedge inside the scribed line.

I—Contour the notch with a series of slicing cuts. Work the far side of the notch, to prevent wood from splintering in your direction.

tried it myself—is actually quite a simple process. Rip a kerf along the length of the log midway between the two scribed lines (photo B). The kerf depth will depend on the distance between the scribed lines and the diameter of the lower log. As the distance between the scribed lines will vary, the depth of this kerf will vary along the length of the log.

Now angle the chainsaw at approximately 60° from vertical, and cut $\frac{5}{16}$ in. to ¼ in. inside the scribed lines to the bottom of the initial kerf (photo C). As you become more proficient with the saw, this margin may be decreased to ⅛ in. Narrowing the margin between the cut and the scribed line reduces the amount of final cleanup. Overcutting, however, will entail at best, complete rescribing; at worst, discarding the log. Gauge the depth of the cut by watching sawdust gather in the center kerf as the two kerfs meet. Rip the entire length of the log in this fashion, remove the waste strip (photo D), and repeat the process on the other side of the groove, working in the opposite direction. This produces a shallow V inside the scribed lines (photo E).

The remaining wood between the cut and the lines must now be removed and the V groove shaped to conform as closely as possible with the top of the log beneath. Arrange scaffolding so that the chainsaw may be held comfortably at arms' length. Tilt the log slightly toward you and

D—Remove the V-shaped waste strip from the log. Usually, the strip will almost fall out; in this case, however, it took a bit of prying with a bar to loosen it.

E—Working the opposite side of the log, angle the chainsaw and cut ¼ in. inside the remaining scribed line. This forms a shallow, V-shaped groove between the scribed lines.

F—Tilt the log toward you and brace it. Hold the chainsaw at arms' length and, using a lateral, brushing motion, remove the remaining waste and shape the curved groove.

dog it into position. Use the bottom rounded portion of the chainsaw tip to remove the waste and shape the groove (photo F). Be cautious: Do not work too deeply into the groove and risk kickback. As the chain approaches the scribed line, the final sliver will splinter out, leaving a clean, accurate line that requires no further work.

Now tilt the log in the opposite direction and, working from the other side of the log, remove the bulk of the remaining waste. The bottom of the groove may require shaping or cleanup with a third and final pass. The tendency is to remove too little waste from the shoulder immediately inside the scribed line and too much from the center of the groove. This causes a log to seat incorrectly on the inside of the shoulder.

When you have finished cutting, put aside the saw and closely compare the logs to be joined. Inspect the scribed line and the groove; ascertain that the taped areas have been adequately hollowed and conform to the contours of the log beneath. Once this work has been completed to your satisfaction, the rough corner notches can be shaped to conform to the scribed line.

Run a kerf through the deepest part of the notch to within ¼ in. of the scribed line (photo G). Then make two angled cuts that meet at the bottom of the initial kerf (photo H). Rough-in the round shoulders with a series of slicing cuts, keeping one edge of the blade level with the sur-

face of the wood. This technique permits rapid shaping. The tendency of the wood to splinter out on the near side of the groove can be alleviated by angling the sawblade and always working the far side of the notch (photo I). You can finish contouring the notch with the bottom tip of the chainsaw blade. An alternative method of shaping the notch is to cut the shoulders with a sharp firmer chisel, and waste the remainder with the saw. Hollow out the middle slightly (about ¼ in.) in the same direction as the lateral groove. This leaves room for inserting fiberglass weatherstripping later on.

Finishing—With the notches completed, it is time to test-fit the log. Roll it gently into position and align the vertical lines on the log ends with those on the log below. If the work has been accurate, the initial fit will be perfect. An imperfect fit will require a close scrutiny of both notches and groove to find the problem area. Run a strip of paper along the groove to detect areas of contact that cannot be seen.

Once the fit is acceptable, turn the log groove-up and apply a non-sealer preservative to the notches and groove. Staple a thin strip of fiberglass into the groove and notches to prevent air infiltration. Don't apply fiberglass to the grooves outside the corner notches: Air circulation is beneficial there. I use 2-in. R-8 fiberglass without

the vapor barrier. It can be readily cut into strips with a knife or shears, and it compresses to less than ¼ in., ensuring an airtight seal between logs. After rolling the log into its final position, trim extraneous fiberglass with a razor blade to prevent absorption of water. I prefer to spike the logs into position at the notches so that they won't move when I install the next round.

This method of log joinery demands of the builder both dedication and time. I spent approximately seven to eight hours working each 26-ft. log: This included cutting, limbing, skidding, stacking, sorting, peeling, raising, scribing, cutting notches and groove, fitting and fiberglassing. The initial logs undoubtedly took longer until I had learned, modified and perfected the techniques. Today, with approximately 100 logs in place, I view with satisfaction the product of three years' demanding work. □

Alasdair Wallace, writer, teacher and cabinetmaker, has recently completed his first log building near Lakefield, Ontario.

Editor's note: Working with a chainsaw can be dangerous. To reduce kickback, consider buying a chainsaw equipped with a safety chain, chain brake or low-kick bar. Wear safety goggles, ear protection, and never work alone.

Log-Building to Last

With proper siting, roof overhangs, notching, chinking and ventilation, a log home should last 200 years

by Alasdair G.B. Wallace

All things deteriorate over time, and log homes are no exception. Some begin to degrade almost as soon as the walls are raised, and have trouble surviving their creators. But if a log house is built thoughtfully, it can have a useful lifespan of 200 years. And it will look no different to the casual eye than one that will rot seriously during its first decade.

The worst culprit in the demise of log homes is moisture. Keep this in mind as you plan each detail of your structure, including siting, foundation and bottom-log detailing; notching and chinking; and landscaping and interior finish.

Site location—Site selection is too important to be determined only by the proximity of schools and shopping centers. You'll naturally be looking for a southern exposure and good natural drainage, but keep in mind that frequent assaults by wind-driven rain or sleet can cut in half the useful lifespan of a log building. Wind sheltering can also greatly reduce the amount of snow you'll have to shovel, and will save money on energy bills by cutting down the wind-chill factor.

However, living on the lee side of a hill or in a hollow can sometimes eliminate a beautiful view, contribute to dampness, or reduce the natural sunlight available to the building. A row of maples or a cedar hedge can provide an equally effective windbreak. In a pinch, a snow fence or 6-ft. slat fence gives you instant protection at a minimal cost.

Wind sheltering should not be confused with eliminating the movement of air around a house. Air circulation is vital. Don't succumb to the rustic appeal of ivy or other vines growing up the walls of a log home. These plants also produce moisture, reduce air circulation and feed on your wood. Landscaping should be kept at least 3 ft. from the exterior of the foundation, even if the foundation is poured concrete and the bottom log is well off the ground. This moves the plants out from the overhang, allowing them to be watered by rainfall, and eliminates the possibility of getting the logs wet with spray from a sprinkler or garden hose. In older homes, where the bottom logs are usually much closer to the ground, you can install a sheet-metal splash guard if the logs are getting soaked. Replacing trees and shrubs right next to the house with a well-drained gravel pathway is an even better solution.

Cedar logs in a cabin built between 1820 and 1830 show extensive weathering but remarkably little rot. The precise dovetail notching on these logs directs any water that falls on them toward the outside of the building and down.

The bottom log—In many early pioneer dwellings, the bottom round of logs was set directly on the soil. The floor itself was packed earth. A surprising number of these buildings are still standing. They are typically homes with cedar bottom logs laid on well-drained, sandy soil. If you're building new, the best way to keep your bottom log dry is to keep it 20 in. above grade. It can be set either on piers or on a perimeter foundation. Using piers makes for a colder floor, but encourages air circulation. The crawl space formed by a continuous concrete foundation requires screened venting to the outside. In either instance, sandwich a vapor barrier or damp course of polyethylene or heavy tarpaper between the logs and the concrete. The pier top or foundation wall should also be formed or tooled with a slight pitch so it sheds water.

Choice of logs—Log builders of the 1980s prefer to use evergreens. These logs are generally less expensive than hardwood. They are also straighter and less tapered, much lighter, and more easily worked. Western red or eastern cedar, Douglas fir, pine, tamarack

(larch), spruce and balsam fir all make excellent building logs. Cedar is the most durable, but also the most expensive.

Unfortunately, woods like balsam, poplar and oak are usually ignored because of misleading information published about their durability. It's true that these species won't last as long as some others under similar conditions, but the many well-preserved 100-year-old houses made with these logs are a strong argument for using available species, and providing adequate protection.

Choice of notch—The best notches shed moisture rather than trap it. Wherever possible, remove wood from the bottom side of the log only. The rounded surface on the top of the log will shed water naturally. If wood must be removed from the top, as in the case of a common lap joint, slope the top of the flat surface slightly to the outside of the building.

The round notch and lateral groove (see pages 176-179) is a popular moisture-repelling, self-draining notch. The saddle notch, square notch, and sheep's head notch are other round-log notches that shed water efficiently. For hewn-log construction, the dovetail notch is best for keeping the joints dry.

The end grain absorbs water more easily than the rest of the log, and this can cause the joint to decay if the log ends aren't extended far enough beyond the corners. In the round-log building that I do, I like log ends to extend at least 15 in. beyond the notch.

Overhangs—Most of the water that shortens the life of log buildings comes as rain, sleet and snow. It's the job of the overhangs at the gable ends and the eaves to protect the walls and foundation. The overhang is measured as the horizontal distance from the fascia board of the eave to the exterior surface of the wall. Overhang widths are nearly always compromises, taking into account the direction of the prevailing wind, windbreaks, door and window placement, the distance of the bottom log from the ground, roof pitch, and aesthetics. An effective overhang will keep rain blowing in at an angle of less than 15° off vertical from striking the foundation or the logs above it. For an 8-ft. wall, this means about a 21-in. overhang. If the gable-end wall of the same house measured 16 ft., then the overhang should be 42 in. at the ends of the house.

Substantial overhangs not only protect the

logs from direct weather, but also shade them in the summer, and direct water run-off away from the foundation and bottom log even if you haven't installed eave gutters. It's tempting to stack firewood, hang canoes and store tools in such a convenient, dry location, but this will discourage air from circulating freely over the walls, and leave the area damp.

Mortar—Perhaps the most obvious sign of a log building's deterioration is cracked and missing mortar. Correctly applied and maintained mortar chinking will act as an effective barrier to any moisture that gets onto the log walls, as well as a seal against the wind. However, even a chinking job that looks well done to the untrained eye can actually trap moisture in the logs and lead to early degradation of the building.

The larger the spaces between the logs, the harder the chinking job. Making mortar hold between round logs is much more difficult than between logs that have been sawn or hewn. There are many ways to persuade the mortar to remain. The most common is to drive nails partway into the logs as a reinforcement for the chinking. Other reinforcements—chicken netting, expanded metal lath, staples and wire—have their advocates; none, however, eliminates the expansion and contraction of the logs with the seasons. The best way to deal with the movement of the wood is to use a mortar that is more pliable than the conventional clay mixture. I like a combination of three parts mason's lime, one part white portland cement and twelve parts fine, clean sand.

Chinking for square or round logs should be inset from the log faces. Flat ledges tend to retain water, and reverse slopes carry water into the logs and hold onto it. With round logs, tooling the mortar so that it's concave or steeply angled is best (drawing, below right), since thin feathered edges will break off in the long run.

The interior—It's easy to become preoccupied with the outside of the logs, and neglect the equally vulnerable interior surface. Since logs will expand and contract with seasonal fluctuations in ambient humidity, it's best to keep the conditions equal on both sides of a wall. If you are planning to oil the outside of your logs, you should oil them on the inside, too, to avoid subjecting the logs to unequal stresses, which can promote checking.

Any log home that is going to be heated in the winter also needs a humidifier. A plate or drum-type humidifier can be attached to the furnace, but a kettle of water on top of the stove works fine too in a small house. Use a humidistat to detect very dry or very wet air.

Be especially careful planning rooms containing bathtubs, washbasins, showers and toilets. Exhaust fans will reduce the humidity in these areas, but exposed logs in these rooms will need a polyurethane finish. Some builders sidestep the problem by covering the logs with a stud wall faced with ceramic tile or vinyl wallpaper. Kitchen areas can also be

treated with a plastic finish or walled over to provide a surface that can be wiped.

All the spaces inside a log home should be well ventilated. Cupboards, closets and the spaces behind counters require particular attention. Set cabinets and their countertops about 2 in. out from the surface of the log walls, and drill a series of holes in the kickplates to allow the natural convection loop from floor to ceiling to circulate beneath and behind the counters.

Finishes—Most of the log buildings that have survived from the last century have done so without the benefit of modern preservatives. But many contemporary log-builders use a preservative of one kind or another, and the relative merits of each are hotly debated. Used correctly, all of them will help preserve the logs. The choice depends in part on the kind of logs you're using, and on the harshness of the weather in your area.

Most *clear liquid preservatives* impart neither color nor visible finish, and are quickly absorbed by the logs. They are usually quite effective if applied every few years. Many of

Chinking and caulking round logs

Laterally grooved round-log construction requires no chinking and is self-draining.

When chinking is used, a concave finish, left, reduces fragile edges. Long feathered edges, right, will break off over time.

Unless caulked and maintained periodically, a lateral check will collect moisture, leading to serious rot in center of log.

Chinking sawn or hewn logs

Rainwater gets trapped here.

Inset sloped chinking drains water.

them contain pentachlorophenol, a highly toxic chemical that is currently being investigated as a possible carcinogen (see *FHB #13*, p. 14). The penta content of these wood-preserving liquids is usually 5%. They should be used only on the exterior and applied with a brush. Other clear penetrating finishes contain chromated copper arsenate (CCA) rather than penta. This chemical is effective, but will stain the logs a grey-green.

Creosote is the same stuff that is used on telephone poles. It is a highly effective water repellent that emits a strong, characteristic, tar-like odor for a long time. It colors logs dark brown or black.

Varnishes seal the logs and help preserve them, but you have to reapply them as soon as ultraviolet light begins to break down the film. If applied incorrectly or when the logs are green or the air is humid, the film is likely to peel right off. Some varnishes bleach out and crystallize with exposure. Resin-base varnishes are likely to blacken some species of wood, notably cedar.

Boiled linseed-oil and turpentine mix. If applied hot—the oil is heated first, then removed from the flame before the highly volatile turpentine is added—the 30% oil/70% turpentine mixture penetrates well and preserves effectively. Some builders recommend that candle wax be added to the mix. This mixture can darken logs under moist conditions. It will blacken cedar.

Paint has been used for centuries to preserve wood. When correctly applied and maintained, it is effective, but most contemporary log builders choose not to obscure the natural color of their logs.

The natural log. Perhaps the most aesthetically pleasing buildings are those which have acquired time's own inimitable coloring and texture. Given adequate shelter, drainage, air circulation and a species known for longevity, the logs will attain the characteristic grey, stratified appearance of glaciated rock.

Maintenance—Keeping the logs dry and promoting good air circulation around them are the priorities of maintaining a log house. Make sure that gutters and downspouts are clear, and that they don't leak. Trees and shrubs near the house should be pruned regularly. Screening over foundation vents may be necessary to prevent animals from moving in. Mortar chinking will require occasional repair, and checks in the logs will need to be caulked. If you're using a wood preservative, regular applications will be necessary.

The resurgent interest in log building is finding expression throughout North America. The caliber of workmanship and design speak to the concern of this new generation to produce buildings of integrity, strength and beauty. Given today's technical expertise, effective design and reasonable maintenance, these houses will long remain as tributes to the industry of their builders. □

———

Alasdair Wallace is a log-builder in Lakefield, Ontario. Photo by the author.

An Island Retreat

A chainsaw and a milling attachment
turn salvaged logs into a simple vacation cabin

by David K. Ford

Our log cabin was built in less than a month on an island in central Ontario where we spend our summers. It's what is known in Canada as a sleep cabin—a single room that serves as bedroom and living room—with a screened porch at the front (south) and a small porch at the rear for firewood and a wash bench. Sanitation? An old-fashioned outhouse back in the woods.

Like the Ojibway Indians who have lived here for generations, we oriented our cabin to the south to catch the warm breezes of Shawandassee, the south wind. Woods to the north screen our site from other buildings on the five-acre island, and also from Keewaydin, the north wind that sweeps off the subarctic tundra around Hudson Bay, and is chilly even in summer.

The wood for our cabin cost almost nothing because we used logs salvaged from the bay behind our island. Over the years, hundreds of them—those with the densest wood—had sunk to the bottom after being pushed into the lake during logging operations. The logs were between 12 in. and 18 in. in diameter, and most had been cut to just over 16 ft. in length. It was easy to raise one end of each log out of the water with a ⅜-in. chain and tow it ashore behind a small aluminum fishing boat with a 10-hp motor. We used pulleys and come-alongs to roll the logs onto the dock. We figured we would need about 50 timbers for the cabin. Working unhurriedly during short summer vacations, it took us three years to collect, cut and stack the logs under another cabin to dry.

Making timbers with a chainsaw—The key to our simplified building technique was a chainsaw milling device called a Lumber/Maker, a simple attachment that uses a 2x4 or 2x6 plank as a track along which the saw rides to make long, straight, ripping cuts. It's possible for a person working alone to saw up a log, but the job goes much faster if one person pulls the saw with a rope while a second holds it upright and guides it along the plank. This way neither person has to work very hard. Any decent chainsaw will do this job. We used a Stihl 015 with a 14-in. bar. Just be sure that the saw you choose has a good anti-vibration system, an effective muffler and an automatic chain-oiling mechanism. Since the cuts are long and sustained, you'll also have to file the teeth regularly for ripping (see the drawing on the next page).

Our first cut was right down the center of the log, halving it lengthwise. Then we flipped each half-log onto its round side, centered a 2x6 on its

The cabin is built of salvaged logs shaped with a chainsaw milling attachment. Each log is cut in half, then squared on two sides (top). Stacked and spiked timbers butt at corners in an alternating pattern (above). Small-diameter logs split lengthwise and nailed under the eaves and gable overhangs reinforce the roof boards. Porch steps and jacks are made of leftover slabs.

new flat side, and ripped along both sides of the plank. These cuts produced a timber with three faces parallel or perpendicular to each other. Because we used a 2x6 plank as a guide, the face between the parallel sides of all the timbers was always 6 in. wide.

To build our cabin, we simply stacked the squared timbers and spiked them together. We further simplified the construction by butting the timbers at corners, alternating joints rather than using the interlocking joints found in most traditional log buildings.

The pioneers used the more complex chinked-log or scribed technique (see pages 176-179) because they had unlimited supplies of logs in any diameter or length they cared to select, and a severe shortage of nails or other metal fasteners. Because they couldn't afford the luxury of drying their logs, they left them long to cope with the inevitable end-checking. Our timbers were dry. Besides, these days it's easy to trim and shape logs with chainsaws, and because inexpensive steel fasteners are everywhere, it made sense to build in a way that gets more mileage out of available logs.

Our Ojibway building crew, who had built many conventional log cabins, were skeptical at first, but by the end of the project they were planning, with the aid of our photographs, to sell similar cabins to others in the area. Even before we were finished, they were bringing around prospective customers.

Preparation for building—We planned the one room to use our full-length timbers, which would yield finished inside dimensions of 15 ft. 4⅞ in. on a side. All doors and windows were planned to have small panes of glass. Large picture windows with double panes are attractive, but we had to carry the materials we used over long distances by boat and by hand. Small panes are easier to transport, and easier to replace if they break.

We drew up working plans, built a small model to check our ideas, and in the spring of 1980, were finally able to tell George Mathias, our friend and advisor, and a highly skilled builder, that he could muster his crew and be ready to go when we arrived August 1.

George cleared the site and poured simple, pyramidal concrete foundation piers, set with their bottoms on the bedrock. We left the rough wood forms on the piers so the bare concrete would not show. Then came the three stringers, 35-ft. full-round, peeled logs. We notched their bot-

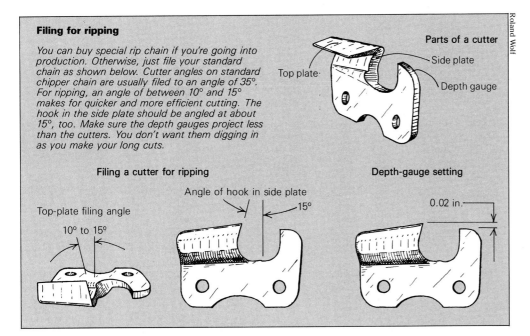

Filing for ripping

You can buy special rip chain if you're going into production. Otherwise, just file your standard chain as shown below. Cutter angles on standard chipper chain are usually filed to an angle of 35°. For ripping, an angle of between 10° and 15° makes for quicker and more efficient cutting. The hook in the side plate should be angled at about 15°, too. Make sure the depth gauges project less than the cutters. You don't want them digging in as you make your long cuts.

Parts of a cutter

Top plate

Side plate

Depth gauge

Filing a cutter for ripping

Top-plate filing angle

10° to 15°

Angle of hook in side plate

15°

Depth-gauge setting

0.02 in.

Concrete foundation piers support three notched full-round stringers (below). Cabin timbers are laid up with their inside faces flush (bottom). Fiberglass between the rounds will fill gaps in case of shrinkage. The outside faces of the timbers are more sharply rounded so snow won't stick and water will run off. Door and window openings were laid up rough, then trimmed to size with chainsaws.

toms and set them on top of the piers (photo center left). In central Ontario, we don't have to worry about termite damage, but in many areas termite shields should probably be installed between the posts and the stringers. We then flattened and leveled the tops of the stringers with drawknife, ax and chainsaw, and were finally ready to start building.

Laying up timbers—We decided not to install the floor until the house was enclosed, because the chipboard subflooring we were planning to use would warp badly if it got wet. So we began to lay the timbers that would make up the walls, nailing each, round side out, to the one beneath it with 12-in. spikes, which we spaced every 2 ft. or 3 ft. With four of us working, three or four rounds went up easily each day. We used full-length timbers for the first round and wherever openings did not interrupt the wall, and shorter ones between doors and windows. Our numerous door and window openings stretched our supply of timbers and also speeded up the work, because short timbers are easier to fit and to spike into place than long ones.

The timbers were thoroughly dry, some having been under cover for as long as four years. They worked beautifully with drawknife, ax and plane when we had to trim their surfaces to achieve a tight fit (not too much work here, because everything had been sawn straight to begin with). We decided to lay strips of fiberglass 2 in. wide by about ⅜ in. thick between the timbers in case shrinkage of the wood opened up gaps. One of the builders said that the last log cabin he had built, using the traditional full-round, green log method, settled 4½ in. the first year. We have a bet with him that our walls won't settle measurably, because the timbers were all so dry and laid so tight.

In addition to trimming the flat surfaces, we also beveled off sections of the gently curving, unsquared outer faces (photo below left). This way the outside wall would present no flat shelves on which rain or snow could accumulate and then seep into the joints. The Ojibways referred to this as "giving it a loggy look."

One of the most satisfying results of constructing a log cabin by this method is the handsome paneled look of the inside walls. We could have planed them smooth and finished them, but we like the texture the chainsaw gave them. As they age, they will take on a pleasing golden color, and when we want to hang anything on a wall, we just drive a nail or bore a hole for a peg.

A change in plans—When the walls reached their full 8-ft. height, we decided to change the details of roof construction. Our plans, made many months before and many miles away, had called for plywood sheathing and drop siding to form the gable ends at north and south, 2x8 roof rafters and ¾-in. plyscore roof sheathing. On the site, we decided first that since we had plenty of logs, we would lay timber walls above the 8-ft. wall heights to form the gable ends. Each course was slightly shorter than its predecessor, but was left with ample stock to allow for the trim. When we reached full height, we snapped chalk lines to mark the final slope, tacked up a straight 2x6 plank and cut the finish line using a chainsaw and the milling attachment (photo right). This went very quickly, and it was an efficient use of the short pieces of timber that had accumulated.

We also felt that the appearance of plyscore roof sheathing on the inside of the cabin would not be in keeping with the rough pine walls, so we bought 1-in. rough-cut pine boards 10 in. to 12 in. wide from a lumber mill. And, at George's suggestion, we decided to save time, material and money by forgetting about rafters. Instead we used a top plate, a ridgepole and a single purlin halfway up the slope on each side of the roof. George trimmed them so the roof planks would lie flat and be easy to nail. We nailed them up diagonally, leaving the planks long at the eaves and the ends of the building, and snapped a line and trimmed them with a chainsaw (photo below right). The final roof was built up of a layer of builder's felt covered by standard asphalt roll roofing. After it was installed, George took a long, slender log and split it lengthwise with his saw. We nailed each half as reinforcement under the board ends along each eave. We did the same at the overhangs on the north and south, tying the splits into the ends of the top plate, purlin and ridgepole. The resulting roof is strong and rustic-looking.

Windows, doors and floors—Once the roof was on, we trimmed the window and door openings square with a chainsaw, cutting them about ½ in. oversize on the sides and the top to allow for any expansion of the frames, contraction and settling in the timbers. When the window and door frames were wedged and nailed into place, we packed these gaps with fiberglass to cut drafts, and covered them on the inside with trim boards. The window frames, the sashes and the two doors were then given two coats of a good sealer.

Although we were building a summer cabin, we decided to insulate under the floor. We started with 2x6 joists, set on edge on top of the stringers and nailed to the walls. The joists were set so that their tops were 1⅞ in. below the level of the finished floor, as determined by the top of the first round of timbers. This required some notching of the joists, but the ubiquitous chainsaw made short work of the job. Furring strips (1x2s) nailed to the bottom of the joist sides made a shelf, on which we laid ¾-in. tarred sheathing board. Batts of 6-in. fiberglass were dropped into the resulting troughs. We used tarred sheathing because it is one of the few materials that the voracious local mice, squirrels and chipmunks won't scavenge when they're building nests for the winter.

On top of the insulation and joists, we nailed a subfloor of ¾-in. chipboard, followed by a vapor barrier of 4-mil polyethylene. Next we nailed 1¼-in. by 4-in. pine tongue-and-groove flooring boards to the chipboard and into joists with 1½-in. finish nails, properly set. As a finish, we applied one coat of penetrating sealer, followed by two coats of polyurethane varnish. All this may seem like a bit much for a simple summer cabin, but a floor finished this way retains its natural wood appearance, is easy to sweep clean and does not absorb dirt stains.

The gable end, laid up of short timbers, is trimmed with a chainsaw and milling attachment using a board as a guide (top). Diagonal roof boards overhanging the eaves are cut the same way (above).

Finishing touches—We framed the screened porch with round log posts supporting the outer corners of the overhanging roof and the ridgepole. The unscreened back porch was too shallow to require supporting posts.

We used a single layer of the 1¼-in. tongue-and-groove pine flooring on both porches. Two coats of a good grey floor enamel protect the porch floors from the weather and from wear.

A few of the logs we cut produced slabs with a 10-in. or 12-in. flat side. These made great stair treads and sides for the porch steps.

Brown creosote stain is the traditional exterior finish preservative around here, but creosote is nasty to work with and doesn't resist the weather very well. The new water-base stains are much nicer to use, but they look unnatural and heavy to us. We liked the light grey patina our long-immersed timbers had taken on, so we decided just to leave them unfinished and let nature take its course. Our liberal roof overhangs shelter the walls from rain, and the space under the building lets air circulate freely. If the timbers begin to show checks and cracks that could collect moisture, or harbor ants and other harmful insects, we'll fill them with a caulking that matches the color of the wood.

Whenever our Ojibway friends got a building

timber, log or other item in place particularly well, they would step back, admire it and comment, "It looks just like it gro'd der." Our oldest son, who was helping on the job, disappeared one day with a nice piece of slab wood under his arm. When he returned several hours later, he presented us with a sign he had carved from it saying, "It Gro'd Der." This so amused the building crew that they used it as the lintel over the screen door on the front porch. Now our island cabin even has a name. □

David Ford is a manufacturers' representative based in Cleveland, Ohio.

The Lumber/Maker is available from Haddon Tool, 4719 W. Route 120, McHenry, Ill. 60050. Other chainsaw mills are made by Granberg Industries, 200 S. Garrard Blvd., Richmond, Calif. 94804, and Sperber Tool Works, Inc., Box 1224, West Caldwell, N.J. 07006.

For information on chain and chainsaw accessories, write for catalogs from Bailey's, Box 550, Laytonville, Calif. 95454, or Zip Penn, Inc., 2008 E. 33rd St., Box 179, Erie, Pa. 16512.

For an authoritative and enjoyable guide to the use and maintenance of chainsaws, try Barnacle Parp's Chain Saw Guide *by Walter Hall (Rodale Press, Emmaus, Pa., $7.95).*

TIPS & TECHNIQUES

Cut-off fixture

I seldom use radial arm saws when I'm working on a job. They are expensive, difficult to move around and constantly out of adjustment. The alternative is to use the simple cut-off fixture shown in the drawing below, a circular saw with a shoe that is parallel to the blade, and a $15 saw protractor.

The fixture can be made from 2x lumber of any width, and should be long enough to sup-

Beveled end stop

2x4 protractor stop

Saw protractor

Line of cut

port the boards you are cutting. Attach a block across the face of the fixture at one end. Make sure that this end stop is perpendicular to the edge of the jig. This block should be back-beveled along the end grain on the working side of the jig to prevent sawdust buildup when it is in use. A 2x4 block can then be nailed to the side of the fixture, with the bottoms flush. This is the protractor stopblock. It should be long enough, with the additions of the protractor arm length and the width of your saw shoe, to produce the cut-off length that you want.

Butt the piece to be cut against the end stop, set the protractor across the board and against its stop, and run your circular saw against the leg of the protractor. This method will produce fast, clean, accurate and repeatable cuts of any angle for everything from joist blocking to finish work without measuring and squaring each board.

If you are production-cutting several lengths, nail protractor stops in several locations along the length of the fixture. Clamping the stop is faster if only one or two cuts are needed at that length.

—Ron Davis, Novato, Calif.

Cutting curves in big beams

Last year I designed and built a house that uses seven cantilevered 6x20 beams to hold up the second floor. The beams are exposed to view on the outside, so I wanted a decorative cut on each one to dress it up. The simple 12-in. radius arc I chose turned out to be a lot easier to draw than to cut with the tool at hand—a Sawzall with a 9-in. blade.

The basic problem with cutting curves in thick stock with a reciprocating saw is that

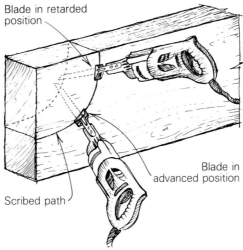

Blade in retarded position

Scribed path

Blade in advanced position

the free end of the blade tends to drift outward, cutting an arc larger than the one that is being followed. After much trial and error, including guiding the protruding blade with a pair of lineman's pliers, I developed the technique explained below.

Begin by scribing each beam carefully on both sides with the aid of a cardboard template, and then place it upside down on a pair of sawhorses. Once approximately ¼ in. of the cut has been made, the operator can advance or retard the blade on the scribed path by leading the cut with either the heel or the tip of the blade. This develops a torsion within the blade that affects its course.

Have a helper stand on the opposite side of the beam and describe the path of the blade as the arc is being cut. If the blade moves outside the pencil mark, the operator needs to advance the blade. If the blade starts to move inside the line, the operator can retard it to pull it back on course. With a little practice, this method works very well.

Incidentally, the helper is actually relief personnel, since each cut on my 6x20 beams took nearly 1½ hours to complete.

—Eric K. Rekdahl, Berkeley, Calif.

Securing electrical tape

Whenever you are using electrical (PVC) tape to wrap wire connections, rope ends or repaired power cords, just swab the end of the tape with PVC pipe cement. This will keep it from coming unstuck and unwinding.

—Bill Hart, Templeton, Calif.

Stair-button rip fence

When I need an accurate rip on a job site without a table saw, I use my stair-gauge fixtures. I clamp them on the front and back of my circular-saw baseplate, equidistant from the blade. This requires a saw with a flat baseplate. Measure the distance from the blade to the stair buttons just as you would with a rip fence, and then make the cut normally. With this quick setup, I get table-saw accuracy.

—Gred Gross, Wooster, Ohio

Homemade scribe

The carpenter's scribes that are sold in lumberyards and hardware stores are so flimsy that they will self-destruct just rolling around with the other tools in your nailbags. Sturdier versions are sold by mail order, but they cost about $20. Rather than spend the cash, I made

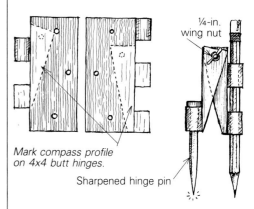

¼-in. wing nut

Mark compass profile on 4x4 butt hinges.

Sharpened hinge pin

the one shown here out of an old 4x4 butt hinge. It's accurate, nearly indestructible and easy to make.

I cut out the basic shape with a hacksaw, filed down the rough edges and drilled holes to accommodate a ¼-in. wing-nut pivot. A sharpened hinge pin makes a good turning point. The pin and the pencil are held in place by friction. If the slots are a little loose, a squeeze in the bench vise will tighten them up in a hurry.

—John H. Sandstrom, Fort Dodge, Iowa

Finish and Woodwork

The Kitchen Cabinet
How to design and build one with basic tools

by Will Hasson

The fundamental building block in every successful kitchen is the below-the-counter cabinet. It supports the work surface above and makes for organized storage beneath. Though this essential built-in can take on a multitude of styles and refinements, at heart the construction requirements are all the same. A sturdy case and framework, easily operable drawers, doors that fit and proper scale are common to good cabinets, whether made of painted plywood or expensive hardwood. In this article, I will describe the basic steps involved in building an uncomplicated, yet handsome, kitchen cabinet unit for about $140 in materials.

The cabinet shown above is the standard 24-in. depth, and is designed for the typical $\frac{3}{4}$-in. thick by $25\frac{1}{2}$-in. plastic-laminate countertop. Its $35\frac{1}{4}$-in. height, also standard, accommodates appliances such as dishwashers and trash

compactors, and presents the countertop at a level on which most people can work comfortably. These dimensions are used widely in the trade, but you can change the height to suit yourself just by adding or subtracting an inch or so. The 54-in. length here is arbitrary, but it makes good use of a single sheet of plywood and has room enough for a small sink to be let into the countertop. Also, a cabinet of this general size is small enough to move easily from shop to site—an important consideration.

Starting out—Constructing cabinets should always begin with a drawing of exactly what you have in mind. The drawing is indispensable for working out proportions, making cutting lists and for seeing the relationships between the different parts. You can use the drawings and cutting lists on the facing page as models for

preparing your own design and material requirements. If you do as much as you can on paper before you cut any wood, you'll save yourself a lot of time and frustration.

Almost all cabinets have the same parts: a case (or carcase), a face frame, doors and drawers. The cabinet shown here is designed to feature wood, and therefore has a carefully joined face frame and panel doors. The face-frame members are doweled together, and the whole frame is then applied to the front of the case. It provides jambs for the cabinet doors, defines the openings for the drawers and generally stiffens the carcase and helps it resist lateral racking. The doors and drawers have $\frac{3}{8}$-in. rabbeted inner edges so they overlap the face frame $\frac{3}{8}$ in. to form a lip on all sides, and so the drawer fronts and door frames project $\frac{3}{8}$ in. beyond the plane of the face frame. This arrange-

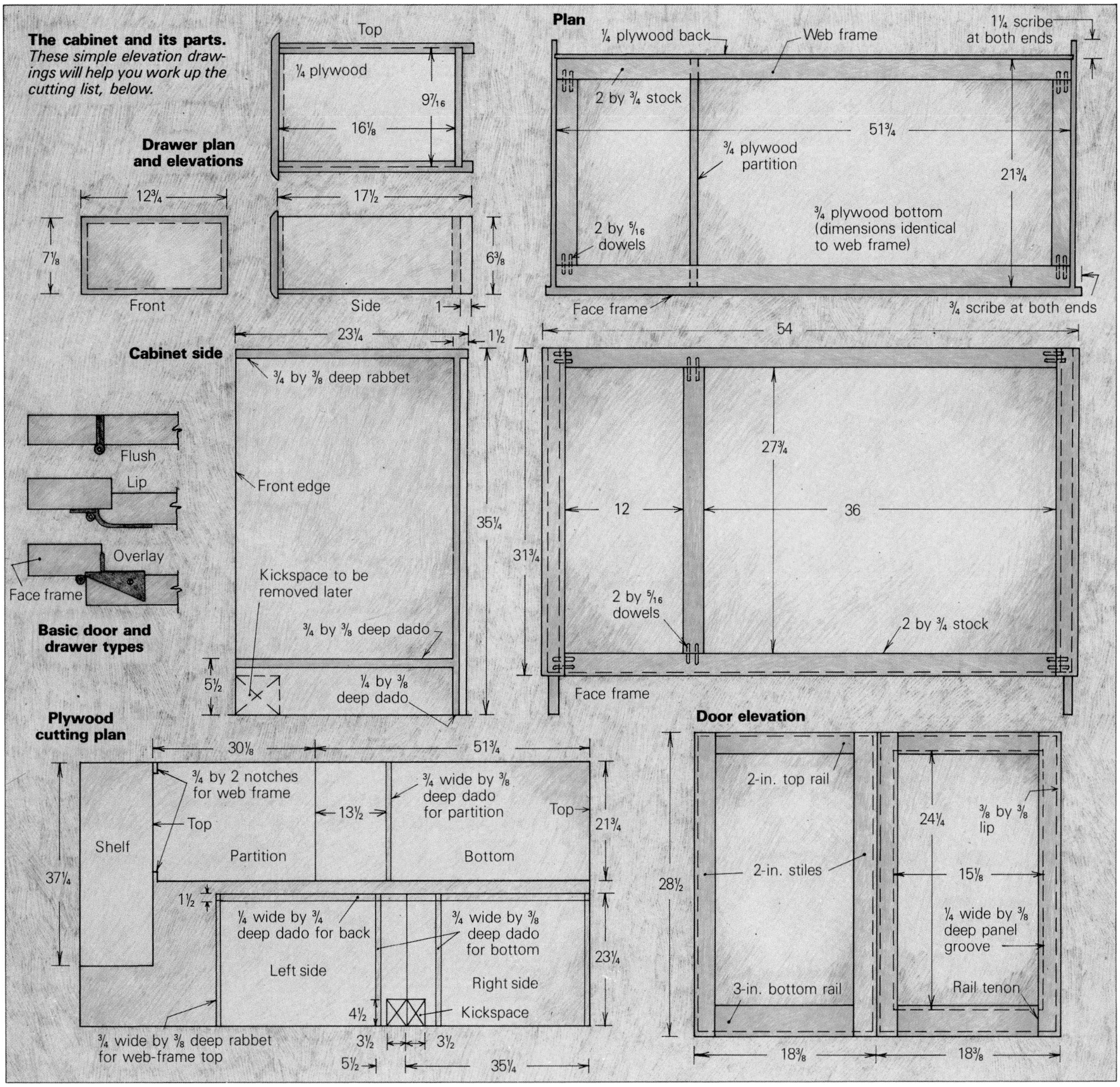

The cabinet and its parts.
These simple elevation drawings will help you work up the cutting list, below.
Top
¼ plywood
9 7/16
16⅛
Drawer plan and elevations
12¾
17½
7⅞
6⅜
Front
Side
1
Plan
¼ plywood back
Web frame
1¼ scribe at both ends
2 by ¾ stock
51¾
¾ plywood partition
21¾
¾ plywood bottom (dimensions identical to web frame)
2 by 5/16 dowels
Face frame
¾ scribe at both ends
54
Cabinet side
23¼
1½
¾ by ⅜ deep rabbet
Flush
Lip
Overlay
Face frame
Basic door and drawer types
Front edge
35¼
31¾
27¾
12
36
Kickspace to be removed later
¾ by ⅜ deep dado
2 by 5/16 dowels
2 by ¾ stock
5½
¼ by ⅜ deep dado
Face frame
Plywood cutting plan
Door elevation
30⅛
51¾
¾ by 2 notches for web frame
Top
Shelf
Partition
13½
¾ wide by ⅜ deep dado for partition
Top
Bottom
21¾
37¼
1½
¼ wide by ¾ deep dado for back
¾ wide by ⅜ deep dado for bottom
Left side
Right side
23¼
4½
Kickspace
¾ wide by ⅜ deep rabbet for web-frame top
3½
3½
5½
35¼
2-in. top rail
2-in. stiles
24¼
⅜ by ⅜ lip
15⅛
¼ wide by ⅜ deep panel groove
28½
3-in. bottom rail
Rail tenon
18⅜
18⅜

ment looks good, and the rabbets allow considerable latitude in fitting doors and drawers into the cabinet. Once the face frame is made and attached, however, you can choose between either overlay or flush detailing, as shown in the drawing above.

In our shop, we make the face frame extend ¾ in. beyond the sides of the case so the cabinet can be scribed to fit snugly against interior walls and against other adjacent cabinet units.

You can build the entire cabinet unit with just a few tools. A table saw or radial arm saw is useful for ripping frame members and drawer sides to width and for crosscutting them accurately to length. The case sides and back, however, are best cut with a skill saw, since in many cases it's almost impossible to wrestle a full sheet of ¾-in. plywood onto a saw table. You will also need an electric drill with a sharp 5/16-in. bit (preferably a brad-point bit) and a doweling jig for joining frame members.

It's possible, but not advisable, to get the job done without a table saw or radial saw. Without standing power tools, you would have to use a circular saw for all the ripping and crosscutting and a router for cutting the grooves, rabbets and dadoes. By hand-planing the frame members and drawer sides to finished width, you could certainly get accurate results, though it would take you considerably longer this way than if you were to use a standing power saw.

Making the face frame—Construction begins with the face frame. Rip ¾-in. solid stock into 2-in. wide strips and cut them to length. A crosscut jig for your table saw can be a great help in cutting accurate 90° angles. One way you can build a jig like this is described in the

Cutting list

Part	How many?	Size (in.)
Case		
Partition	1	30⅛ x 21¾ x ¾-in. ply
Bottom	1	51¾ x 21¾ x ¾-in. ply
Sides	2	35¼ x 23¼ x ¾-in. ply
Shelf	1	37¼ x 14 x ¾-in. ply
Back panel	1	51¾ x 30½ x ¼-in. ply
Web frame	2	51¾ x 2 x ¾
	2	17¾ x 2 x ¾
Face frame	2	50 x 2 x ¾
	2	31¾ x 2 x ¾
	1	27¾ x 2 x ¾
Doors		
Stiles	4	28½ x 2 x ¾
Rails	2	15⅛ x 2 x ¾
Rails	2	15⅛ x 3 x ¾
Panels	2	15⅛ x 24¼ x ¼-in. ply
Drawers		
Fronts	4	7⅛ x 12¾ x ¾
Sides	8	6⅜ x 17½ x ¾
Backs	4	6⅜ x 9 7/16 x ¾
Bottoms	4	16⅛ x 9 7/16 x ¼-in. ply
Kickplate	1	54 x 3½ x ¾
Hanger bar	1	37¼ x 3½ x ¾

The frame. With the doweling jig indexed on the pencil mark (1) Hasson bores out the end grain in a frame member. The hole is bored to the proper depth when the electrical tape on the drill bit is flush with the top of the jig, which is a self-centering dowel-guide made by Dowl-It (Box 147, Hastings, Mich. 49058).

For joining face-frame members, a single pencil line (2) marks the intersection of an upright. Lines to mark bore centers have been extended with a combination square down the edges to ensure accuracy in locating holes.

2

1

Crosscut jig

A simple table-saw attachment is especially useful for cutting frame members to length. We have four different versions for cutting various angles and sizes of work. The one drawn below is for making precisely square cuts in stock up to 2 ft. wide.

The guide rails on the bottom of the jig fit into the two miter grooves milled into the saw table on either side of the blade. We use oak or maple for the rails because these woods are hard and slippery. The runners should slide easily without any side-to-side movement.

The rails must be exactly perpendicular to fence A, which is rabbeted on its underside. In our latest version, I placed the rails in the table grooves and put a dab of glue atop their ends. Then I positioned fence A on the glue points and adjusted it perpendicular to the rails with a combination square. When the glue had set up, I applied glue to the edge of a ½-in. plywood panel and carefully slid the edges into the rabbet at the bottom of fence A, clamped it in place, and screwed the plywood to the runners, without moving the assembly from the table saw. Later I glued fence B to the top of the plywood. It keeps the ends of the plywood from flopping around and gives ballast to the front of the jig.

Passing the jig over the sawblade a few times will cut a workable slot in the plywood table. If you use the jig with a dado blade, make several shallow cuts to form the blade slot.

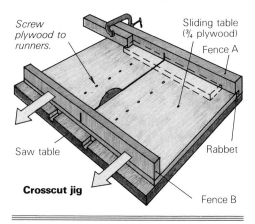

Screw plywood to runners.

Sliding table (¾ plywood)

Fence A

Saw table

Rabbet

Crosscut jig

Fence B

sidebar at left. Lay the frame members on a flat surface in the positions they belong in and label both sides of each joint. Holding one piece tightly in position with your fingers or a clamp, draw two lines across the joint. These lines mark the centers of the $\frac{5}{16}$-in. dowel holes, so don't put them too close to the edges of the stock—about ½ in. in is good. Next, carry the marks over the edges and ends of the strips with a combination square. A doweling jig is crucial to this operation because it ensures that the dowel hole is square to the surface and in just the right spot. Center the indexing mark of the doweling jig (mine is self-centering) on the lines and bore $\frac{5}{16}$-in. dia. holes slightly more than 1 in. deep for 2-in. dowels (photo **1**, above). Drill a test hole to the right depth, and mark the drill shank with a piece of electrical tape where it's flush with the top of the jig. The tape will mark the depth for all the holes.

Before you join the frame members, think which parts should go together first. Generally, the rule is to work from the inside out. With large, complicated frames it is sometimes useful to glue up the frame in sections, letting the glue in each section dry before assembling it into a complicated whole. After drilling all the holes, baste the dowels on one end with a thin coat of yellow (aliphatic-resin) glue, and drive them home (**2**). To assemble each section, apply glue first to the end-grain edges, letting it soak in and reapplying as necessary. Then coat the dowels and close the joint.

Clamp the frame together with bar clamps, and use wooden pads between the clamp jaws and the face frame to avoid crushing the edges and to direct the clamping force to the joint area (**3**). Small pieces of paper between the clamp and the frame will prevent the glue from reacting with the metal and staining the wood black. Don't rush. Work steadily and carefully, making certain that everything is flat and square. After the glue dries, tool off the squeeze-out with a sharp chisel or paint scraper. Plane or sand the face frame with a belt sander to even out the joints; then set it aside.

3

When gluing up the face frame (3), check for square, and position clamp blocks to direct clamping pressure across joint lines.

The basic case (4): ¾-in. plywood sides, bottom and partition are glued and nailed in place with 6d cement-coated nails. Notches in the top of the partition will receive the web-frame top.

Apply the face frame (5) with the cabinet laid on its back. Before gluing face frame to carcase, check the scribe depth (the amount by which the frame overlaps the sides). Re-check periodically while tightening the clamps. The web-frame top has already been glued and nailed into notches in the partition and into rabbets in the case sides.

4

5

The case—The case or carcase determines the interior space of the cabinet. The sides, bottom and interior partition and shelf in this cabinet are all ¾-in. plywood, and the back is ¼-in. plywood. We use lauan plywood for these parts because it's easy to work, and has an unobtrusive color and grain pattern that doesn't clash with the solid woods on the face of the cabinet. A web frame, similar to the face frame, is fitted into rabbets and holds the carcase together at the top. The bottom of the case is housed in dadoes in the sides.

Now lay out the cabinet sections on the plywood in the most economical fashion (drawing, p. 189). Cut the plywood for the sides, bottom and partition, and mark the positions of the dadoes and rabbets. Remember there are two opposing sides, one the mirror image of the other. Avoid making a pair of identical sides by accident, and having to scrap one.

Set your table saw or router to cut a groove ¾ in. wide by ⅜ in. deep. Now rabbet the tops (inside) of the case sides and dado the bottoms

as shown in the drawing on p. 33. Then cut ¼-in. wide by ⅜-in. deep grooves for the plywood back.

Next, dowel-join and glue up the web-frame top just as you did the face frame. It has the same outside dimensions as the plywood bottom panel, and is held in the rabbets cut into the top of the case sides and in two notches cut into the partition top. The web frame holds the top of the case together, and serves as a cleat for attaching the countertop.

When the carcase components are cut to size, nail and glue them together from the bottom up (4). Use 6d cement-coated nails, and draw a centerline opposite the dadoes to aid in locating the nails. Lay the partially assembled case on its back and align the side and partition while nailing the web frame in place. Make sure the distance between the partition and the sides is equal at the top and bottom.

After assembling the case, attach the face frame (5). If the cabinet is to be painted or if nail holes are not objectionable, the face frame

can be glued and face-nailed to the body with 6d finishing nails. A neater way is to glue and clamp the face frame to the case, though this requires a good supply of bar clamps and C-clamps. Whichever method you use, leave an equal overlap at each side for scribing. When the frame is glued in place, saw out the kickspace notches.

Drawers—Doors and drawers fall into one of three categories, depending on how they relate to the face frame. Lip doors and drawers are the traditional choice. Overlay doors and drawers are used almost exclusively in mass-produced kitchen cabinets, mainly because they are so quick and easy to install. But they also give a kitchen a clean, solid look. Flush doors and drawers (they fit flush with the face frame) make the cabinets look like furniture, and are by far the most difficult to install, as each one must be hand-fitted.

The details for drawer construction depend upon the type of drawer (lip, overlay or flush)

6

7

8

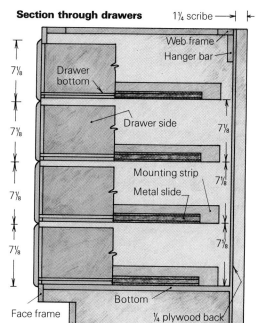

Section through drawers

1¼ scribe

Web frame
Hanger bar
Drawer bottom
Drawer side
Mounting strip
Metal slide

7⅛
7⅛
7⅛
7⅛

7⅛
7⅛
7⅛
7⅛

Face frame
Kickplate
Bottom
¼ plywood back

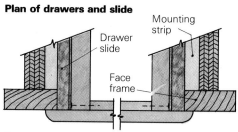

Plan of drawers and slide

Mounting strip
Drawer slide
Face frame

Drawers. After cutting the components (drawing, above), assemble the drawer by gluing and nailing the dadoed sides to the back (6). Slide the drawer bottom into its groove, and cap the box with the drawer front. Bore pilot holes for nailing the sides into the rabbeted front to avoid splitting the wood. The front edges of the drawer front have been shaped with a router and a ⅜-in. rounding-over bit.

Screw the slides to the drawer sides (7). The wheel flange should be flush with the side's bottom edge, and the track parallel to the edge.

Inside the cabinet (8) drawer slides are fixed to a mounting strip that holds them flush with the face frame. The mounting strips are glued and screwed to the case sides.

Door-frame assembly

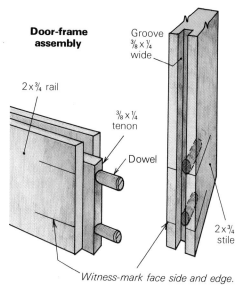

Groove ⅜ × ¼ wide

2 x ¾ rail

⅜ × ¼ tenon

Dowel

2 x ¾ stile

Witness-mark face side and edge.

A bottom rail, with doweled tenon and panel groove. The stubby tenon adds strength and lets you plow unstopped grooves in the stiles.

and the kind of drawer-slide mechanism you use. We use metal slides with nylon rollers for mounting the drawers because they are easy to install and because drawers glide smoothly and quietly on them. The slides come in a number of sizes, based on the dimensions of the drawer and the expected weight of its contents. Each drawer requires a pair of slides, which fit into a corresponding pair of tracks mounted on the inside of the cabinet. The slides in this cabinet are Grant #336 (Grant Hardware Co., High St., West Nyack, N.Y.; and 16651 Johnson Dr., City of Industry, Calif. 91745).

Begin laying out the drawers by considering the inside dimension of the opening in the face frame into which they will fit. In this cabinet, the drawer opening is 12 in. wide and 27¾ in. high. To keep things simple, we'll construct four identical drawers (drawing, facing page, top). Add the opening height plus the width of two lips (27¾ in. + ⅜ in. + ⅜ in. = 28½ in.) and divide the total by four to determine the height of each drawer front (7⅛ in.). The length of the drawer front is the width of the opening plus the width of two lips (12¾ in.).

Cut out the drawer fronts and calculate the width of the rabbet on the inside of each drawer front required to overlap the drawer sides, the metal slide and the face frame. The side thickness is ¾ in.; almost all metal slides require ¹⁷⁄₃₂ in. between the drawer side and the face frame; and the lip width is ⅜ in. This adds up to 1²¹⁄₃₂ in.

Cut the rabbets in the drawer fronts, and set them aside while you make the other drawer parts. The width of the sides will be the drawers' front height less the width of the two ⅜-in. lips (in our case, 7⅛ in. − ¾ in. = 6⅜ in.). The drawer length of 17½ in. in this cabinet is arbitrary. Drawers can be as long as the depth of the cabinet will allow.

The drawer sides, back and front are grooved along their bottom inner edges to receive the ¼-in. plywood bottom. Plow the groove ¼ in. deep and ⅜ in. from the bottom edge. The sides have a ¾-in. wide dado cut 1 in. from the rear edge to house the drawer back. Nail and glue

the drawers together, making sure they are square (6). Apply glue to the joints but not to the bottom panel, which should float freely in the groove. This is especially important if the drawer bottom is made of solid wood rather than plywood, because gluing it will cause it to crack eventually.

Half of the drawer slide is screwed to the drawer sides (7); the other half is attached to the sides of the cabinet on a spacer strip that brings it flush with the face frame. Locate the bottom edges of the spacers the same distance apart as the height of the drawer fronts. This will give you a guide for attaching the slides (8). All metal slides come with slotted holes to allow up-and-down and in-and-out adjustments. Use the slotted holes until the drawers run properly, and then drive the rest of the screws to lock the slide in place. You'll probably have to shave the top and bottom edges of the drawer fronts with a plane to regularize the spaces between them.

Panel doors—Frame-and-panel construction is common to most cabinet doors because it can be varied in so many ways. The inner edges of the frame may be molded with a router or shaper, be treated with applied molding or just be left square. The panel may be flat or raised, or glass may be used instead of wood.

Construction is basically the same regardless of the details. The first step, again, is to make a cutting list. Door stiles should be equal to the height of the door opening plus ¾ in. for the lips top and bottom. Calculate the width the same way. For a pair of doors, the opening plus ¾ in. is divided by two to get the width of each individual door. The length of the rails equals the width of the door less twice the width of the stile plus twice the depth of the panel groove for cutting a short tenon on the rail ends. In this cabinet, rail length is 15⅛ in. (18⅜ − 4 + ¾). Typically stiles and top rails are 2 in. wide. Bottom rails are 1½ times as wide to add visual weight and to increase the amount of wood involved in the joint.

After ripping the stiles and rails to width and

cutting them to length, arrange the pieces on a flat surface and mark the joints for dowels. Keep the holes far enough from the edges to prevent cutting into the dowels when rabbeting the outer edge for the lip and grooving the inner edge for the panel (drawing facing page, bottom). Mark the inner edge of each piece so rabbeting, grooving and drilling get done in the right spots. Then bore for dowels.

Make a ¼-in. wide groove in a scrap piece of wood for testing the thickness of the rail tenons. Now cut the tenons ¼ in. thick by ⅜ in. wide on both ends of the rails. Do this on a table saw or with a router, or clamp the rail in a vise and cut the tenon with a good backsaw. Make sure the tenons fit snugly, but not too tightly in the test groove.

The next step is to groove the inner edges of the frame members to receive the panels and tenons. Be certain to mark the outside face of each member and hold that face against the saw fence when plowing the grooves for the panels. Cut the ¼-in. plywood panel to fit, and then assemble the frame around it. Again, don't glue the panel into the groove. Be sure the doors are square and flat; otherwise they won't lie flat against the cabinet when closed.

Once the glue has dried, cut the ⅜-in. by ⅜-in. rabbet all around the door's inner edges (9), and you're ready to put on the hinges. Hinges are made for all three kinds of doors, and each style requires a specific type of hinge. It is best to install the hinges for lip doors on the doors first, spacing them one hinge length down from the top and the same up from the bottom. Next, set the cabinet on its back, put the doors in place and mark the position of the hinges (10). Drive one screw into each hinge and see if the door swings freely and lies flat. If it does, drive in the rest of the screws. If the fit is skewed, remove the screws, readjust the hinges as necessary, and try again. Now install catches to keep the doors flat against the face frame. Mount the catches opposite the door pulls. We use the common roller variety. Screw the male half to the top of the door stile, engage the two halves and screw the catch plate to the mount-

9

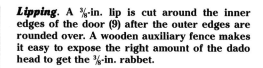

Lipping. A ⅜-in. lip is cut around the inner edges of the door (9) after the outer edges are rounded over. A wooden auxiliary fence makes it easy to expose the right amount of the dado head to get the ⅜-in. rabbet.

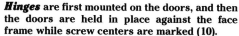

Hinges are first mounted on the doors, and then the doors are held in place against the face frame while screw centers are marked (10).

10

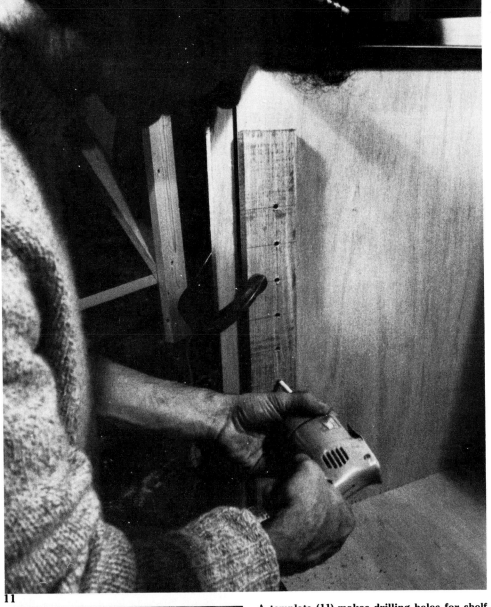

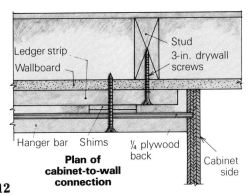

A template (11) makes drilling holes for shelf rests quick and accurate.

The hanger bar for attaching the cabinet to the wall is glued and nailed to the bottom of the web frame (12). The kickplate overlaps the side by the same amount that the face frame does.

Stud
3-in. drywall screws

Ledger strip
Wallboard

Hanger bar   Shims   ¼ plywood back

Plan of cabinet-to-wall connection

Cabinet side

12

Attaching adjacent cabinets

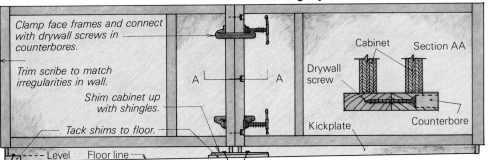

Clamp face frames and connect with drywall screws in counterbores.

Trim scribe to match irregularities in wall.

Shim cabinet up with shingles.

Tack shims to floor.

Level   Floor line

A ⌐   A

Cabinet   Section AA

Drywall screw

Kickplate

Counterbore

ing block with the door closed. Magnetic catches are also good, but best of all are self-closing hinges, which let you do away with catches all together.

While the back is still off, bore holes for the shelf rests. Make a drilling jig like the one shown in photo **11** with ¼-in. dia. holes, 2 in. on center, for press-in shelf rests. Use a piece of tape on the drill bit to regulate the depth of the hole. I drill the vertical rows of holes about a foot apart toward the back of the cabinet. The shelf should be about 14 in. wide, rather than span to the front of the cabinet. This narrower shelf lets you get at pots and other stuff stored in the rear of the cabinet.

A 1x4 hanger bar for installing the cabinet should now be nailed and glued to the bottom of the web frame and to the sides of the cabinet (**12**). Finally, slide the ¼-in. plywood back into its groove and nail it in place. Give everything a good sanding, and protect the wood with the finish of your choice. For wood finishes, we prefer Watco Danish Oil or lacquer, but for cabinets that are used frequently, polyurethane is the best choice.

Installation—Chances are good that the space your cabinet will occupy is neither level, square nor bordered by perfectly flat walls. These imperfections are usually slight and of no structural consequence, but they can drive the meticulous installer crazy. A few tips:

First, measure the highest and lowest points in the room and determine the differential. If the difference is an inch or less, I shim up the bottom of the cabinets, starting at the lowest point in the room. If the difference is more than 1 in., I split it by adding shims at one end and cutting the bottoms off the cabinets at the other. When the cabinets are level, pull them away from the wall and attach 1x4 ledger strips to the wall opposite the hanger bars.

Return the cabinets to their intended positions and check any counter-to-wall intersections for fit. If the wall is irregular, plane the scribing strip to match the wall contour (for more on scribes, see the next article).

When the cabinets are scribed and level, clamp the neighboring face frames together (if you've got more than one cabinet) and attach them with three counterbored drywall screws. Don't worry about plugging the counterbores—they won't show, and you can easily remove the cabinets if you decide to take them with you. Once the face frames are joined, screw the hanger bars to the wall ledgers, filling any gaps between the two with shims. It doesn't take a lot of screws for a solid connection; two per cabinet should do it.

Cut the kickplate to fill the gap at the back of the toe-space, and attach it in place with 6d finishing nails. We recommend finishing the juncture between the flooring and the kickplate with dark-colored vinyl coving. The coving makes this busy but hard-to-reach kitchen cranny easier to clean, and protects the kickplate from scuff marks.   □

Will Hasson is a partner in Fourth St. Woodworking, Berkeley, Calif.

Counter Intelligence

Some tips on working with plastic-laminate countertops

by Jesus Granado

For the past 25 years, plastic-laminate surfaces have been the most popular countertop finish in the kitchen and bathroom. They come in a wide variety of colors, patterns and textures, and the material is affordable, durable and easy to clean.

Plastic laminates are manufactured by several large companies (see box, p. 197) and sold under a variety of names, the best known of which is Formica, but they are all basically the same. Five to seven layers of kraft paper, with a final colored or patterned top layer, are saturated with melamine plastic and subjected to heat and pressure in excess of 1,000 psi in a large press. The result is a hard, thin panel, about $\frac{1}{16}$ in. thick. A final skin of melamine is applied to the top, and the product is ready for use.

The standard countertop is $25\frac{1}{2}$ in. wide, with a $3\frac{1}{2}$-in. high backsplash. Typically, the substrate is made of $\frac{3}{4}$-in. high-density particleboard, although plywood is sometimes used. On commercially constructed tops, the plastic laminate is bonded to the core under heat and pressure. This allows one piece of plastic to wrap the entire counter from backsplash to nosing and eliminates dirt-catching junctures of horizontal and vertical surfaces. The tops are

designed to cover the common 24-in. wide cabinet, leaving a 1-in. overhang. The remaining $\frac{1}{2}$ in. is a scribe allowance at the top of the backsplash for fitting the counter to irregularities in the wall.

Custom cabinet shops and large building-supply centers frequently carry post-formed countertops from 6 ft. to 12 ft. in length, or can have them made to order. These stock countertops come in a limited range of colors, and cost $5 to $7 per running foot. Your supplier should also have matching sidesplash and open-end trim kits for finishing an installation. For more money, you can get an extraordinary selection of colors and patterns from the specialty shops, even textured and metallic finishes.

Installing a post-formed countertop—Putting in your own counter isn't very difficult, and it's a good way to save between $100 and $200. The easiest top to install is the straight section, with one open end and no miter (more about miters later). Begin by measuring the distance from the end of the cabinet to the wall, and have the counter cut to this length.

If your dealer can't cut the top to length, use a circular saw and cut from the bottom. This

will keep the upward-cutting circular-saw blade from chipping the brittle laminate as it passes through it. If you use a handsaw to make your cuts, use a crosscut with at least 10 teeth per inch, and cut from the top. For added protection against chipping, no matter which saw you use, run a strip of masking tape over the cut line, and saw through it. The laminate should be fully supported to within a few inches of the cut line. Make sure the blades are sharp on any tools used for cutting plastic laminates.

At the wall end of the counter, attach the sidesplash to the countertop with four 2-in. drywall screws in pilot bores and a bead of silicone caulk between the counter edge and the sidesplash (drawing, below). The $\frac{3}{4}$-in. thickness of the sidesplash will extend the open end of the counter to the necessary overhang for the open-end cap. If you aren't using a sidesplash, remember to add $\frac{3}{4}$ in. to the counter length for the open-end overhang.

When the unfinished sidesplash is screwed in place, position the counter on the cabinet and check for fit against the back wall. Chances are, the backsplash will touch the wall in places, with ugly gaps in between. This is where the scribe allowance comes in handy. With the

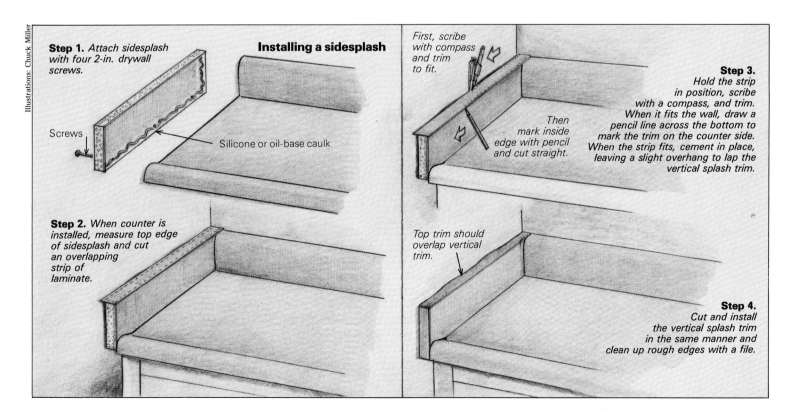

Illustrations: Chuck Miller

Step 1. *Attach sidesplash with four 2-in. drywall screws.*

Screws

Installing a sidesplash

Silicone or oil-base caulk

Step 2. *When counter is installed, measure top edge of sidesplash and cut an overlapping strip of laminate.*

First, scribe with compass and trim to fit.

Then mark inside edge with pencil and cut straight.

Step 3. *Hold the strip in position, scribe with a compass, and trim. When it fits the wall, draw a pencil line across the bottom to mark the trim on the counter side. When the strip fits, cement in place, leaving a slight overhang to lap the vertical splash trim.*

Top trim should overlap vertical trim.

Step 4. *Cut and install the vertical splash trim in the same manner and clean up rough edges with a file.*

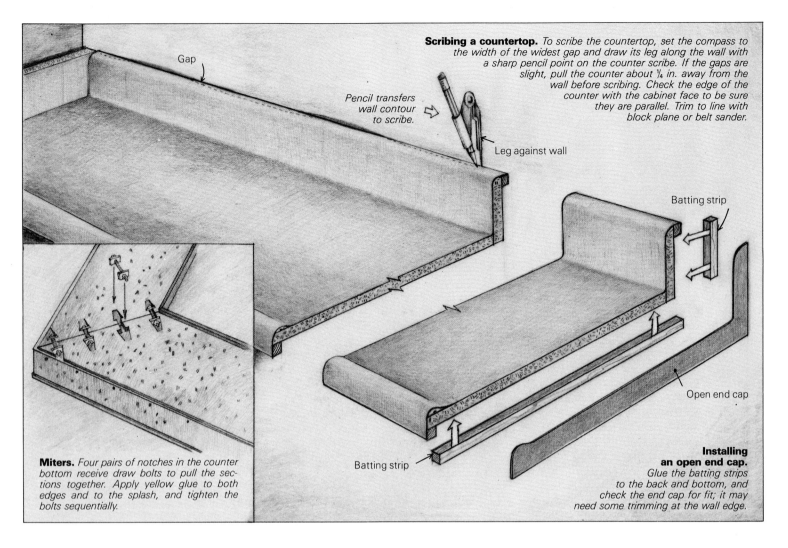

Scribing a countertop. *To scribe the countertop, set the compass to the width of the widest gap and draw its leg along the wall with a sharp pencil point on the counter scribe. If the gaps are slight, pull the counter about ¼ in. away from the wall before scribing. Check the edge of the counter with the cabinet face to be sure they are parallel. Trim to line with block plane or belt sander.*

Pencil transfers wall contour to scribe.

Leg against wall

Batting strip

Open end cap

Batting strip

Miters. *Four pairs of notches in the counter bottom receive draw bolts to pull the sections together. Apply yellow glue to both edges and to the splash, and tighten the bolts sequentially.*

Installing an open end cap. *Glue the batting strips to the back and bottom, and check the end cap for fit; it may need some trimming at the wall edge.*

counter against the wall and parallel with the face of the cabinet, run a compass, set to the width of the widest gap, along the top of the backsplash (large drawing, above). As the compass is drawn along, the wall's contour is transferred to the scribe allowance. Trim to the pencil line with a block plane or a belt sander until the counter fits tightly against the wall.

Counters often need to be cut out for sinks, which are usually centered below windows or above cabinet doors and set back 2 in. from the counter's front edge. Mark the shape of the opening in the right place on the underside of the counter. Cut straight lines with a circular-saw pocket-cut (see page 110), and connect them with saber-saw cuts at radiused corners.

Reposition the counter on its cabinet and make sure it's flat. Sometimes a top will be slightly twisted and one corner will not touch the cabinet, and then you'll have to use screws. Usually the cabinet's web frame for attaching the counter is ¾ in. thick. If so, use 1¼-in. screws to avoid penetrating through the ¾-in. thickness of the counter. You don't want screw tips poking through your nice new top. Place the screws 10 in. on center, and at all corners. If the top rests flat, use panel adhesive to secure it to the cabinet—it's quicker than screws. Spread a bead of adhesive on all the cabinet surfaces that will touch the bottom of the counter and put the counter back on for the last time. In 24 hours, the adhesive will have cured enough for you to use the counter.

When the top is secured, finish the sidesplash

trim and the end cap. The splash kit will include a strip of laminate for covering the raw edges. Cut a strip long enough to cover the top, hold it in place and scribe it with the compass. Trim it along the pencil line with a pair of tin snips. When it's tight to the wall, run a pencil along the underside, against the inside edge of the splash, for trimming the outer edge. Finished, this piece should conform to the contour of the wall and to the straight edge of the sidesplash, covering any gap between the two. To glue it in place, coat both surfaces with contact cement and press it into position. Bevel the rough edges with a file.

A strip of laminate cut to the cross-sectional shape of the countertop is supplied by the dealer to finish the open end. Glue the batting strips to the bottom of the counter and the back of the splash (drawing, above right), and cement the end cap in place. Some end caps are precoated with heat-activated glue; use an electric iron to install them.

The length of a wall-to-wall counter, plus the thickness of its two end splashes, should be ½ in. shorter than the space it will occupy. The ½ in. will give you some room to maneuver the top into position. Center the top on the base cabinet, with equal gaps at both side walls, and trim as shown, letting the scribed laminate cover the gaps.

Miters—An L-shaped or U-shaped counter arrangement is more complicated than a straight one. A 90° corner in a countertop is the same

as in a picture frame, but the miter cut is nearly 3 ft. long and has to be dead on. It's impossible to make this cut without special equipment, so buy sections of counter with the 45° miter already sawn. Some suppliers have a random selection in stock; others will have to order counters cut to your specifications. You can also take your counter to a custom shop that will make cuts on a miter saw for a fee.

To provide a bearing surface for draw bolts at the miter joints, cut tapered notches with a router in the counter bottom. The draw bolts, along with yellow glue in the joint, will hold the sections together (drawing, above left). It's best to scribe a mitered top to the walls before the sections are joined. When both counter pieces fit their respective walls, and the miter joint is tight, you can glue the sections together.

Gluing can be done in place, or the sections can be moved to a bench and glued upside down. I prefer gluing in place. It's a little more cumbersome, but the assembled top doesn't have to be moved around. Whichever method you choose, apply glue to both edges and align the surfaces of the two counter sections at the miter before the glue dries. Lightly tap a block of wood up and down the seam until the surfaces are perfectly flush.

Rolling your own—Even without a high-pressure press, you can make a durable countertop, and by ordering your laminates through a custom shop, you can get the color and texture you want. A counter you make yourself won't have the rounded nosing and curved backsplash like the post-formed ones, but you do have the option of varying the width of the counter and the height of the backsplash.

The substrate should be high-density particleboard. Don't use ordinary floor underlayment; it's just too soft. Glue and nail batting (boards that thicken the edges) and scribe strips to the underside of the show edges (drawing, right). Next cut the plastic-laminate to cover the counter surface so that it overhangs the visible edges about ⅛ in., and fits flush with edges that butt into walls. With the top surface applied, cut the laminate strips to cover the edges. Do this on a table saw, and use a veneer-cutting blade (carbide-tipped blades cut smooth and chip-free). Contact-cement the strips into place, butting them hard against the underside of the overhanging top. Finally, trim the overhang off with a bevel or flush-cutting plastic-laminate bit in your router. The bit you use must have a ball-bearing pilot. A bit with a solid pilot can burn and otherwise mar the plastic surface it bears against.

Another method is to glue on the edge strips first, trim them flush with the top and bottom edges with a 90° laminate-trimming bit, apply the field (countertop) laminate second and finally trim the overhang with a bevel-cutting laminate bit.

When you're bonding the laminate to the particleboard substrate, take care to apply the contact cement evenly to both surfaces. Covering large areas with a brush leaves tiny ridges in the cement and hinders bonding. Instead use a cheap, short-napped paint roller to get complete, uniform coverage. Once the cement is dry to the touch, press the two surfaces together to activate bonding. Be certain that the laminate is properly aligned before you lower it onto the substrate, something best done by two people. Once contact is made, that's it; there's no ungluing. Use a linoleum roller to flatten the countertop and to bring the two surfaces into intimate contact. A rolling pin or veneer roller will do for pressing the edges. To avoid air bubbles, start in the middle and work toward the edges.

Maintenance—I tell my customers to treat their countertops the way they treat their furniture. Don't use abrasive cleansers on plastic laminate; the top layer of melamine will erode and eventually it will stain and peel. Use a liquid cleanser like 409 or Fantastic and a sponge to clean the top, and then treat it to a coat of liquid furniture wax every four to six months. The wax will protect the plastic and make it easier to clean. Don't use the top as a cutting surface, and don't put a pot straight from the stove on the counter. The temperature of boiling water is the hottest the surface can get without blistering. There are also several chemicals that will harm plastic laminates; hydrogen peroxide is the most common. Drain cleaners shouldn't come in contact with your countertops either. □

Jesus Granado makes and installs countertops in Richmond, Calif.

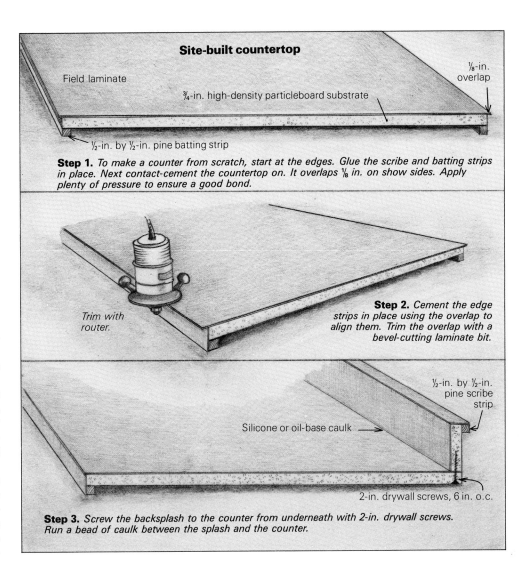

Site-built countertop

Field laminate

¾-in. high-density particleboard substrate

⅛-in. overlap

½-in. by ½-in. pine batting strip

Step 1. *To make a counter from scratch, start at the edges. Glue the scribe and batting strips in place. Next contact-cement the countertop on. It overlaps ⅛ in. on show sides. Apply plenty of pressure to ensure a good bond.*

Trim with router.

Step 2. *Cement the edge strips in place using the overlap to align them. Trim the overlap with a bevel-cutting laminate bit.*

½-in. by ½-in. pine scribe strip

Silicone or oil-base caulk →

2-in. drywall screws, 6 in. o.c.

Step 3. *Screw the backsplash to the counter from underneath with 2-in. drywall screws. Run a bead of caulk between the splash and the counter.*

Buying plastic laminates

Plastic laminates were first used as an electrical insulating material in the early 1900s. They were a lighter and less expensive alternative to mineral and glass insulators. The addition of colors and designs printed on the top layer of paper, underneath the transparent resin surface, produced the first decorative laminates. Radio manufacturers, who were using laminates as insulators, began making imitation wood-grain cabinets.

The National Electrical Manufacturers Association (2101 L St. N.W., Washington, D.C. 20037) sets the standards for this industry. These include tests for thickness, dimensional change, formability and resistance to wear, scuff, impact, heat, staining, light and blistering. NEMA publishes Standards Publication #LD 3-1980, *High Pressure Decorative Laminates* ($12.55 plus 20% for postage and handling). This overpriced 60-page pamphlet has an excellent section on fabricating and installing plastic laminates. However, this same information is free of charge in the advertising literature of all the major manufacturers.

Nearly all high-pressure decorative plastic laminates in the U.S. are made by seven companies—Consoweld Corp. (700 DuraBeauty Lane, Wisconsin Rapids, Wis. 54494); Formica Corp. (10155 Reading Rd., Cincinnati, Ohio 45241); Lamin-Art (6430 E. Slauson, Los Angeles, Calif. 90040); Nevamar Corp. (8339 Telegraph Rd., Odenton, Md. 21113); Pioneer Plastics (Pionite Rd., Auburn, Maine 04210); Westinghouse Electrical Corp. (Hampton, S.C. 29924); and Wilsonart (600 General Bruce Dr., Temple, Tex. 76501).

Decorative plastic laminates are produced in several grades. Standard grade can be used for most applications. Some manufacturers make plastic laminates for use on countertops that are thicker and more durable than the material used for walls or other vertical-surface applications. Post-forming grade is specially made to bend under pressure, and the maximum bending radius and other data are available from the maker. Fire-rated plastic laminates can also be purchased. Sheet sizes for all these grades vary from one manufacturer to another, although 24 in., 30 in., 36 in., 48 in. and 60 in. are standard widths, and 6 ft., 8 ft., 10 ft. and 12 ft. are standard lengths.

Other variables are color, pattern, texture, and finish, and these determine price. Sold by the square foot, laminates with deeply textured surfaces, detailed patterns, and exotic (called "decorator") colors cost more. Solid white is the least expensive, in the $1.10 to $1.50 range. Shopping around will definitely pay off. —*Paul Spring*

Greenhouse Shutters

Insulated panels and rigging that folds them away overhead

by Stan Griskivich

When my wife and I built our house in 1978, I made and installed thermal shutters that fold back away from the windows (see *FHB* #2, p. 52). When we added a greenhouse last year, I knew we'd need insulation for its sloping south-facing glass, too. But shutters hinged on sloping studs would be awkward to operate, and would get in the way of people and plants when folded back. Insulated quilts don't work well on sloping windows larger than about 2 ft. by 6 ft. either, because their

Stan Griskivich is a carpenter and cabinetmaker who lives near Yarmouth, Maine.

own weight pulls them out of their tracks or away from their magnetic strips. So I developed an overhead bifold arrangement operated by a line-and-pulley system adapted from nautical rigging. It takes just a few minutes each day to raise and lower the shutters.

The shutters, which fit into the slope of the wood-frame wall, incorporate a Thermax core and two dead-air spaces. They give double-glazed windows an R-value of about 14, as opposed to the R-8 of the house shutters, which have no Thermax core, and only one dead-air space. Because so much of a greenhouse is glass, it is important for greenhouse shutters

to have a high R-value if it can be done cost-effectively. With my shutters it can be. Owner-built with the tools and materials listed on the facing page, they cost about $3 per sq. ft.

Measuring and cutting—Measure the width and height of your windows. Subtract ½ in. to determine the size of your shutter sets. These fold horizontally, so divide the set's height in half to get the length of each shutter's stiles (vertical frame members). I use a rabbet joint at the corners, so I subtract 1½ in. from the shutter set's width to find the length of the rails (horizontal frame members).

Rip 5/4 by 6-in. pine boards into 1¾-in. wide rails, and then cut them to length. If your south wall slopes, as mine does, you'll have to cut the bottom end of the stiles and the two sides of the rail on the lower shutter at an angle to match that of the slope. I used a sliding bevel to find the angle, and ripped the rails on my table saw and crosscut the stiles on my radial arm saw. Plow a ¼-in. deep by ½-in. wide groove down the inside face of all stiles and rails. For the groove in the bottom rail of the lower shutter, set the table saw just as you did to cut the edges of the rail to the proper angle. The ½-in. Thermax core fits into this slot. Now cut the rabbet ¼ in. deep by 1 in. wide in the ends of the stiles.

Cut the Thermoply panels to the same size as the shutters, two panels per shutter. I use a fine-tooth veneer blade on a table saw. A utility knife and straightedge also work fine.

Cut the ½-in. Thermax board 1½-in. shorter and narrower than the shutter. Be sure that the Thermoply and the Thermax panels are cut squarely and accurately.

Assembly—Put yellow (aliphatic-resin) glue on the ends of one rail and fasten stiles to rails with 2-in. drywall screws. Slide the Thermax board into the grooves of the stiles, then install the other rail. You now have a wood frame the same size as a shutter with a Thermax core recessed ⅝ in. deep within it.

The upper shutters require a pulley in each rail. To support them, glue ⅝-in. thick blocks to the center of their window sides, as shown. Thermoply will cover the blocks.

Run a bead of panel adhesive around the

Shutters prevent radiant and conductive heat loss through glazing. Here the author lowers a pair of his folding greenhouse shutters.

perimeter of the completed frame and lay down a sheet of Thermoply, foil face in. Fasten it with ½-in. to ⅝-in. staples set flush with the surface, about 8 in. o.c. Turn the shutter over and install the other sheet of Thermoply the same way. If necessary, sand the edge of the Thermoply flush with the surface of the frame. Then paint or seal the edges of the frame to prevent moisture absorption. The white surface of the Thermoply can be left as is, painted, stenciled, or covered with fabric or wallpaper.

Hinges and rigging—When you have completed both shutters for a single folding unit, join them with a pair of 3-in. by 3-in. loose-pin hinges. Locate the hinges with their barrels facing the window, and make sure that you leave 1-in. to 1⅛-in. clearance for the pulley, which would otherwise keep the shutters from folding completely. Once the hinges are in place, separate the two shutters by pulling out the hinge pins. Now install hinges on the top horizontal rail of the upper shutters—this time with their barrels away from the windows. The shutter pair will hang from the top plate of the south wall.

Measure across the ceiling a distance equal to half the height of the windows, and screw a 1x3 or 1x4 along the length of the greenhouse to support ceiling pulleys.

The rigging (drawing, top right) is easy to understand if you have ever been sailing. Begin by installing a pulley (I use Nicro Fico NF 414) on the ceiling strip, directly behind the vertical center of the shutter set. Now install another NF 414 pulley in the upper shutter's top-rail screw block, and an NF 219 pulley in the screw block on the bottom rail.

This part is easier with two people. Mark the location of the hinges on the top plate by holding the upper shutter against the stops with its pulleys facing the windows. Leave equal clearance on both sides of the shutter.

After the upper shutter is hinged to the top plate, re-connect the lower shutter. Drill a ¼-in. hole through the bottom rail of the lower shutter. Run a ³⁄₁₆-in. braided nylon line through the hole from the window side, and tie a stop-knot on the inside surface. Run the line to the NF 414 pulley in the upper shutter, then down to the NF 219 pulley at the meeting rail. Drill another ¼-in. hole through the shutter and run the line to the second NF 414 pulley, mounted on the ceiling. Once this rigging is complete, pulling on the line will simultaneously fold the shutters and draw the pair against the ceiling. A cleat fastened to the back wall secures them out of the way.

When you want to lower the shutters, let them down from the ceiling until you can reach up and begin to unfold them. From this point they will unfold by themselves as you pay out the line.

Install 3M V-Seal weatherstrip on the 1x4 glazing stops and the top, bottom and meeting rails. Install 1½-in. by 3½-in. turn buttons at the midpoint and bottom of all the sloping 2x6 studs; they will hold each set of neighboring shutters closed. □

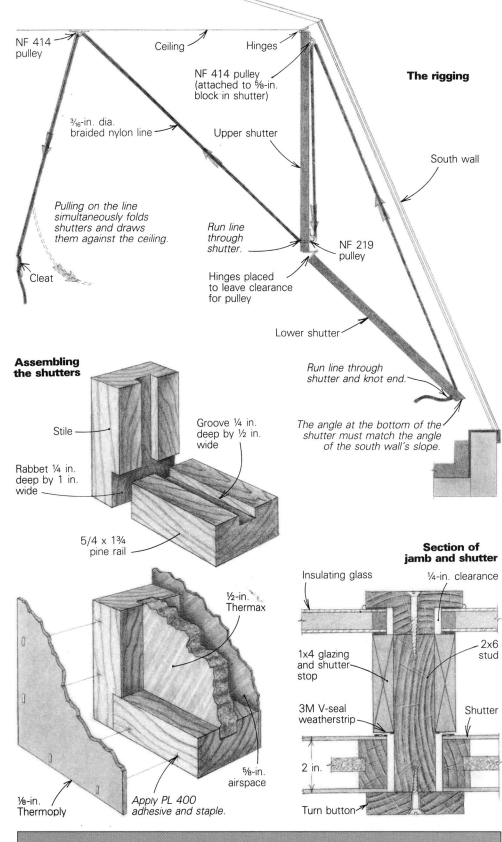

The rigging

NF 414 pulley

Ceiling   Hinges

NF 414 pulley (attached to ⅝-in. block in shutter)

³⁄₁₆-in. dia. braided nylon line

Upper shutter

South wall

Pulling on the line simultaneously folds shutters and draws them against the ceiling.

Run line through shutter.

NF 219 pulley

Cleat

Hinges placed to leave clearance for pulley

Lower shutter

Run line through shutter and knot end.

The angle at the bottom of the shutter must match the angle of the south wall's slope.

Assembling the shutters

Stile

Groove ¼ in. deep by ½ in. wide

Rabbet ¼ in. deep by 1 in. wide

5/4 x 1¾ pine rail

½-in. Thermax

⅝-in. airspace

⅛-in. Thermoply

Apply PL 400 adhesive and staple.

Section of jamb and shutter

Insulating glass   ¼-in. clearance

1x4 glazing and shutter stop

2x6 stud

3M V-seal weatherstrip

Shutter

2 in.

Turn button

Materials
⅛-in. Farmline Thermoply (Simplex Products, Box 10, Adrian, Mich. 49221). Also known as White Face Thermoply, it comes in 4x8 sheets with one foil face and one white.
½-in. foil-face Thermax insulation board (Celotex Corp., Tampa, Fla. 33622), available in 4x8 and 4x9 sheets.
5/4 x 1¾-in. pine (ripped from 6-in. boards)
Contech PL 400 adhesive (caulking-gun cartridges)
3-in. by 3-in. loose-pin hinges
3M #2743 V-seal weatherstripping
Nicro Fico NF 414 and NF 219 pulleys (Nicro Corp, 2065 West Ave. 140th, San Leandro, Calif. 94577)
³⁄₁₆-in. dia. braided nylon line
4½-in. aluminum cleats
½-in. to ⅝-in. staples
2-in. drywall screws
Yellow (aliphatic-resin) wood glue

Tools
Table saw or radial arm saw with dado head
Utility knife and straightedge or fine-tooth plywood blade
Electric or hand stapler
Electric or hand drill and ¼-in. bit
Chisel or router (for hinges)
Screwdriver, awl, caulking gun, tape measure

Cornice Construction

Building the return is the tough part

by Bob Syvanen

All gable-roofed houses need to have a cornice of some sort. Functionally, the cornice fills the voids between roof and sidewall. It extends from the shingles to the frieze that covers the top edge of the siding. There are two basic kinds of cornices: one includes a gutter (drawing **1A**, below); the other is a simple cornice molding (**1B**). Traditionally, each calls for a different method of roof framing. For the gutter, rafter tails are combination-cut (plumb and level) and bear on the top plates. For a cornice without the gutter, a double plate is nailed across the tops of the joists to support the rafters. There are a vast number of possibilities in cornice construction and detailing (photos, facing page), but they are all variations on the basic, step-by-step sequences shown and discussed here.

The cornice return at the juncture of eave and gable is the most noticeable and most intricate part of this architectural detail. Sometimes called a boxed return, it adds a delicate touch to the large, repetitive detail of clapboard or shingle siding.

Building the cornice itself is simple; building its return is a bit more challenging. Let's start with the eave corner, assuming that you've already framed and sheathed it. The rafter ends and ceiling joists should still be exposed, and the first step is to nail up the fascia board. For the guttered cornice (**2A**), the fascia will go against the rafter tails; otherwise, nail it to the ends of the ceiling joists (**2B**). For the fascia, I use #2 pine boards ripped to width on a table saw. The top edge of the fascia usually needs to be beveled for a tight fit.

At the corner of the house, the fascia meets the ear board—the broad, flat backing board at the gable base that holds the cornice return. The joint between the fascia and the ear board has to be mitered, and unless the fit is

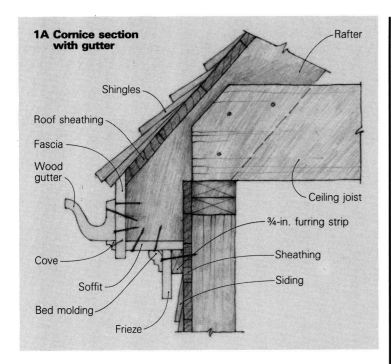

1A Cornice section with gutter

Rafter
Shingles
Roof sheathing
Fascia
Wood gutter
Ceiling joist
¾-in. furring strip
Cove
Sheathing
Soffit
Siding
Bed molding
Frieze

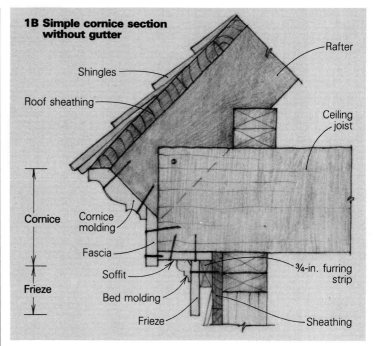

1B Simple cornice section without gutter

Shingles
Roof sheathing
Rafter
Ceiling joist
Cornice
Cornice molding
Fascia
Soffit
¾-in. furring strip
Frieze
Bed molding
Frieze
Sheathing

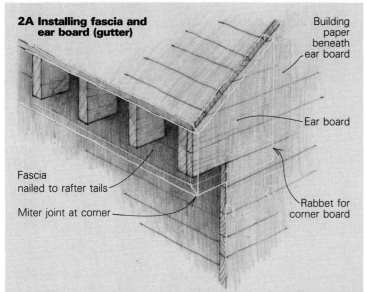

2A Installing fascia and ear board (gutter)

Building paper beneath ear board
Ear board
Fascia nailed to rafter tails
Miter joint at corner
Rabbet for corner board

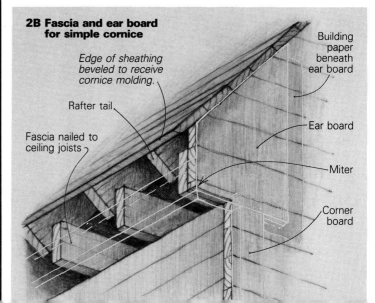

2B Fascia and ear board for simple cornice

Edge of sheathing beveled to receive cornice molding.
Rafter tail
Building paper beneath ear board
Fascia nailed to ceiling joists
Ear board
Miter
Corner board

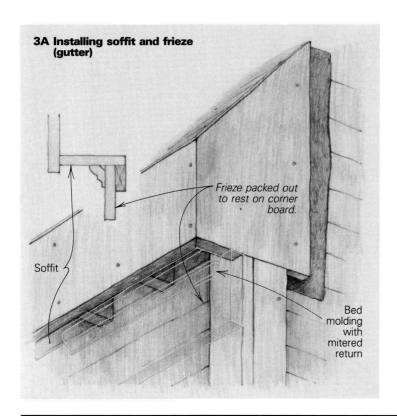

3A Installing soffit and frieze (gutter)

Frieze packed out to rest on corner board.

Soffit

Bed molding with mitered return

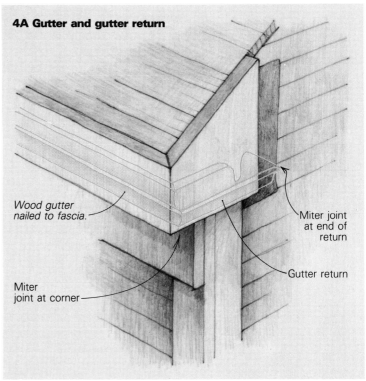

4A Gutter and gutter return

Wood gutter nailed to fascia.

Miter joint at corner

Miter joint at end of return

Gutter return

tight, it's a good idea to use a weatherproof adhesive caulk. Phenoseal (made by Gloucester Co., Box 428, Franklin, Mass. 02038) and other marine adhesives work well in this type of joint. I also staple a piece of building paper behind the ear board.

If you want the ear board to be flush with the corner board, which is installed before the siding is nailed up, the ear board needs to be rabbeted at its bottom edge so that the siding and a tongue at the top of the corner board can be tucked up underneath it. The ear board should also cover the gable edge of the roof sheathing if you're not planning to use rake molding, which is shown in the inset drawing **6B**.

Now install the soffit, frieze and bed molding, in that order. I cut the soffit and frieze boards from pine or fir, and fur out the frieze a full ¾ in. so that the siding will fit underneath it. Cut the frieze to extend over the corner board, as shown in **3A** and **3B**. You can use something fancier than bed molding to cover the joint between soffit and frieze, or simply leave this juncture plain.

The return—The next piece to go on is the cornice (crown) molding (**4B**), or the wood gutter that replaces it on a guttered cornice (**4A**). Here again, the corner is mitered to make the return. In fitting the return, I like to tack a temporary guide strip to the ear board, on a level line where the bottom of the cornice molding or gutter will fit. This makes test-fitting, trimming, and retesting the miter a bit faster.

A good tight joint here is important. Corners are seldom perfectly square, and you can either adjust the cut in the miter box or just keep fitting and trimming with a chisel or block plane. The small triangular piece at the end of the return is delicate, and I usually cut

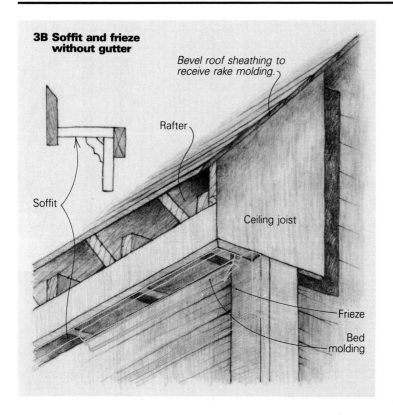

3B Soffit and frieze without gutter

Bevel roof sheathing to receive rake molding.

Rafter

Soffit

Ceiling joist

Frieze

Bed molding

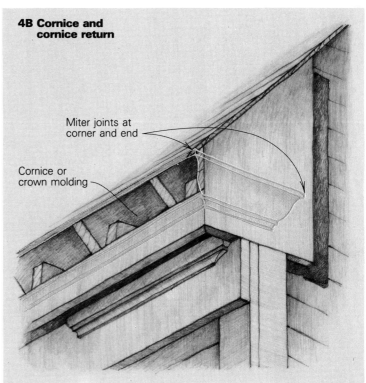

4B Cornice and cornice return

Miter joints at corner and end

Cornice or crown molding

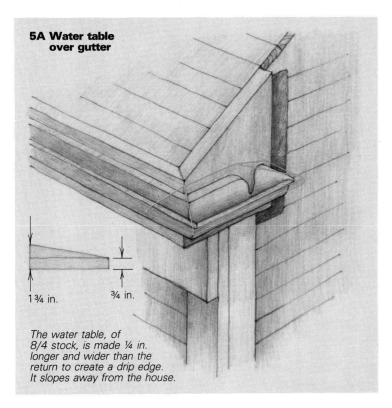

5A Water table over gutter

1 ¾ in.   ¾ in.

The water table, of 8/4 stock, is made ¼ in. longer and wider than the return to create a drip edge. It slopes away from the house.

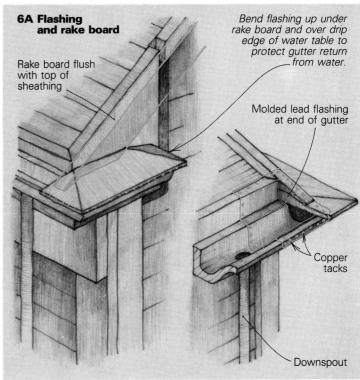

6A Flashing and rake board

Rake board flush with top of sheathing

Bend flashing up under rake board and over drip edge of water table to protect gutter return from water.

Molded lead flashing at end of gutter

Copper tacks

Downspout

and fit it on the ground after the corner miter is done.

What you have now is the mitered return with an opening on top. This opening must be covered with a water table (**5A**, **5B**), which is a piece of 8/4 pine, beveled away from the side of the house to shed water. You can cut a single bevel, or bevel the board downward on all open sides for a fancier look. Size the board a little longer and wider than the return to create a drip edge. The water table for the simple cornice return has to be trimmed to fit beneath the roof sheathing (inset, **5B**).

Once you've nailed the water table in place, flash it before installing the rake board along the corner between the roof and the gable wall. The flashing has to extend up beneath the rake board, so if you're working on an old house, pry the rake board up, tuck the flashing underneath it, and then renail the rake. If you are going to use rake molding (**6B**), nail the rake board to the gable end, furring it out away from the sheathing at least ¾ in. so that the siding can be fitted underneath. To provide good bearing for the rake molding, bevel the gable-end sheathing. If you don't use a rake molding, leave the sheathing square-cut so that the rake board and its furring strip

(**6A**) can be nailed to it. In either system, the rake board and molding die onto the water table, and must be scribed accordingly.

If you're building a guttered return, there's one final, important step: sealing off the corner. Water that gets into the return side of the gutter will soon rot the wood. Lead is the best flashing material because it can be formed to the gutter contours more readily than copper or zinc. Use the end of your hammer handle to form the lead flashing so that it covers the corner and fits over the water table, rake board, and roof sheathing. Then nail it down with copper tacks in a bed of caulking.   □

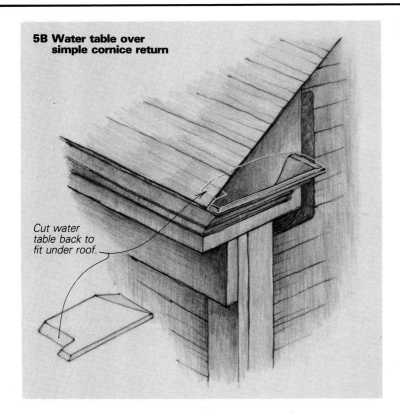

5B Water table over simple cornice return

Cut water table back to fit under roof.

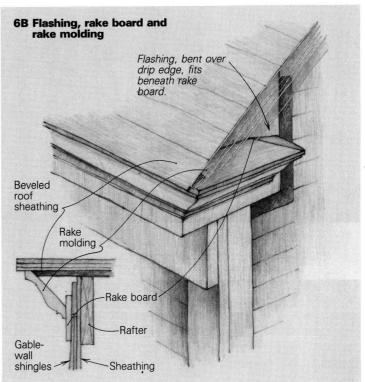

6B Flashing, rake board and rake molding

Flashing, bent over drip edge, fits beneath rake board.

Beveled roof sheathing

Rake molding

Rake board

Rafter

Gable-wall shingles

Sheathing

Table-Saw Molding

The secret is in the order of cuts

by Bruce Andrews

When the landmark Winooski Block was finished in 1862, the builders festooned it with all manner of ornamental moldings and wooden filigree. But by the time we (Moose Creek Restorations Ltd.) got the repair contract, 117 Vermont winters had weathered, cracked and split all of its remaining woodwork. Three-fourths of the building's cornice moldings were either rotten or missing. We were to replace 10,000 linear feet of various moldings, not one of them a type manufactured today, and we didn't even own the usual tool for milling moldings, the spindle shaper. We still were able to complete the job, relying on our table saw and a lot of careful planning. We found that the table saw could handle most any profile—it could even scoop out concave curves—but we also learned that every profile required its own sequence of cuts. Figuring out that sequence is the heart of our method.

The first thing we worried about was getting enough good stock. Molding stock must be the highest quality, close grained and knot free. We were still short of stock after several deals to obtain a couple thousand board feet of Vermont pine in varying widths, thicknesses and lengths—all rough cut and in need of finish planing, dimensioning, and in some cases, drying. We were bemoaning our plight when two young entrepreneurs wandered into our office. They asked if we knew anyone who could use several thousand board feet of redwood and cypress beer-vat staves from the old Rheingold Beer brewery that was being dismantled in Brooklyn, N.Y. Well, yes, we probably knew someone. The wood reeked of stale beer, but it was superb for our purposes. It was straight, close grained and of course, well seasoned.

Before any shaping could be done, we had to prepare our stock. We thought that the wood might have nails hidden in it, but we found none. We did find metal flecks where the vat bands had deteriorated, but with wire brushes and large paint scrapers we removed almost all the rust. On our 16-in. radial-arm saw, we ripped the lumber to the rough sizes we needed, about ¼ in. thicker and ½ in. wider than the dimensions of the finished moldings. Next we prepared the stock on a jointer and a thickness planer. Once we had dressed down the old surfaces ¼ in., the wood was perfect and unmarked. As we worked,

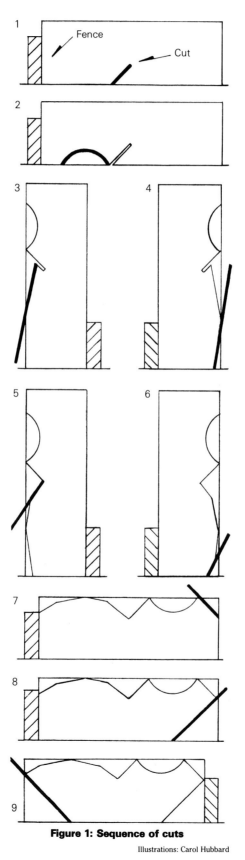

Figure 1: Sequence of cuts

Illustrations: Carol Hubbard

we checked our cutters for sharpness. Our stock was as straight and as square as we could make it; we were ready to begin shaping.

Setting up—Milling complex moldings on a table saw requires precision. Begin with an accurate template of the molding, to which you can adjust the sawblade's settings and against which you can compare results. The best template is a short piece of the molding you want to copy. If you must create a template from molding in place, you'll have to use a profile gauge. (See Figure 2 on the next page.) Many exterior moldings are too large to be handled with one application of the gauge. If this is the case with your trim, you'll have to take a series of readings, transfer them to paper and combine them for the complete profile. In fact, it's a good idea to sketch all molding profiles on site, for the gauges may get distorted before you return to the shop. Fashion your template out of a rigid material such as Masonite or plywood.

Before any cutting, even before setting the sawblade, scrutinize the template or molding cross section. The question is how to determine the order of the cuts. You don't want to take out a piece of stock you'll need later to run against the fence for making another cut. Think things through on a piece of paper. Certain cuts simply have to be made before others.

In Figure 1, for example, cut 1 is crucial because it is a dividing line between two curves: If its angle is incorrect or its cut misaligned, the proportions of both curves will suffer. If it is too deep, it undercuts the convex curve; if too shallow, material in the notch will have to be cleaned out later—a waste of time.

Cut 2, which creates the concave curve, must meet precisely the high point of cut 1. Because the stock is fed into the sawblade at an angle, this is a delicate cut.

Cuts 3 through 6, creating the convex curve, must be made after 2. If they had been made before 2, the convex curve would have made subsequent cuts a problem. (The stock could easily roll on that curve as it is fed into the sawblade.)

Cut 7 is delayed so that the point it creates with cut 2 won't be battered as the stock is maneuvered over the saw. Cuts 8 and 9 are made last, because leaving the corners of the stock square

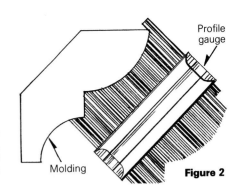

The Winooski Block (left) is capped by a cornice assembly over 6 ft. wide; it consists of 14 elements, including 7 moldings that were reproduced on a table saw. The milling of the molding is described on the facing page.

You need a template to mill new moldings. Use a piece of the original, or transfer readings from profile gauge to paper on site and cut a template later in the shop.

Figure 2

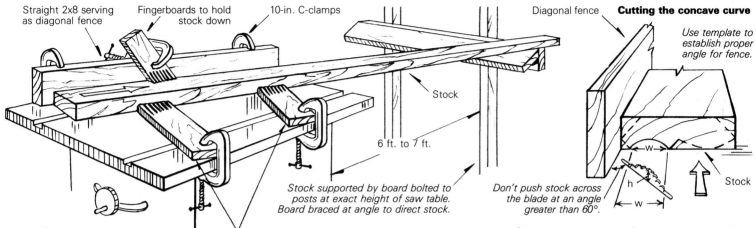

**Figure 3
Cutting setup**

Straight 2x8 serving as diagonal fence — Fingerboards to hold stock down — 10-in. C-clamps — Stock — Diagonal fence — **Cutting the concave curve**

Use template to establish proper angle for fence.

6 ft. to 7 ft.

Stock supported by board bolted to posts at exact height of saw table. Board braced at angle to direct stock.

Don't push stock across the blade at an angle greater than 60°.

Fingerboards to hold stock against fence

Concave curves may require several passes, starting with the blade set low. On the last cut, saw points should just touch curve outline.

ensures the stability and accuracy of preceding cuts. (Cuts 3 through 6 would have been almost impossible if cut 9 had preceded them.)

To save time, pass all molding stock through a given saw setting; be fastidious about such settings, making practice cuts on scrap work. Cut more molding stock than you'll need at each setting, so you'll always have waste stock with the necessary previous cuts. In other words, to get an accurate setting for cut 5, you'll need stock with cuts 1 through 4 already made.

Cutting—We used a 10-in. Rockwell Unisaw with a 48-point carbide-tipped blade for all molding cuts. For most cuts we used the rip fence provided by the manufacturer. For cut 2 however, we needed a diagonal fence, so we trued a 2x8, used a template to carefully set it at the proper angle for the desired cove, and clamped it to the table with 10-in. Jorgensen C-clamps (Figure 3). To reduce stock flutter we used fingerboards, pieces of wood with a series of parallel kerfs cut in one end. Two fingerboards clamped to the table held the stock against the fence, while one fingerboard clamped to the fence held the stock down. The kerfs allowed enough play to let the wood slide through, but maintained enough pressure to ensure a straight cut. Using fingerboards and extension tables, you could cut all the molding unassisted, but you may prefer to have a helper to pull the stock gently through the last few inches of a cut. Several times a day, wipe the tabletop and sawblade clean with turpentine, to minimize binding.

Except for cut 2, all cuts were made with the

rip fence running parallel to the blade on one side or the other. As shown in Figure 1, cuts 1 and 2 were made with the stock face down on the table, while cuts 7, 8 and 9 were made with it face up. The stock stood on edge for cuts 3 through 5. (When cutting some symmetrical convex shapes, you can leave the sawblade at the same angle, and after one pass, turn the stock 180° to get the cut whose angle mirrors the first.) Each cut was preceded by carefully adjusting blade height and angle against the template. The last cuts on a molding (cuts 8 and 9) should be slightly larger than 45°—if your sawblade will tilt just a little more—to avoid gaps where the building surfaces are not quite perpendicular.

We cut the concave shape (cut 2) into the molding by passing the stock diagonally across the table-saw blade. (See Figure 3, at right.)To set the blade and fence correctly, you'll need a piece of the old molding. (A template is less effective.) Holding the high point of the curve over the blade, slowly crank up the blade so that the tip of the highest tooth just grazes that curve's apex; lock the setting and try a few cuts. To create the width of the curve, angle the piece of molding until all the teeth of the exposed blade lightly touch the arc of the molding. Another person should snug the fence against the angled molding and then clamp the fence to the table while you hold the molding in place. You'll have to tinker a bit to get the exact angle you need.

You can create almost any symmetrical curve with this method. Pushing the stock across the blade at a wider angle will result in a wider curve. However, the widest angle at which we

would push wood across the blade is 60°; with wider angles, not enough of the sawteeth are gripping and the blade will bind. (I'm not sure why, but a 48-tooth carbide blade binds up less than an 82-tooth one. It may be that the chips clear more easily.) If the blade is binding, make several passes to get the curve, starting the blade low and cranking it up ¼ in. for each pass. Don't get so wrapped up in your calculations that you become careless. Keep fingers clear of the blade. The speed at which you feed the stock must be determined on the job: Too fast and the blade will bind, too slowly and the wood will burn. The greater the angle of feed, the more often you should clean the blade.

The quality of the wood greatly affects the complexity of cuts you can make. Hardwoods are more difficult to mill without proper equipment. If concave curves are possible at all on hardwood, you'll have to make many gradually increasing cuts; the angle of the stock to the blade will be limited. Fortunately the grain in our cypress varied less than ¼ in. in 15-ft.

To refine the shape of our convex curves, we used many tools, including jack planes, curved shavehooks and spokeshaves. Among power sanders, Rockwell's Speed-block was the favorite; we clamped the finished molding to benches and sanded it using 50-grit pads.

Using these techniques we milled 1,000 linear feet for each of nine molding types, some more complex than the one described above. ☐

Bruce Andrews is a partner of Moose Creek Restorations Ltd., in Burlington, Vt.

Batten Doors

Building a solid door from common lumber

by Bruce Gordon

A fine-looking batten door can be made from materials sold at any building-supply house, and can be built with limited funds and equipment. In the years when our business had no shop and little machinery, we produced custom batten doors at job sites, using only a table saw, an electric drill and a few clamps. They looked great, and were also competitive in price with factory-made doors.

Batten doors do have some inherent problems, though. Wood moves. A 36-in. door can vary as much as ⅜ in. in width between a dry winter and a humid summer. This will show on the side opposite the hinges, and the door that fits perfectly this winter may need to be planed down next summer and have its latch mortise reworked. The problem can be minimized by accommodating wood movement in the construction. Sealing the wood also helps, but if you use an oil finish, the door will move more than if you use varnish.

As a rule, batten doors do not stay perfectly flat and straight. They tend to bow across their widths and sag away from their hinges. The severity of these problems will depend on the species, grading, dryness and thickness of the lumber that you use, and how carefully you put the door together.

Boards 1 in. thick are best for interior doors, as are 1⅝-in. boards for exterior doors, although you can use ¾-in. tongue-and-groove stock for interior doors and 1½-in. stock for exterior doors. The batten should be 1¼ times as thick as the door body, and 6 in. to 8 in. wide.

To begin, select the stock and cut it to approximate size. If it is roughsawn, joint one face, thickness-plane, joint one edge, rip to width, cut to length, tongue or groove (or half-lap) each edge and do any decorative milling.

I edge-join the boards with dowels to keep the door from sagging away from its hinges and use a Stanley self-centering doweling jig to drill two holes for 2-in. long hardwood dowels in the mating edges of all the boards. The holes should be drilled level with the eventual position of the hinges—about 13 in. from the top of the door and 7 in. from its bottom. Dowel diameters vary with the thickness of the stock, but I use ⅜-in. dowels with ¾-in. boards, and ½-in. dowels with thicker stock. Once the holes are drilled, I apply a sealer (usually Watco or tung oil) and a first coat of finish or stain to the boards. On a door that will be painted, a coat of primer will do.

On a flat surface, assemble the door, inserting the hardwood dowels in the drilled holes. The dowels should not be glued, nor should the boards be pulled up tightly in the clamps. Instead, I insert strips of Formica between the edges to produce uniform gaps between the boards (photo facing page, top). The resulting gaps allow the wood to expand.

Being careful to keep the eventual location of hardware in mind, lay out the battens on the back of the door. Several pattern possibilities are shown in the drawing below. Check to be sure the door is square and flat, then clamp the battens in place. Battens should never be glued to the body. Attach them with metal fasteners so the wood can expand and contract freely.

I use four types of fasteners: rose-head clinch nails for their old-style look and ease of application, drywall screws for speed when I'm not concerned with the looks of the batten side of the door, wood screws, countersunk or counter-bored and plugged, and carriage bolts for the substantial look they give the face of the door. In any case, the batten should have an oversized hole, to allow the body of the door to move without bowing the door. Be careful to predrill even for clinch nails to avoid splitting the wood where the nail breaks through on the opposite side. Ideally, the clinch nail should be bent twice so that it penetrates back into the wood (drawing, facing page), ensuring a tight fit. However, it is common practice simply to fold the clinch nail flat. When the battens are secure, remove the Formica spacer strips.

A few tips regarding hardware may help you avoid frustration. Interior batten doors are usually too thin to take either a mortised latchset or a cylinder latchset. Consequently, you should plan on either a thumblatch or a rim lock. Batten doors are most often installed with a strap hinge, an H-L hinge or an H hinge.

If you don't want the traditional look such hinges give a door, you can use butt hinges. They should be sized so that the screws fasten to the edge of the door itself, not to the end grain of the battens where they won't hold. There are also offset hinges that can help you work around batten placements. Size the hinge so that its throw and the length of the batten allow the door to open 180° without hitting the casing. In some instances it may be best to hang the door from the casing, not the jamb, or use a half-surface hinge, one that combines a strap across the surface of the door with a butt plate mortised into the jamb. □

Bruce Gordon is a partner in Shelter Associates, a design and building firm in Free Union, Va.

Double batten   Z   X   Double Z   Double X   Triple batten

Illustrations: Susan Karzenski

Photos: Susan Mortell

Tongue-and-groove stock cut to approximate length is laid out across two sawhorses and clamped together flat and square. Two hardwood dowels inserted in holes in the mating edges of each board prevent sagging (detail, below). Above, battens are clamped to the back of the door. Formica strip spacers produce gaps that will permit the inevitable swelling. Battens should be fastened with either drywall screws, clinch nails or wood screws (left to right in bottom photo) or carriage bolts. Battens are predrilled for oversized holes, right, to allow for wood movement.

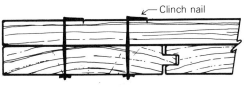

← Clinch nail

Making an Insulated Door

This handsome entry keeps heat in while it keeps breezes out

by Irwin L. Post

Well-insulated, tightly fitting doors reduce fuel bills and make a house more comfortable. I recently built one for a client in Weston, Vt., who wanted a weathertight door that would complement her post-and-beam house. Her interior doors are made from knotty eastern white pine tongue-and-groove paneling and have custom-forged strap hinges. We decided that the inside of the new door should have the same appearance. For the exterior, we settled on five vertical boards surrounded by a border, all of knotty pine (photo right). The hinges, latch and knocker were made by C. Leigh Morrell at the West Village Forge in West Brattleboro, Vt. For a door this thick, you'll either have to modify standard locksets or special-order commercial models. I silver-soldered an extension to the rod that turns the deadbolt.

The heart of the door is a core of pine spacers and 1-in. thick beadboard polystyrene foam (R-value: 3.3), sandwiched between two pieces of ½-in. plywood. The plywood facing stiffens the core and makes it dimensionally stable and resistant to warping. One advantage of using a structural plywood core is that it allows great freedom in designing the finished surfaces on both sides of the door. You can glue stock of any shape or thickness to the plywood skin, creating, for example, freeform or geometric patterns. The facing boards on both sides of the door are splined together. As the drawing shows, the exterior face is made from nominal 1-in. paneling, bordered with 1x stiles and rails. I applied ¾-in. trim strips to the vertical edges, and these are overlapped by the facing materials on both sides. I chose to make the core height equal to the full height of the door and not to cover the top and bottom edges of the core with trim strips, since they are not visible.

Construction sequence—First, determine the size of the door. I made mine ¼ in. shorter and ³⁄₁₆ in. narrower than the finished opening. Unless the door fits the opening precisely, air infiltration will render any amount of core insulation useless. Check to see that the opening is square. You can do this by measuring the diagonals—if they are the same length, then the opening is square. If it's not, you can either make the door fit the irregular opening or rework the jambs and casing to square things up.

Next, decide on the design for the door's inside and outside faces. It helps to make scale drawings to check the overall proportions of the various elements. Knowing the finished di-

Irwin Post

mensions of the door, use your drawings to calculate the dimensions of the beadboard core.

Now, cut the two pieces of ½-in. plywood to the same dimensions as the core (the full height of the opening less ¼ in., the width of the opening less 1¹¹⁄₁₆ in.). I used underlayment grade A-D ply with exterior glue. Cut 1x3 pine for spacers as shown in the drawing, step 1. For 1-in. beadboard, the spacers need to be a full 1 in. thick. Edge spacers can be made from short pieces if necessary, but they should be mitered and tightly butted to prevent air leaks. The diagonal spacers keep the door from twisting. The small blocks along the long sides are for extra reinforcement around the lockset. I glued a block on each side so I didn't have to keep track of which side would have the lockset and handle.

Glue the spacers to one piece of plywood. I used yellow glue. Be sure your worktable is absolutely flat, because a door built on a warped surface will be a warped or twisted door. A few 1¼-in. brads will keep the spacers from shifting while you apply clamps.

With a sharp knife or a razor, cut the beadboard to fit tightly between the spacers. Be sure

not to leave any gaps. Now glue the second piece of plywood onto the spacers to complete the core (step 2). Clamps on the edges and cauls (cambered strips of wood which, when clamped at their edges, will exert downward pressure along their lengths) across the width of the door are a good way to get a bond. Use brads to keep the plywood from shifting as it's being clamped.

Attach the trim strips to the vertical edges (step 3). On this door, the strips were ¾-in. thick pine about 2¼ in. wide. I glued them on one at a time, and used a 2x4 to distribute the clamping pressure, as shown in the drawing. After both edges are glued on, trim them flush with the plywood on both sides. I used a ball-bearing pilot flush-cutting bit in my router, but a hand plane would work just as well.

While the glue is setting, prepare the paneling for the faces. I used splines and grooves to join adjacent knotty pine boards (step 4). You can use a dado head in a table saw to plow the grooves, or a slotting cutter in your router. The splines were cut on the table saw. The V-grooves that show on the outside of the door were made by beveling the boards with a hand plane.

Glue the paneling to the plywood core. I used no nails or screws because I wanted to fasten the panels in a way that would keep the gaps between the boards constant and minimize any distortion, such as cupping, due to seasonal changes in moisture content. To do this, I glued the boards along the outer edges of the door on their outer edges, and the others with a narrow bead of glue along their centers. I glued one board at a time, starting from one edge, and I used cardboard spacers between the boards during assembly to create space for expansion. The unglued splines allow for movement.

The door is most easily finished with it lying flat on a table. I stained the outside of this door, and then gave all exposed surfaces three coats of polyurethane varnish.

This door is 3½ in. thick overall. Its feeling of mass and sturdiness goes well with the post-and-beam construction. The total R-value is 7.25, calculated by the methods in the USDA Forest Products Lab's *Wood Handbook* ($10 from the U.S. Government Printing Office, Washington, D.C. 20402). By comparison, a standard 1¾-in. solid-core exterior door has an R-rating of about 2.5, and double-glazed windows have an R-value of about 1.9. □

Irwin Post is a forest engineer. He lives in Barnard, Vt.

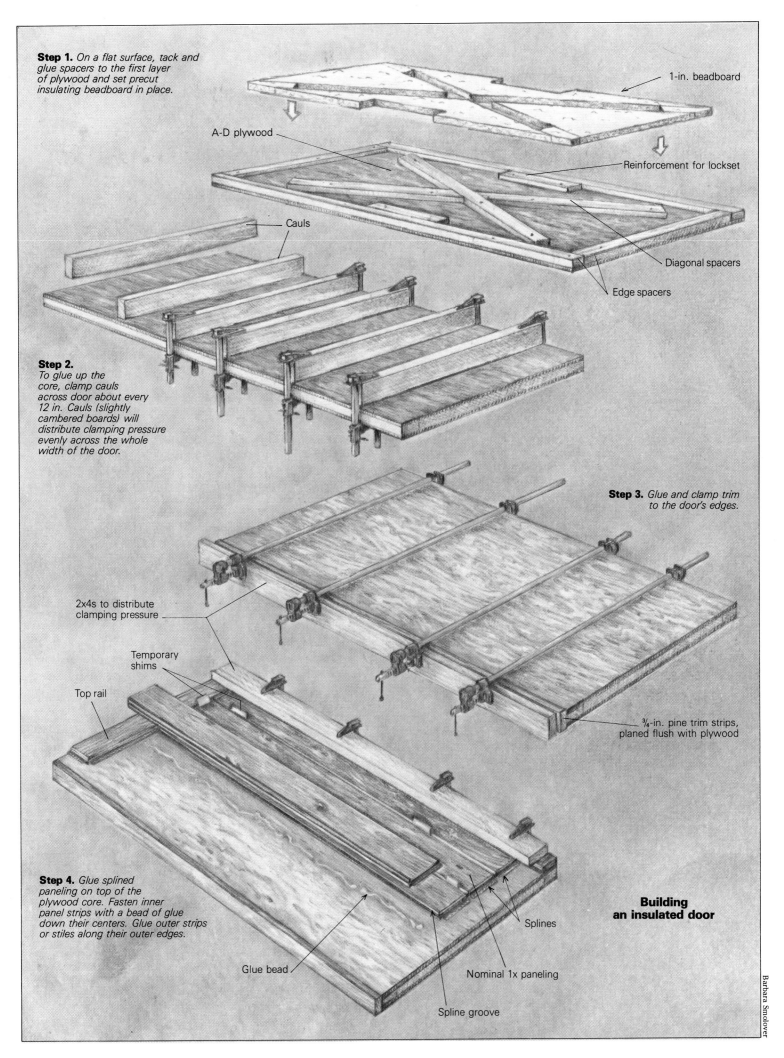

Step 1. On a flat surface, tack and glue spacers to the first layer of plywood and set precut insulating beadboard in place.

1-in. beadboard

A-D plywood

Reinforcement for lockset

Cauls

Diagonal spacers

Edge spacers

Step 2.
To glue up the core, clamp cauls across door about every 12 in. Cauls (slightly cambered boards) will distribute clamping pressure evenly across the whole width of the door.

Step 3. Glue and clamp trim to the door's edges.

2x4s to distribute clamping pressure

Temporary shims

Top rail

¾-in. pine trim strips, planed flush with plywood

Step 4. Glue splined paneling on top of the plywood core. Fasten inner panel strips with a bead of glue down their centers. Glue outer strips or stiles along their outer edges.

Splines

Building an insulated door

Glue bead

Nominal 1x paneling

Spline groove

Barbara Smolover

209

Hardwood Strip Flooring
How to lay out and install it

by Don Bollinger

Putting down a hardwood floor is one of the few kinds of woodworking that can be classified as both rough and finish carpentry. Anyone who has installed a ¾-in. oak floor will attest to the brute force required to drive home the nails, and to the telltale evidence of every mistake. And since all the work happens at ground level, a floor mechanic spends half his time on his knees and the rest stooped over like a field hand.

Still, the bundles of hardwood flooring that get broken open for each job contain some of the most carefully graded and highest-quality building materials available today. It's a pleasure to work with this stuff, and when wood this good is properly installed and cared for, it makes for a durable floor that can add a lifetime of color and warmth to any home.

Strip flooring—There are two basic types of strip flooring: tongue-and-groove (T&G) and square-edged. Some T&G strips are end-matched (drawing, above right), which means they're T&G on their ends as well as along their lengths. Strip flooring ranges from ⁵⁄₁₆ in. to 1½ in. thick, and from ¾ in. to 3¼ in. wide. Here in the Northwest, most strip flooring is red oak, white oak or maple, and the most common size, as it is throughout the country, is ¾-in. thick by 2¼-in. wide T&G. I like end-matched strips because they give the floor a bit more strength, and reduce bouncing where strips butt end to end between joists.

Hardwood strip flooring is further subdivided into random lengths and shorts, depending on quality. The better the grade, the

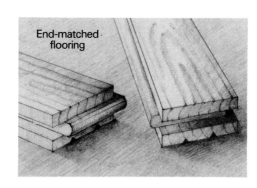

End-matched flooring

fewer the knots, sapwood and pinworm holes, and the longer the strips. For instance, the best grade of oak is called *clear*, and the strips are at least 1¼ ft. long, averaging about 3¾ ft. They are usually flatsawn, but the more attractive quartersawn strips can be special ordered. The poorest grade is *#2 common*. These strips, which average 2¼ ft. in length, have knots, checks and worm holes. When you order any grade, add 5% to 10% to allow for waste.

Precautions—Have your flooring delivered at least three days before you install it, and break the flooring bundles open so the boards can adjust to the ambient moisture. If the floor is going into a new house, make sure the place is sealed up, and that all masonry, plaster, drywall and other wet-process work is completely dry. Three weeks before the flooring is to be delivered, turn up the thermostat to your typical setting to help dry things out. If there is too much moisture in the house when

the floor is installed, the floorboards will shrink when the humidity drops, and you'll get ugly gaps between them. Moisture can also come up through the subfloor, so make sure the basement is dry and well ventilated.

Subfloors—A hardwood floor is no better than what's under it. Almost all squeaky floors, cracking finishes and other signs of early aging can be traced directly to inadequate underlayments. I've found it best to install a subfloor that's heavier than those generally specified in the building code (see the chart below). You will need an especially heavy subfloor if you're planning to lay your strip flooring parallel to the joists.

Plywood is the most common underlayment for strip flooring—I don't use particleboard because nails don't hold in it and water deteriorates it. For ¾-in. thick strips (like the ones we're talking about in this article), use ¾-in. plywood blocked under all edges, with a ⅛-in. expansion gap between each panel. Glue the plywood sheets down with panel adhesive and nail them with either hot-dipped galvanized or ring-shanked nails. This combination will reduce movement and squeaks.

If you use planks for your subfloor, they should be 6 in. to 8 in. wide and at least ¾ in. thick. They can be either square-edged or shiplapped, but shouldn't be T&G unless they are at least 1½ in. thick—anything less will flex and squeak.

Without a moisture meter, it's hard to tell whether boards will expand or contract. To play it safe, leave ⅛-in. gaps between your

Orientation of flooring	Recommended underlayments for tongue-and-groove hardwood flooring			
	Random-length strip		Shorts (18 in. or less)	Square-edge strip
	up to 3¼ in. wide, ¾ in. thick	4 to 8 in. wide, ¾ in. thick	up to 3 in. wide, ¾ in. thick	1 to 3 in. wide, ½ to ⁵⁄₁₆ in. thick
Strips running at 90° to joists set 16 in. o.c.*	¾-in. T&G plywood ¾-in. square-edge plywood ¾-in. by 6 to 8-in. plank or shiplap laid diagonally or two layers of ½-in. square-edge plywood with staggered seams	⅞-in. T&G plywood 1-in. square-edge plywood ⅞-in. square-edge plywood ¾-in. by 6 to 8-in. plank or shiplap laid diagonally or two layers of ½-in. square-edge plywood with staggered seams	1-in. T&G plywood or square-edge plywood Two layers of ½-in. square-edge plywood with staggered seams	1⅛-in. T&G plywood or square-edge plywood Two layers of ⅝-in. square-edge plywood with staggered seams
Strips diagonal or with joists set 16 in. o.c.*	1⅛-in. T&G plywood, or two layers of ⅝-in. square-edge plywood with staggered seams Bridging between joists			
* For joists set 24 in. o.c., add ¼ in. underlayment; for joists set 12 in. o.c., subtract ⅛ in. underlayment.				

subfloor planks. A standard ¾-in. plank sub-floor should be installed diagonally across joists spaced 12 in. to 16 in. o.c. Thicker T&G material may be laid at right angles to the joists. In either case, be sure that all underlayment planks butt end to end over a joist or blocking. Otherwise you'll get soft spots, fractured finishes and squeaks.

Whatever underlayment you use, be sure that its moisture content has stabilized before you install any hardwood over it. Don't use for underlayment plywood that has been used in concrete forms, because it has absorbed a lot of moisture and could delaminate.

Once the subfloor is installed, any differences in height between adjoining pieces of underlayment should be sanded flat with a belt sander using a coarse-grit belt. Scrape away any plaster or joint-compound lumps, then sweep or vacuum the subfloor. Prowl it one more time with a handful of 2-in. ring-shanked nails, and fix any humps and squeaks. Finally, mark the location of each joist at the base of the wall.

It's usually best to run the flooring in the direction of the longest dimension in the room. This will reduce the area over which the flooring will do most of its seasonal expansion and contraction. A vapor barrier between subfloor and finish flooring will keep out dust and moisture from below, help prevent squeaks and stop unwanted airflow through plank subfloors. I usually use three-ply resin paper, but 15-lb. to 30-lb. asphalt-saturated felt works just as well. Roll it out in the direction you want to run the strip flooring. If the floor is directly over a furnace or uninsulated heating ducts, add an extra layer of paper or felt to help keep the finish flooring from drying out.

Cover the length of the room where the floor-laying will begin with a few courses of VB paper or felt stapled to the floor. The baseline (see next page) is chalklined onto the felt. Subsequent courses needn't be stapled. Overlap the edges of the strips 3 in. to 6 in., and lay more down as the floor progresses, to keep traffic from tearing up the paper.

If the total thickness of the subfloor plus underlayment is ¾ in. or less, nail ¾-in. strip flooring into and between all joists. Snap chalklines between the marks on the wall so you know where the joists are. You'll have to restrike this line each time you lay a new strip of felt.

Moisture and movement—Like any wood, each piece of hardwood flooring expands across its grain as it picks up moisture, and contracts as it loses it. Given a large lateral area, the cumulative effect can cause a floor to

Installing a hardwood strip floor requires planning, especially in places like hallways, where misaligned boards are easy to see. Once the first few rows are nailed in place, the floor is racked, or loosely laid out, several rows at a time. When the flooring is extended into the kitchen (at right), a hardwood spline will connect the grooves of adjacent boards, allowing a change in the direction of the leading edge.

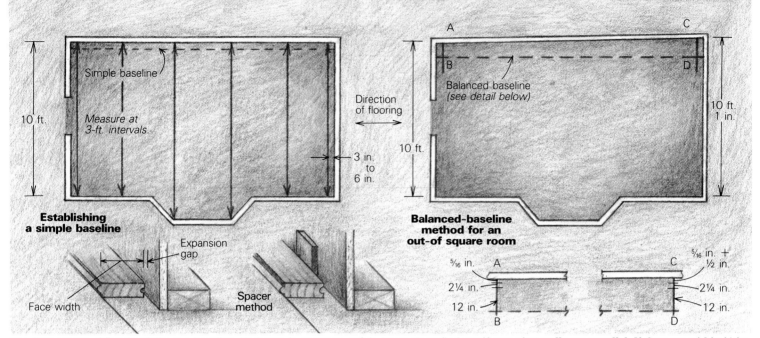

The first course of flooring follows a baseline. To locate it, measure at 3-ft. intervals to find out if opposing walls are parallel. If they are within ½ in., use a simple baseline, left. The balanced baseline, right, averages the error in a room whose opposing walls are more than ½ in. off parallel.

buckle. I've seen entire walls uprooted at the plate as a result of a leaky roof.

A little seasonal movement is normal. To allow for expansion, leave gaps between the edge of the floor and the wall. Since a baseboard or shoe molding will conceal this gap, many installers simply allow ¾ in. around the entire perimeter. A rule of thumb is to expect hardwood to expand and contract ¹⁄₁₆ in. for every running foot of cross-grain flooring.

Wood movement along the grain is negligible, but we leave about a ½-in. gap at each end wall. This space makes it easier to fit the boards together and allows air circulation.

Layout—Assume there is no such thing as a square room. I know I've never found one. The most carefully constructed rooms are usually out of square by ½ in. to ¾ in. These inaccuracies are easy to mask in the typical strip floor just by allowing for them in the expansion gaps around the edges of the room. Severe cases of out-of-square can call for a tapered floorboard hidden under a cabinet kickspace or a counter overhang, or a mask in the form of some type of border arrangement. In one terribly out-of-square room, I finally installed the flooring diagonally, after checking to make sure the subfloor would support it.

A good installer will spend as much as a third of the job sizing up the layout, installing the first few boards and anticipating and avoiding the glitch that would foul up the floor. Remember that some wall-to-floor intersections are in more visible locations than others, and that the baseboards and floorboards should be as close to parallel as possible. Converging floorboard and baseboard lines look especially awful in hallways. The error caused by out-of-square walls has to be dispersed gradually. Don't try to deal with it at a single intersection, where it will smack you in the eyes.

When you're installing strip flooring you need to worry only about the alignment of walls parallel to the flooring. Run a tape measure between these walls every three feet along their lengths. Start 3 in. to 6 in. away from the corners. They tend to flare out because of the way wallboard is taped. If there are more than two facing walls, make at least two measurements for each additional wall. If the opposing walls are less than ½ in. out of parallel, you can use a simple baseline to begin laying the floor.

Baselines—The baseline is a starting point for the flooring in a room, and is used as a reference device to keep straight, even runs of flooring. Since every installation is different, the location of the baseline is never quite the

same, but there are some guidelines to determine where to put it. It usually parallels the longest wall or longest uninterrupted run of flooring in the room. Given a choice between two walls of equal length, establish the baseline near the most visible wall.

In the simplest situation, you go to the end of one wall and pencil a mark on the vapor barrier. The distance from the wall to the mark is equal to the face width of one floorboard plus the expansion gap, as shown in the drawing, above left. Repeat the procedure at the other end of the wall and snap a chalkline between the two marks. This is the baseline. The first course of flooring is then face-nailed in place with the baseline as a guide. Begin nailing halfway along the length of each board, then work toward each end.

If the opposing walls are close to parallel, an easier way to start the first strip is to place removable shims equal to the thickness of the expansion gap against the wall. The first strip butts these shims, and is nailed in place. If you are using this method, be sure to sight along the leading edge of the flooring to make sure any dips or wows (depressions or bulges) in the wall aren't being telegraphed into the strips.

The balanced baseline—If opposite walls are more than ½ in. out-of-parallel, the baseline has to average out the discrepancy and distribute the error to both sides of the room. This is called a balanced baseline.

In the drawing above right, one wall is 10 ft. long and the opposite wall is 10 ft. 1 in. Start at point A and measure out 14 ⁹⁄₁₆ in. to find point B. This dimension is the total of the expansion gap (⁵⁄₁₆ in. per side for 10 ft. of cross-grain floor) plus the flooring face width (2¼ in.) plus a 1-ft. workspace. At point C, add ½ in. to the measurement to account for half of the 1-in. discrepancy. Mark point D. This total measurement is 15¹⁄₁₆ in.

Now snap a chalkline between point B and point D, and forget about any other reference

Chart is adapted from the National Oak Flooring Manufacturers Association line. Measure back 1 ft. to locate the leading edge of the first piece of flooring.

Sometimes a balanced baseline alone won't solve a special problem. We once had a job in which a stairwell, in a run of only 2 ft. 4 in., wandered ⅜ in. out of parallel. We decided to compensate for the error with two different steps. We added ½ in. to one end of the baseline measurement, bringing the stairwell and baseline a little closer to parallel, then we took up the rest of the error by slightly skewing the nosing around the stairwell. This left that opening slightly out of square, but no one without a tape will ever notice.

Installing the first boards—Sort through your longest bundles and find enough straight pieces for two starter and four finishing rows. Check them by sighting down the boards as though you were looking for a good pool cue. Using bent or crooked pieces to finish out a floor usually results in ugly gaps in the last few rows. It's hard enough to pull the flooring up tight next to a wall without having to fight recalcitrant boards.

For right-handers, it's easier to work from left to right, so most floors begin at the left-hand corner (drawing, top right). Set the first strip squarely on the baseline ½ in. out from the wall; the tongue side should face you, with the end-matched groove to your right.

Align the leading edge of the groove side with the baseline. Starting in the middle of the board's length, about ½ in. from the grooved edge, face-nail ¾-in. boards with 8d cut nails or 2-in. power-driven fasteners (the chart on the facing page gives other nailing specs).

We use a power nailer to set the boards (photo right). It fits over the leading edge of the board, and with one blow of the mallet to the nailer's plunger, the tool draws the board tight against its neighbor as it drives a barbed fastener through the top of the tongue and into the subfloor at a 45° angle. We use a power nailer designed for face-nailing in tight spots near walls.

Power nailers are fast and easy to use. They lessen the chance of dings from misguided hammer blows. You can rent one from a tool-rental yard or a flooring-supply center, or buy your own. Two companies that make them are Porta-Tools (Box 1257, Wilmington, N.C. 28402) and Power Nailer (Power Nail Co., Rte. #22, Prairie View, Ill. 60069).

Place the nails 8 in. o.c., and make sure every other nail sinks into a joist. At board ends, drill ¹⁄₁₆-in. pilot holes to keep the boards from splitting. Blind-nail the same number of fasteners through the board's tongue, taking care to avoid knocking the board out of alignment. Set all nails below the face of each board.

Select strips for the second row that won't end within 3 in. of the butt joints in the first row (drawing, center right). You should stagger butt joints this way throughout the floor. Tap the boards in place using a buffer

One blow to the power nailer is enough to pull the flooring tight against the previous course and set a barbed nail.

Illustrations: Frances Ashforth

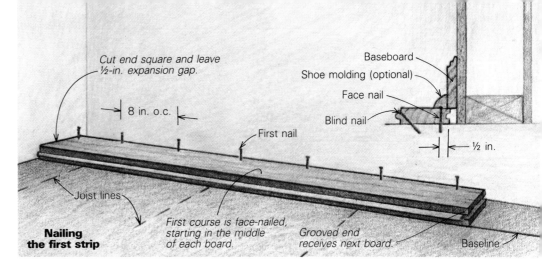

Cut end square and leave ½-in. expansion gap.

8 in. o.c.

First nail

Joist lines

Baseboard

Shoe molding (optional)

Face nail

Blind nail

½ in.

Nailing the first strip

First course is face-nailed, starting in the middle of each board.

Grooved end receives next board.

Baseline

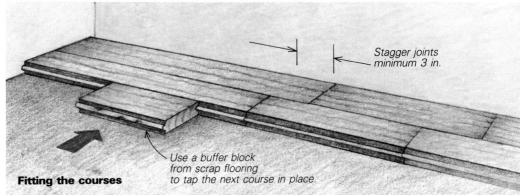

Stagger joints minimum 3 in.

Fitting the courses

Use a buffer block from scrap flooring to tap the next course in place.

Two header boards intersect in a mitered corner. The door casing has been trimmed off at the bottom to allow the flooring to slip under for a finished look. The tape measure registers the distance from the board's face edge to the baseline.

At a doorway, a header board is screwed in place 8 in. o.c. with #10 2-in. square-drive screws. Hardwood plugs, with their grain turned at 90° to the grain of the header board, will fill the counterbores. The first course of flooring will be mitered to meet the edge of the header board.

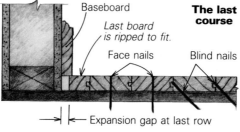

The last course

Baseboard

Last board is ripped to fit.

Face nails   Blind nails

|← Expansion gap at last row

The last course. Strips are levered into place, above, and face-nailed with a power nailer.

At row ends, left, boards are marked for length by eye and cut off at 90° with a power miter box. Trimming this board will remove a flaw, making it a good candidate for a row-ending piece. Another common practice is to pick end pieces that leave cutoffs at least 6 in. long, which are used to start subsequent courses.

board—a short offcut used to protect the board from hammer blows while positioning it—and blind-nail them in place. In most cases you'll get a snug fit with little effort, but occasionally a crooked piece will require levering with a prybar or a screwdriver. Before you exert too much force, check to see if a board is hung up on a cracked tongue, on an unset nail or on a small bit of wood.

Once you've gotten this far, you can move right along. I like to keep my power miter box just behind me and to my right as I'm nailing. It's easy to reach there, and the sawdust it makes is away from my clean vapor barrier.

Once the first few rows are nailed down, we "rack" the floor. This means laying out several loose rows of flooring a few inches ahead of the last row installed (photo p. 211). We place the pieces so the butt joints are as far apart as possible, and so that short pieces get evenly distributed. We use long strips across doorways and down halls, and save the shorts for closets, bathrooms and other confined areas.

Unless a precise cut is needed, we don't measure cutlines at the end of a row. Instead, we hold the board against the wall so that its tongued edge faces the nailed-down rows and eyeball a cutline (photo bottom left). When the board is drawn tight into its end-matched groove, there is a ½-in. gap at the wall.

If the flooring butts something that isn't already in place, such as a threshold, hearth or finished trim, we work to a temporary straightedge, start the flooring at it and face-nail the ends of each board.

Header boards—At doorways and passages, we install header boards for decorative effect. They're wider T&G planks than the rest of the flooring, and they're screwed to the subfloor rather than nailed (photos facing page, top). Instead of blending into the flooring with a butt joint, they join at a 45° angle. Their countersunk screw holes are plugged using the same wood, with the grain turned 90° for a subtle highlight.

You can't leave an expansion gap when you install header boards. Instead, we add 50% to the minimum gap at the opposite wall.

Check the alignment of the rows as they march across the room to make sure the occasional dip or wow isn't becoming amplified. Usually they can be handled with a buffer board and a small sledge, but sometimes a board will need a little wood removed. I use a plane to take material off the groove side of the strip to match fluctuations in neighboring boards. Cull out the really distorted pieces—they're just too much trouble to deal with.

When you reach the opposite wall, you'll need to face-nail the last few rows while they're held tightly in place. Don't do more than two rows at a time. Get out those straight boards you've been saving and use a piece of flooring held against a block on the wall to lever them into position (photo facing page, bottom right). Position the force of the lever over a wall stud, as shown, and use a block long enough to span two studs to keep from damaging the wallboard.

Once the last full row has been nailed down, you'll probably need to rip one last course to finish the floor (drawing, facing page). Occasionally a ripped piece will be too narrow to nail without splitting. In such a case, I glue the narrow piece to the last full strip with yellow glue and temporarily wedge it in place until the glue dries.

Stairwell nosing—When hallway flooring is nailed off, we take another set of measurements to see how close the strips and the stairwell are to being parallel (photo above right). If things are a bit off, we pencil-mark the necessary adjustment on the subfloor and adjust the nosing strips to compensate for the error. Then we temporarily screw them in place. When it's time to install the notched floorboard (photo below right), we remove the nosing so that we have room to tap the grooved edge of the flooring into place. If there's no nosing to adjust, we dress a slight taper into a number of strips with a plane to hide the error.

Finishing up—We wait two weeks before we go back to a floor to sand and finish it. Here in the humid Northwest, this allows the newly installed floor to acclimate itself to its new surroundings, and to adjust to the stresses brought on by being nailed and screwed to a flat surface. After sanding, we stain the floor and finish it. And that's another story. □

Don Bollinger owns the Oak Floors of Greenbank in Seattle, Wash.

The distance from the advancing floorboards to the stairwell nosing is measured to check for parallel, above. Once this is established, the nosing is cut to length and screwed in place. In the finished floor, below, the strip next to the angled nosing piece required notching. The nosing was removed to let in the strip, then replaced. The countersunk screw holes were plugged.

Floor Sanding

Follow a pattern with powerful machines for best results

by Don Bollinger

Sanding a floor is nasty work. It's noisy and dusty, and it takes a fair amount of practice to do it right. It's one of the most frequently subbed-out jobs. But a lot of people are willing to take it on for the first time, as owners of tool-rental shops will readily attest. Here are some tips, and a few precautions.

Preparation—Take everything that isn't nailed down out of the room. Cover the built-ins with plastic sheeting, and tack sheets of plastic across any doorways. Rolled-up towels laid against the bottom of the far side of doors in rooms being sanded will also help to contain the dust.

Sweep the floor clean and set any protruding nails at least 1/16 in. below the surface. Repair loose boards or squeaks with nails driven into joists. If there's room under the floor, I like to fix squeaks by driving screws through the subfloor to draw the hardwood tight against the underlayment. These screws have

Don Bollinger owns Oak Floors of Greenbank in Seattle, Wash.

to be 1/4 in. shy of the total thickness of the subfloor and the flooring.

You'll need a dust mask while you're sanding. You may want to wear some ear protection, too. Wear shoes that don't have crevices in the soles that can pick up grit. Sneakers or running shoes are good, but avoid the ones with black soles—they can leave scuff marks that are hard to remove.

The tools—It takes two basic kinds of sanders to finish (or refinish) a floor: a drum sander and a power edger. Both are available at tool-rental shops, and both require some muscle and practice to use correctly.

The drum sander, or floor sander, is used to sand most of the floor (photo facing page, top). It is a formidable machine. Even the smaller versions weigh about 125 lb., and they look like a cross between a lawn mower and a steamroller. They need their own 15-amp circuit to operate an 8-in. drum that rotates at about 5,000 rpm. When this drum is fitted with coarse-grit paper and lowered onto the floor, it wants to take off like a dragster. If you

hold it in one place, it is inclined to eat its way through the floor and into the basement. It's not a machine to be taken lightly.

But it's the only tool for the job, and with some practice, an operator can develop the required light touch. The first-time user should practice on a section of floor that won't be in direct view. Try a bedroom floor or part of a room that will be covered with a rug. And sand only with the grain.

As in any sanding job, you start with coarse-grit paper and work up to the fine grit. It takes quite a few sheets of at least three different grits to do an entire floor. Sheets are sold at the rental shops, and you can generally return the ones you don't use. Take plenty.

If you are refinishing a floor that's covered with paint, begin sanding with a very coarse paper—12 to 20 grit. For a new floor, start with 24 to 40 grit.

The drum sander is designed to make a slightly deeper cut on the left side of the drum (drawing facing page, top). This delicate angle allows you to feather the edge of the cut on the right side. To benefit from this feature, you should start sanding on the right side of the room, and work toward the left.

Begin about a third of the way up the floor (drawing facing page, center), and gradually lower the drum to the floor by letting up on the handle. It's important that you walk forward as you do this so the drum won't dig in in one spot. The weight of the machine will do the cutting. You want to make sure that the drum smoothly engages and disengages with the floor.

As you near the wall (about 1 ft. away), begin lifting up the drum, and then lower it again as you back up over the same path. You're towing the sander now, and this is when it does its best cutting.

Move to the left in 2-in. to 4-in. increments, making a forward and a backward pass over each section. When you've covered two-thirds of the floor, go to the left wall, turn around and sand the remaining third in the same manner. Take care to feather the slight ridge where you changed direction.

Sometimes a floor will be so uneven that it has to be sanded diagonally to the strips. Do this very carefully, and only with the coarsest paper. Start in one corner and move from right to left until two-thirds of the floor is covered. Go to the opposite corner, reverse direction and finish the remaining third. Then

sand the entire floor in the direction of the grain with the coarsest paper.

Since the drum sander can't reach in close to walls, corners and other tight areas, you'll have to sand these surfaces with an edger (photo bottom right). This powerful disc sander has grips for both hands built into the body. When it's tilted back on its wheels, the disc is lifted off the floor. When allowed to tilt forward, the machine begins its work. Like the drum sander, it is a difficult tool to use correctly without some practice. Try it out where you can't do too much damage.

The edger has a light mounted on the front of its chassis that helps you see what you're doing. If you're renting one, check to see that the bulb works—frequently they don't.

Unlike the drum sander, the edger makes its deepest cut on the right-hand side (at about one o'clock), and should be moved across the floor from left to right. A standard pattern for moving the tool is the semicircular path shown in the drawing, bottom right. There are other ways to operate an edger, too. Do whatever works best for you. Sand the areas missed by the drum sander with the same or a slightly finer grit.

When the entire floor has been sanded with coarse-grit paper, fill any holes in the floor. I use a lacquer-base filler blended to match the species of wood that I'm finishing, and I spot-fill nail holes and cracks between boards with a putty knife. If it's a top-nailed floor with a lot of nail holes or a parquet floor with numerous gaps, I trowel on the filler with a concrete trowel and wipe away the excess with a burlap rag. When the filler has dried, the floor can be sanded with medium (40 to 60-grit) paper. When you've finished with the drum sander and edger, check for shiners—nailheads turned silvery from being sanded. Set them, and refill the holes.

Final sanding should be done very carefully with 80 to 120-grit paper. Feathering is most important now because any ridges will show in the finished floor.

When the floor has been completely drum-sanded with the last paper, clean up the corners. For this task, the professionals I know use a common paint scraper, sharpened to a razor edge with a file or stone. Most of the time you should pull the scraper with the grain, but for hard-to-reach crannies you may have to work it at a 45° angle. This is acceptable practice, but never pull across the grain. When the scraping is done, hand-feather the perimeter with a sanding block wrapped with the final grit.

Cleanup—Sweep the ceiling, walls and floor as clean as you can with a good broom. Next, lightly dampen a medium-sized towel with paint thinner, lacquer thinner or alcohol, wrap it around the broomhead, and go over the entire floor with it. This is called tacking the floor, and it will collect most of the fine dust that still remains. Thoroughly vacuum the edges and corners using a crevice attachment, and you're ready to apply stains, sealers and finishes.   □

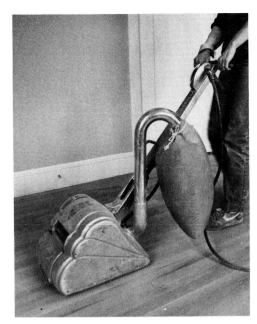
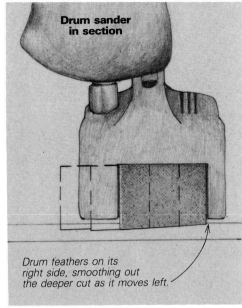

Drum feathers on its right side, smoothing out the deeper cut as it moves left.

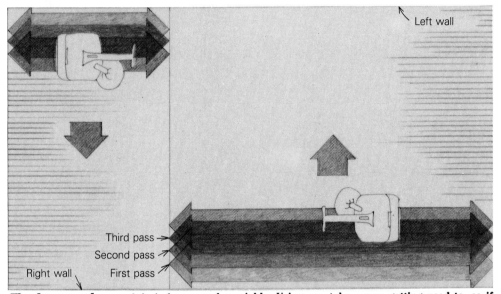

The drum sander, top left, is heavy and unwieldy. Using one takes some getting used to, so if you'll be doing your own floor, practice in an area that will be out of sight. When you're ready, begin sanding a third of the distance up the floor, as shown in the drawing, above. Push and pull the sander over the same area before advancing 2 in. to 4 in. for the next pass. When you've done two-thirds of the floor, reverse direction and sand the rest of the floor in the same manner.

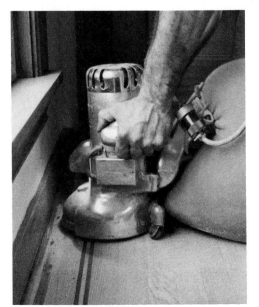
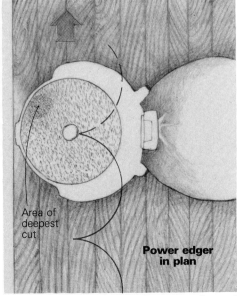

The power edger sands the spots you can't reach with the drum sander. Its heavy metal guard keeps the disc from touching the baseboard. A looping, semicircular motion, as shown at right, will keep the edger moving in the correct direction and prevent it from gouging the floor.

Connecticut River Valley Entrance
Reproducing a famous 18th-century doorway

by Gregory Schipa

This past spring we were moving into the last stages of the restoration of the Burley house, a beautiful Georgian structure built in New-market, N.H., around 1710. It had been dismantled, moved to Riverton, Vt., and re-erected on a new site.

Although we had found a Federal entrance on the building at its original site, framing mortises in the structure left no doubt that the house had once boasted double, 24-in. wide doors. As part of the restoration, the owner wanted us to recreate a Connecticut River Valley broken-pediment entrance. Since these beautiful entrances became fashionable around 1750 in the Newmarket area, she surmised that the old Burley house might have gotten one as part of a facelift to reflect the family's increasing affluence. The proportions of the building were right, and I was easily persuaded. After all, a job like this is a joiner's dream. We decided that it would be best to join such an entry by hand in the original manner. We had only to do a little research and then find our favorite example.

Development of the style—In 1636, settlers of the Massachusetts Bay Colony began to move westward from Boston to the Connecticut River Valley. Preferring the isolation and dangers of the wilderness to the overcrowding and religious intolerance in the Boston area, they first settled Springfield, Mass., and then Hartford, Windsor and Wethersfield, Conn. The migration continued for another 100 years, extending up the Connecticut River to New Hampshire and Vermont. It was not long before these pioneers began adding distinctive touches to the houses they built.

Most of the early work in the Colonies had been Jacobean in style, and was executed by English craftsmen, but the Connecticut River Valley towns soon developed their own unique details and embellishments, while staying with the same basic house plan. With the start of the Georgian Period, change accelerated. It can most easily be traced in the treatment of the front entrance, which rapidly evolved from a simply framed batten door through embellishment with the common architrave, transom and crown to the introduction (about 1700) of pilasters, double doors and the classic three horizontal members: architrave, frieze and cornice. Connecticut River Valley craftsmen lavished ever-increasing creativity on these entrances, apparently in celebration of the lessen-

The historic Burley house was moved from its original location to central Vermont, and its new entry was modeled after the one on the Rev. John Williams house in Deerfield, Mass.

ing dangers and growing wealth and freedom of the period. Soon cornice members evolved into full pediments, and before the middle of the 18th century, these elaborate pediments were further embellished by being broken and ended with volutes or rosettes.

Finding our model—Some of the very best examples of the Connecticut River Valley entrance can be found at Old Deerfield Village, Mass. Perhaps the best known of these is on the Rev. John Williams house (see p. 221). We chose to reproduce this particular entrance because its door opening was the same size as the Burley house rough opening. Also, the dimensions of the two houses were almost the same, and the window placements were simi-

lar. The Williams entrance also seemed to represent the best of the Connecticut River Valley pilastered doorways, with an early Jacobean feeling reflected in its steep, handcarved moldings (very few of which are classical), and its high raised-panel tombstone pedestals. Its massive broken-scroll pediment is very steep and abrupt, and its double pilasters are deeply fluted, with carved rosettes in the necking. It is perfectly balanced and well detailed. We hoped we could match the work of the 18th-century craftsman who built it.

Getting started—My foreman, Richard Tintle, and I first went to Deerfield to measure, draw and photograph the entryway. We figured it would take the two of us at least four weeks to do the job. We built the entrance primarily of 5/4 and ¾-in. clear local white pine. The stock had to be clear because we were cutting all moldings by hand. We could usually get two moldings from each 1x4 board, but cost was very much a factor, so we tried to use every extra strip of pine for the smaller moldings.

Very little was necessary to prepare the building to receive the work. The size of the rough opening was already dictated by the double stud mortises in the frame, and at 51 in. wide by 80 in. high, it was almost exactly the same as the one in the Williams house. The windows in the Burley house were far apart, and we had sheathed with ⅝-in. plywood, so we had a large, flat, blank space on which we drew the outline of the entire entrance. This gave us an immediate feel for its impressive dimensions, and made it easy for us to figure how much wood we would need for our first step, the underlayment. We also rigged a tarp so the entry would be under cover until it was completed and primed.

The underlayment, more properly called the clapboard catch, consists of the boards to which the entire built-up entrance detail is attached. We used two 8¾-in. wide boards of nominal 1-in. stock on each side, planning to hide the seam under the pilasters. Two horizontal 11½-in. wide boards across the top extended above the eventual level of the drip cap (the horizontal member of the cornice), and high enough to accept the bottom of the broken-scroll pediment. Two more 1x12s were later used to back the pediment and match its curve. All the boards used for the clapboard catch were joined by hand and jackplaned. The Williams entrance had been further embellished with quoin work (carved V-grooves imitating masonry joints). On our project, we carved these grooves with a V-gouge.

Pedestals and pilasters—Once we were done with the clapboard catch, we began working from the ground up. The base of a pilaster is the plinth block. Because it is close to the moist ground, it is usually the first part of an entrance to rot out. This had happened on the Williams house, where the original plinth blocks had been replaced very shabbily with old rough-cut studs. We used solid sugar-pine blocks on the Burley house, hand-dressed to 7 in. high, 12 in. wide and 2¾ in. deep.

The pedestals for the pilasters rise immediately from the plinth blocks. They are doubled, the under-pedestal being 2½ in. wider than the one on top. They both called for carved quoin work, and the surface pedestal on each side is a tombstone panel (photo right). We were surprised to observe that on the Williams house these little panels were not joined in the customary manner, with stiles and rails, but were actually carved 1-in. boards. We followed suit on our bench, using butt chisels and a V-gouge.

The double pilasters themselves represented quite a challenge. Both the pilasters and under-pilasters were fluted, and their entasis, or taper, had to be reflected in the fluting. To do this, we gradually raised the plane to reduce the depth of each flute toward the top. This also had the effect of reducing its width of cut. We found the Stanley #55 multiplane (photo below right) handy for this job, but its fence could not be used. We had to fabricate one that would follow the taper. The final 16 in. or so of each flute had to be carved with a U-gouge after every few passes, or the plane would ride up. I also clamped a small tab of sheet metal under the stop-block, extending from it to the end of each flute. This protected the flat area beyond the flutes from being scarred as the plane's bottom passed over it.

The fluted part of the pilasters is about 56 in. long, but the pilasters themselves extend to about 77 in., past the capital molding group and carved necking, under the architrave and frieze, finally ending at the bottom of the cornice, where the molding of dentiled corona steps around its top. The pilasters taper from 7⁷⁄₁₆ in. at the bottom to 6½ in. at the top. The under-pilasters are proportionately wider, reflecting the same shadow-line as the pedestals. Pilaster and under-pilaster each project 1¹⁄₁₆ in., and they are both fluted on their sides as well as on their fronts. Before finishing the pilasters, we laid out and carved the 6-in. dia., six-petal rosette on the necking of each. When we finished this part of the project, we had already worked 70 man-hours.

The moldings of the pedestal cap and those of the pilaster base form a single, solid group. They are quite primitive, mostly a series of hollows and rounds—early Jacobean rather than classical. Nevertheless, they have considerable impact because of the depth to which they're cut and the distance they project. We cut out these and other moldings with antique planes (photo top right) and the Stanley #55. We found

The tombstone panel of one of the pilaster pedestals is nailed up, below. The under-pedestal extends 1¼ in. out from each side of the face panel, and both sit on the solid sugar-pine plinth block. Two 8¾-in. boards make up the underlayment, or clapboard catch, to which everything else is attached. All the elements have been gouged to resemble stonework.

Antique planes, like the one above, are profiled to shape a specific molding. They came in handy on this job, where so many traditional shapes were required. A Stanley #55 plane (below) was used to cut the pilasters' fluting and, as shown here, to shape many of the entryway's moldings. This tool, designed for joiners, accepts blades of many different types and patterns.

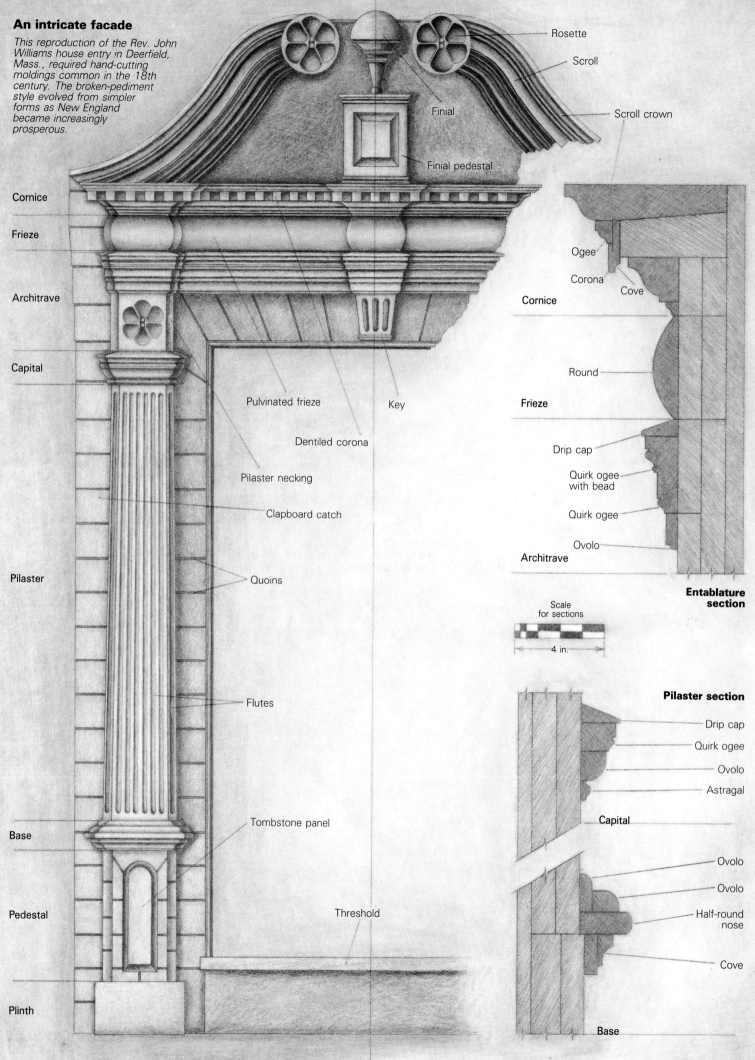

An intricate facade

This reproduction of the Rev. John Williams house entry in Deerfield, Mass., required hand-cutting moldings common in the 18th century. The broken-pediment style evolved from simpler forms as New England became increasingly prosperous.

Rosette

Scroll

Scroll crown

Finial

Finial pedestal

Cornice

Frieze

Architrave

Capital

Pilaster

Base

Pedestal

Plinth

Pulvinated frieze

Key

Dentiled corona

Pilaster necking

Clapboard catch

Quoins

Flutes

Tombstone panel

Threshold

Cornice

Ogee

Corona

Cove

Round

Frieze

Drip cap

Quirk ogee with bead

Quirk ogee

Ovolo

Architrave

Entablature section

Scale for sections

4 in.

Pilaster section

Drip cap

Quirk ogee

Ovolo

Astragal

Capital

Ovolo

Ovolo

Half-round nose

Cove

Base

Illustration: E. Marino III

The perilous history of the Rev. John Williams and his house

In histories of Deerfield, Mass., the Rev. John Williams is known as "the redeemed captive." His house was burned to the ground and several of his children were killed during the French-fomented Indian raid of 1704, and Williams was carried off a prisoner to Canada. His wife was murdered on the march, and his 10-year-old daughter Eunice was adopted by the Caughnawagas. She converted to Catholicism and eventually married an Indian named Amrusus—a process that one partisan chronicler called "her lapse into barbarism." In 1706, Williams himself was exchanged for a Frenchman held prisoner by the English. His loyal congregation enticed him back to the wilderness from the safety of Boston by building him a new house in 1707.

When John Williams died, the house passed first to his second wife, and then, upon her death in the 1750s, to their son Elijah. A subsequent owner of Elijah's house bequeathed the structure to Deerfield Academy in 1875.

In the late 19th century, old houses were often destroyed with little regard for their historic value, and the school planned to raze the house to make room for a new building. But George Sheldon, president of the Pocumtuck Valley Association, saved the Williams house—and its glorious entry—by mounting a campaign that linked it closely with the Rev. Mr. Williams, the most romantic figure in the town's history. As researchers Amelia F. Miller and Donald L. Bunce put it, "With skillful omissions and careful wording, he allowed his readers to believe that this house was the one built in 1707." It wasn't. Miller and Bunce say that the 1707 house was pulled down, and a new house built in 1759 or 1760. Elijah had become a man of means and wanted a grander residence.

Miller and Bunce also found evidence that the great broken-pediment entryway was built by Lieutenant Samuel Partridge in the summer of 1760. Partridge had been with General Wolfe at Quebec, and was related to Elijah by marriage. He must also have been a master of his trade who could demand top dollar. Elijah's account book for September shows a disbursement of £39/0-2 (39 pounds, two pence) in an era when journeyman carpenters were making only a few shillings a day.

—*Mark Alvarez*

that applying molding around a double pedestal and pilaster requires three times the miter work as on single elements.

The pilaster capital molding group of the Williams entrance lines up just below the head jamb. It is steep and simple, which is appropriate to the period, but its moldings are more recognizably classic, with a ⅞-in. astragal, a 1¼-in. ovolo and a ⅝-in. ogee. The group has its own beveled drip cap, cut from a ¾-in. board and shaped with a bench plane.

The architrave—We thought we had done a few miters up to this point, but as we worked on the 95 pieces of the architrave band of moldings, the word began to take on a new mean-

ing. The band has five components (including a spacer), and it steps around both double pilasters and the tapered key. The key, with its own handcarved flutes, extends from the door head to the bottom of the cornice, and matches the size and taper of the quoins. The architrave band is 5⁹⁄₁₆ in. wide, made up of a 1⁹⁄₁₆-in. bottom piece cut to a ¼-in. ovolo molding, followed by a 1¾-in. piece cut to a ⅝-in. ogee, and then a 1½-in. ogee with a bead. The full band has its own ¾-in. tapered drip cap, like the ones on the pilaster capitals. The architrave is ½ in. deep at its bottom and 1⅝ in. at its top—considerably less projection than the pilaster base and cap moldings.

Immediately above the architrave, and sit-

ting flush upon its drip cap, the craftsman at the Williams house had fashioned a rounded, or pulvinated, frieze. We found that we could fabricate this element with a large hollow, a tool that was specifically made for the pulvinated molding, or with a bench plane, scratch-stock and sandpaper. The inaccuracy inherent in the handmade process could easily be worked out with a little additional sanding after the frieze had been mitered and fitted.

Sitting upon the frieze of the Williams entrance, and fitted against the bottom of the cornice like a bed molding, we found the classic 18th-century crown molding. Consisting of a small ogee on the bottom, with a bead above and a large reverse ogee on top, this molding

The mitered moldings of the entry required painstaking sanding. At left, Richard Tintle works on the crown molding beneath the cornice. The clapboard catch, grooved to resemble quoins, is visible to the left of the pilaster.

The scrolls (bottom left) were glued up from sugar-pine planks, and the curves were smoothed with an antique compass plane. On the facing page, the completed pediment.

The finial between the scrolls sits on a paneled pedestal, built as a sturdy box. The side panels are flat, but the front panel, below, is carved. The box will get a hipped top.

(photo above left) was one of 18th-century New England's most consistently popular architectural embellishments. We nailed it in place before we put up the frieze. We were very lucky to own the appropriate antique plane, because its use was the key to cornice and pediment detail on this particular entrance. The crown molding is repeated on the broken scroll, where it is mitered into the small ogee on the cornice fascia.

The pediment—A hundred and twelve man-hours after leaving to measure the famous entrance in Old Deerfield, we were ready to start the pediment. We had made and fitted 270 pieces up to this point. But the craftsman who worked on the original had saved his most creative efforts for last. The massive cornice, rising in steep reverse curves to form the broken pediment, and terminating in bold carved rosettes, is one of the most impressive details of 18th-century domestic architecture. We knew it would be a challenge to reproduce.

The cornice structure begins with the usual horizontal element, which serves as the drip cap for the entrance below. On our entry, we reproduced this with a 2-in. thick, 5-in. wide sugar-pine plank, which we beveled with a jack plane to shed water. We added 1-in. blocks to simulate the step-outs of the pilasters and the key below. The fascia applied to this plank has a $\frac{5}{16}$-in. reveal on the bottom, and a small $\frac{1}{4}$-in. cove cut into its back below that. Its front bottom, also called the corona, has small dentil-

Jim Eaton

like cuts, 1¼ in. square. These can be cut with a fillister plane, or a rabbet plane with a fence clamped to it. It is then topped with a ⅝-in. ogee molding, which we mitered into the bottom ogee of the crown molding, on the return of the broken-scroll pediment to the house.

We laid out the scrolls using a string and a pencil on a very large table. This was hard because we had only our photographs as guides, and we wanted to match the proportions of the original builders. To miss here would sacrifice a lot of very careful work. Once we felt we had proper proportions, we drew out templates on plywood, including all of the cove and thumbnail bed moldings that would follow the reverse curve pediment immediately below it. Then, as we cut them off, each snake-like strip became a different template. The scrolls themselves we cut from laminated sugar-pine planks. We used a beautiful old compass plane to smooth them after an initial rough cutting (photo facing page, bottom). On the rough scroll we attached a ½-in. fascia (again with the reveal and cove on the bottom), but here we discovered a real quirk of the craftsman who built the Williams entry. He had used a different crown on the scroll than on its returns to the house. Although he used a large reverse ogee with a bead and a small cove as his crown molding cut on the scroll, he used the very typical large reverse ogee, bead and small ogee for the returns. The large reverse ogees were the same size, however, and were mitered at their meeting. The small ogee was mitered into the ogee above the

corona. It was a tricky resolution. We made the curved moldings by simply carving them and then working them hard with scratch-stock and sandpaper. We used the large reverse curve of our old crown molding plane as the template for the profile of the scroll crown.

The resolution—Large carved rosettes resolved the scroll tops. They were very similar to the rosettes on the pilaster necking, but were 2 in. larger, and our carving was in bas relief. We did all the rosette carving with U-shaped gouges. The pediment, from clapboard catch to the outside edge of the crown molding, is the same 8-in. depth as the block the rosettes are carved in, so we could terminate all moldings on the circular barrel of the rosette.

Between the rosettes, the pediment on the Williams house had been further embellished with a turned finial on a carved, paneled pedestal. Here we again struggled with our layout, for the finial had been too high on the real entrance for us to measure. Our drawing had to be our template, and when we finally settled on the proportions, we turned it out easily on the lathe—just as it had been done originally, if with a different power source. We left a large round tenon to fit into a mortise in the roof of the pedestal. The pedestal itself we made by first creating a box 8 in. wide by 12 in. high by 4 in. deep (small photo, facing page). Then we hand-carved a raised panel on its front, and flat panels on its sides (another oddity of the Williams entry). We fashioned a hipped top, flat

only on the small portion meeting the bottom of the finial. A deep mortise accepted the finial's substantial tenon. We flashed the cornice, then set the pedestal and finial in place.

So there we were. Three-hundred and fourteen handcarved pieces, 154 man-hours of labor, and we had ourselves an entrance—and a lot of respect for those Connecticut River Valley craftsmen. In hopes of its lasting half as long as the original, we used handmade rosehead nails where they would show, and galvanized nails everywhere else. We also used almost two quarts of glue and untold pounds of lead for flashing. In addition to the horizontal cornice, we also flashed the scroll pediment, overlapping the edge of the crown molding by almost ¼ in. We even fitted the finial with a small leaden skull-cap.

After the awesome entrance, we built the doors, almost as an afterthought. They were real beauties, though. The client chose as our model a fine set of early four-paneled doors, again from Old Deerfield. They were fitted with beaded vertical battens on the inside, large strap hinges, and a huge bar and staples lock. The battens were fastened with clinched roseheads, the stiles and rails with pegs. The client was lucky enough still to have the huge granite step from the original house. We had only to move it gently up under the plinth blocks to make the job complete. □

Gregory Schipa is president of Weather Hill Restoration Co. in Waitsfield, Vt.

Staircase Renovation

Loose, tilted flights signal trouble underfoot, but even major problems can be fixed with screws, blocks and braces

by Joseph Kitchel

There seems to be a great deal of reluctance among renovators to tackle any kind of stair rebuilding. I have seen many beautifully renovated homes with sagging staircases, loose balusters and makeshift railing supports. People who wouldn't think of leaving cracked plaster cracked or sagging floors unshored will put up with staircases so crooked they would make a sailor seasick, and squeaks so loud they wake the whole family at night.

Procrastination and lack of understanding of the mechanics of stair construction are at fault here. No one seems to know quite where or how to start, or who to call to do the job. To be sure, attacking an ailing stair takes courage, resolution and perseverance. It is one of the dirtiest and most disruptive renovating jobs.

The stairway is the spine of the multistoried row house. It links the sometimes minimal area of individual floors into what can be a spacious and accommodating floor plan. Because changing the stairway can seriously affect the physical flow of people and spoil the aesthetics of the whole interior design, it is far better to renovate than redesign or reposition.

Parts of a stair—Each step consists of a riser, the vertical portion that determines the height of each step, and the tread, the part on which you step. The dimensions of steps must be consistent within flights, though they may vary from one flight to the next. Variation within the flight will break the stair-user's physical rhythm and cause tripping. Tread widths may vary in the bottom two or three steps of the main flight for aesthetic purposes, and at the top of a flight, where pie-shaped steps are needed to negotiate a curve. Riser height, however, must not vary.

The newel post, usually decorated with paneling or carving, supports the railing at the bottom of the flight. Though it gives the impression of being heavy and sturdy, after years of being swung around by ebullient children, it has probably loosened enough to sway from side to side or lean out toward the hall. The vertical supports of the railing, aligned more or less behind the newel post and marching upward with each suc-

cessive step, are balusters or spindles, which are dovetailed into the treads.

Keeping the balusters from slipping out of their joints on the open side of the staircase are the noses, continuations of the molding that forms the front edge of the treads. These noses are removable and not part of the tread itself.

On each side of the stairs are the stringers, which hold the risers and treads. The stringer attached to the wall is usually routed or plowed out to receive the steps; this is called a housed stringer, and it produces a strong, dust-tight stairway. The outside stringer may be open or closed. Stringers may be simple, laminated with other pieces to form the curves of the stairs, or partially concealed by decorative filigree.

Were you to remove the plaster under the staircase, you might find two or three 4x4s or 4x6s running the length of the flight—one nailed to the inside of the outside stringer, one centered beneath the steps, and perhaps an-

other nailed to the inside of the stringer attached to the wall. These are the carriages, the members that add extra support. At the bottom, the carriages ideally rest on top of the stairway header, a joist that frames the stairwell end. They are sometimes attached to the inside face of the header. (This was the cause of failure in one flight I repaired. The weight of the stair forced out the toenails holding the carriage.) At the top, the carriages attach to the face of the upper-story header joist or, in the case of a curved flight, to angled braces running from notches in the wall to the upper floor joists.

Other elements of stair construction visible from beneath the flight are the wedges or shims driven between steps and stringers. Tapping these wedges tight or adding wedges made from shingles or building shims can do a great deal toward tightening up a staircase.

Diagnosis and dismantling—Problems with stairs fall into two categories. The simpler ones concern the railing, balusters, newel post and railing supports—the superstructure. Trouble here, though relatively easy to repair, can be symptomatic of more serious problems in the steps, carriages and stringers—the substructure.

Before you remove any plaster, explore the failings of the flight. If the steps are loose and tip toward one side, then the carriage or stringer on that side has weakened. The cause may be rotting, breaking, warping and splitting, or the carriage may have separated from the stringer.

Another symptom of major deterioration is a series of gaps or cracks along the ends of the treads or risers where they fit into the stringer. These rifts may occur on either side of the stairs, but are most often on the wall side, suggesting that a center or outside carriage has shifted downward, skewing the flight toward the center of the stairwell.

Large cracks in the plaster at the top or bottom of a flight are a good sign that the carriages have come loose from the headers. However, cracks generally parallel to the carriages or crisscrossing their length may indicate that vibration has caused the plaster to loosen and the keys to

Photos: Jeff Fox

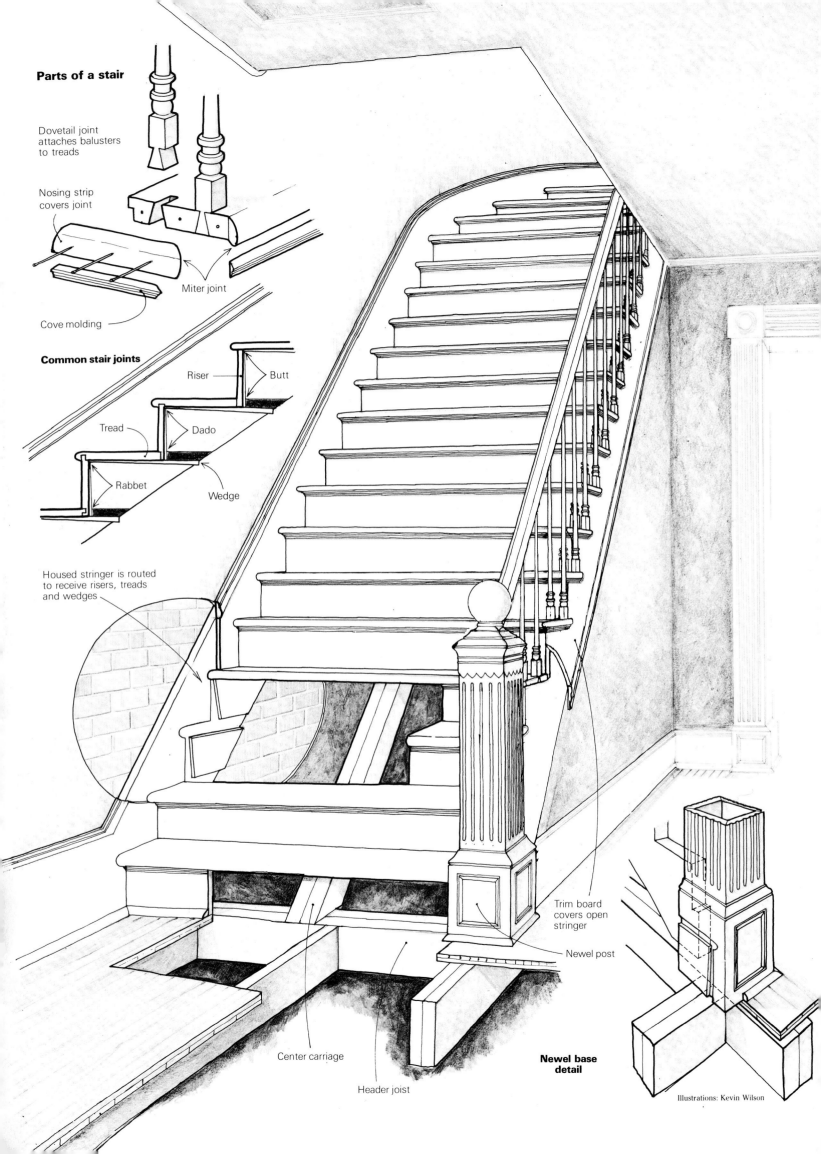

Parts of a stair

Dovetail joint attaches balusters to treads

Nosing strip covers joint

Miter joint

Cove molding

Common stair joints

Riser

Butt

Tread

Dado

Rabbet

Wedge

Housed stringer is routed to receive risers, treads and wedges

Trim board covers open stringer

Newel post

Center carriage

Header joist

Newel base detail

Illustrations: Kevin Wilson

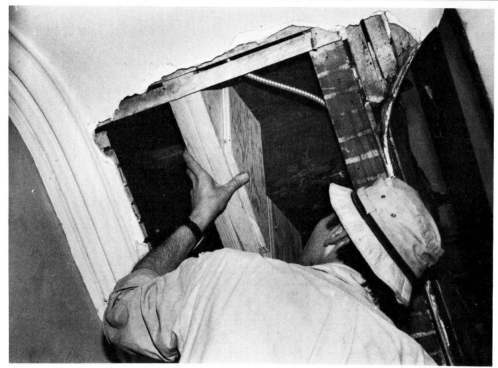

Substructure repairs are major problems and should be tackled first. To correct a distorted stringer, author Kitchel, above left, strips stairs of treads and risers. Plaster has been removed from beneath the stairway previously, and the resulting debris cleared away for a dust-free work area. The jack between wall and stringer provides lateral support until the stringer has been repositioned and reattached, and all steps have been renailed. Above, a carriage pulled away from the joist can be reattached with an angle iron made from $\frac{1}{8}$-in. plate attached to the header with $\frac{3}{8}$-in. bolts. The carriage usually rests on top of the stairway header, but occasionally it is attached to the inside face. Left, support blocks of $\frac{3}{4}$-in. plywood fastened with Sheetrock screws minimize deflection.

break off, which is a much less serious repair.

If your stair has a decorative plaster molding running along its wall side, a continuation of the ceiling cornice, don't despair. This can be saved if it is not badly damaged and is still attached securely to the wall. To dismantle, carefully cut through the plaster and lath along the molding with a masonry blade in a circular saw, cutting parallel to the stair edge of the molding, leaving the molding intact. Plan this cut so that plaster or Sheetrock can be rejoined to this edge.

To further assess causes of stair failure, remove nearby plaster and lath. You may not have to remove all of the ceiling covering if the plaster isn't bad and if the problem is localized. Take time to clean up all the resulting plaster dirt. Debris allowed to accumulate on the flight below makes moving your stair platform difficult and dangerous, and will also worsen the

spread of dust throughout the rest of the house.

If necessary, remove noses and balusters, but label all parts first. Assign each step a number (I usually start at the bottom) and tape this number to the outside of the step and to its adjacent nose molding and spindle.

Using a screwdriver or chisel, gently pry the noses away from the stringer. They're usually finish-nailed in two or three places. You'll see the dovetail joint that connects the spindle to the tread. Tap the spindles out at the bottom and pull them down out of the railing.

Examine the joint between riser and tread to judge how to handle repair of squeaks or gaps. The joint may be a dado, a rabbet or simply butted and nailed. Past repairs, such as wedges driven into the gaps between riser and tread, fillers in the spaces above the treads along the wall stringers, or braces along the railing or

spindles, may have been purely cosmetic. Remove them so that all parts of the stairs may be properly aligned.

Substructure repair—You should now know the causes of your stair failure; tackle the bigger ones first. If a carriage is rotted out or cracked, tear it out and replace it. If it is intact but springy, it may be undersized; laminate a new beam or a steel reinforcing plate onto it. Solutions at this point must be as individual as the problems. But by far the most common failings are bolts and nails which have worked loose, allowing carrying elements to pull free from walls and header. In this case, mending plates made from $\frac{1}{8}$-in. steel work well.

If you must raise or remove carriages, first free them from each riser and tread; otherwise, the attached superstructure will loosen as the skewed members are jacked up. To remove treads and risers, pry them apart at joints or cut through the nail shanks if pulling is impractical.

For major repairs, jack the stringers or carriages into place in a manner that won't damage the superstructure and will keep it from damaging you. To give the jack or brace a level bearing surface, attach angle blocks cut to the slope of the stairs with clamps or screws.

If the wall stringer has come loose from a masonry wall, reattach it by raising it to its proper position and nailing it with cut-steel masonry nails of sufficient length to go well into the wall. Wear goggles. A better method is drilling through the stringer and into the masonry with a carbide bit, and tapping in a lead sleeve; a lag bolt and washer expand the sleeve and tighten the stringer to the wall. If it's a frame wall, lag screws alone will hold stringers to studs.

If the outside stringer has twisted or warped, and is pulling the treads out of the wall stringer opposite, remove the treads and risers and force the stringer in toward the wall. Get the necessary leverage by temporarily bracing from the outside of the stringer toward the partition wall opposite. Screwing or bolting the stringer to the accompanying carriage will correct matters.

With the supporting members repositioned, now attend to the steps. Repair split treads and risers by removing them, lapping a piece of plywood over the back of the split, and gluing and clamping overnight. If you glue and screw the lapping piece you may forego clamping, and can replace the piece immediately. If the very edge of the nose is split, insert dowels from the edge to reattach it, being careful not to split the riser's dado or rabbet joint.

After you've corrected stringer and carriage problems and each tread and riser is back in place and renailed to the outside stringer, strengthen the steps by nailing step blocks to the center carriage. Why this wasn't done originally has always puzzled me. From a piece of ¾-in. plywood cut a step block to fit under each tread, and place it firmly against the back of each riser. Nail or screw the blocks to the side of the carriage, and then nail through the face of the tread and riser into the edge of the blocks. Trim the bottom edges of the blocks to conform to the angle of the carriage. I usually alternate the blocks on opposite sides of the carriage, but they all may be attached to the same side if the staircase is narrow and you can't get between the center carriage and the wall. Nailing blocks on both sides of the carriage is overkill, but add them wherever extra support is needed.

At this time a test run up and down the stairs will tell you where additional nailing and bracing are needed. For all face nailing in treads and risers, I use 6d or 8d finish-head, spiral flooring nails. For nailing where it doesn't show, I prefer 6d or 8d rosin or cement-coated box nails. I find 1½-in. or 2-in. Sheetrock screws driven with a variable-speed drill useful where hammer space is limited. Screws often add more strength than nails, because they pull things together and don't require pounding, which may disturb the alignment of nearby areas.

If risers, treads or stringers are to be refinished separately from the spindles, consider doing this now. Stripping, sanding and painting are easily done with the upper parts out of the way. Before finish is applied, set and fill all nails.

Superstructure repair—Newel-post problems are best dealt with after all other structural problems are solved. Although removing or loosening the newel post may be necessary to work on the carriages or stringers, it can usually be left in place to support the railing.

Newels are usually attached to the bottom step with a threaded rod, and to the railing with a hanger bolt. (Hanger bolts have wood-screw threads on one end and machine-screw threads on the other.) On the machine-screw thread of the hanger bolt is a star nut, which can be turned through the access hole with a screwdriver or needlenose pliers after the bolt is in place. (A plug fills the access hole later.) For extra strength screw into the railing from the inside of the newel post.

Straighten or tighten a shaky newel post by shimming under its bottom edges and renailing or screwing it to the floor. For greater support, try one of these repairs using a threaded rod. First, remove the newel post. Bolt a threaded rod to a bracket or wooden block so that the bolt fits flush with the bottom of the block. Screw the block and rod assembly to the floor, then slip the newel post over them and reattach it. To repair the post without removing it, attach two brackets under the tread of the first step as

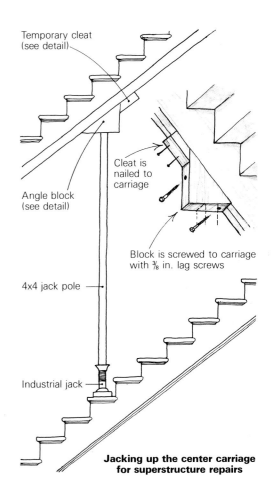

Temporary cleat
(see detail)

Cleat is nailed to carriage

Angle block
(see detail)

Block is screwed to carriage
with ⅜ in. lag screws

4x4 jack pole

Industrial jack

**Jacking up the center carriage
for superstructure repairs**

Building a Stair Platform

Before you renovate your staircase, you'll need to build a stair platform. Part of the reason stair repair is so difficult is that there is no place to stand, no way to get up to the job.

The exact dimensions of the platform depend upon your stairs. To make the platform, set a straightedge, level, on the fourth or fifth step from the bottom of your main staircase. The height of the platform is the distance from that step to the floor; its length is the distance from the back of that step to the front of the bottom step. Check these dimensions at different locations on various flights of your

staircase. Sometimes rise and tread dimensions or angles of incline vary from flight to flight, but not usually. Then cut and attach legs and braces as shown in the drawing.

The platform should be wide enough to hold your stepladder comfortably, but narrow enough to allow passage on the stairs when it's in position. The platform will probably fit your neighbor's stairs, and you can use it when repainting your own stairwell; it is therefore a tool to retain after renovaton. You might consider making it collapsible for easier storage.

—J.M.K.

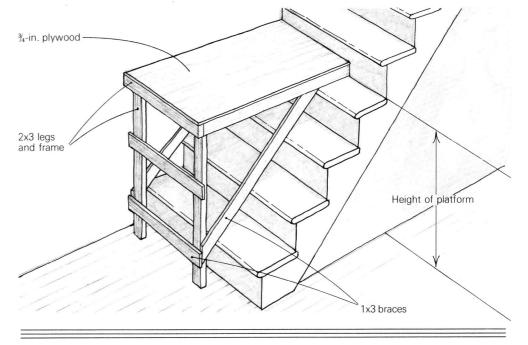

¾-in. plywood

2x3 legs
and frame

Height of platform

1x3 braces

Newel-post attachment

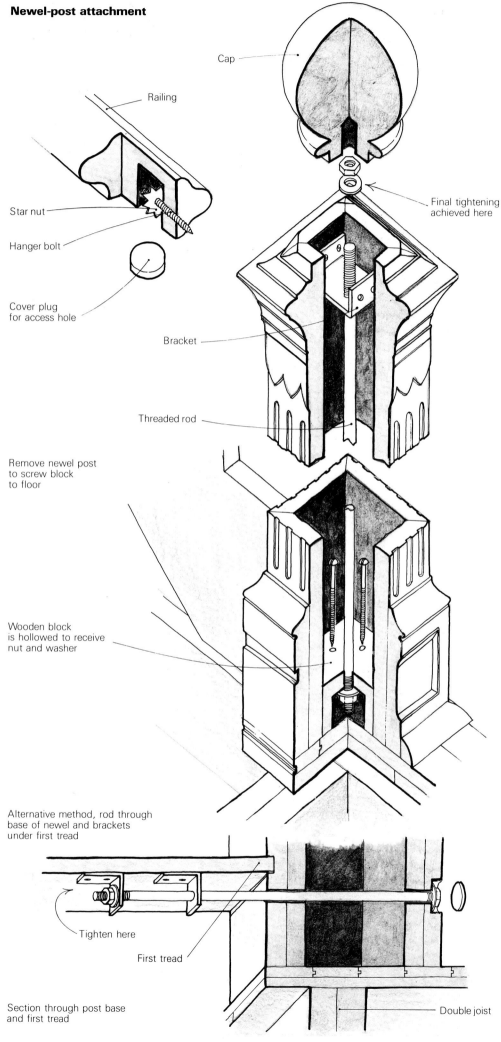

Cap

Railing

Star nut

Hanger bolt

Cover plug
for access hole

Final tightening
achieved here

Bracket

Threaded rod

Remove newel post
to screw block
to floor

Wooden block
is hollowed to receive
nut and washer

Alternative method, rod through
base of newel and brackets
under first tread

Tighten here

First tread

Section through post base
and first tread

Double joist

shown at left. Drill a hole through the base of the newel post to align with the holes in the brackets. Insert a threaded rod through the holes, bolt at both ends, and fill the access hole with a plug.

To replace the spindles, first coat both ends with white glue. Insert the top end of the spindle into the railing underside first, then slide the dovetailed end into its tread slot. Shim the joint wherever necessary, from underneath or from either edge, securing it with a finishing nail through the dovetail into the end of the tread. I find a rubber mallet useful when replacing spindles, because it will not mar the finished surface of the wood.

As you proceed from bottom to top, occasionally check the alignment of the railing and make adjustments by trimming or lengthening the spindles. Temporary braces hold the railing in place until all spindles have been installed and the glue is dry.

You can lengthen spindles (or adapt spindles from another stair) by adding a short piece of dowel. Drill into the top end of the spindle, and glue and nail the piece of dowel in. When the glue is dry, rasp and sand the dowel to the contour and taper of the spindle tip. Stain to complete the match.

When the final spindle has been inserted, the small bracket that originally connected the top of the rail should be reattached, or a new one made to fit. This prevents lateral movement of the rail and will hold the spindles in their correct positions as the glue dries.

Refinish the noses before you attach them. If you remove old nails by pulling them through from the back with a pair of nippers, you'll avoid the splitting that usually occurs when the nails are pounded back through and pulled from the face. Glue the noses and nail them twice along the side and once through the miter where nose meets tread molding.

Replacing the ceiling under the flight is the final step; use lath and plaster or Sheetrock. If your stairs curve, use short sections of wire lath to recreate the original curve.

Take care to keep all nailing surfaces in the same plane. This can be done with two straight-edges; one the width of the ceiling area, from the outside of the stringer to the wall, and the other as long as possible to run the length of the flight. Use building shims where necessary to keep furring strips in the correct plane. Determine your plaster line in relation to any plaster molding along the wall and to the bottom edge of trim pieces that adorn the outer carriage. I usually let the beaded or molded edge of this outer trim protrude below the plaster line. Existing pieces of plaster or plaster molding to be replaced may be drilled and secured with screws before touch-up spackling or painting. □

Joe Kitchel is a prop builder, cabinetmaker and renovator from Brooklyn, N.Y.

Making Curvilinear Sash

How to lay out and assemble a semi-elliptical window

by Norman Vandal

In fine Federal and Greek Revival structures, curvilinear windows are common—above pilastered entrances, in gable pediments and incorporated with Palladian windows. Built under roof gables or in other confined areas where conventional rectangular windows couldn't fit, the curvilinear window was a decorative means of providing light to upstairs rooms and attic space. Since I'd recently been commissioned to build such a window for a nearby restoration project, I began to notice many more of the type I'd been asked to create. I became critical, noting some of the small details that separate the fine from the not-so-fine. Fortunately, there are many semi-elliptical windows to be found around Roxbury, Vermont, where I live. I was able to examine different styles and to develop a taste for the most desirable features.

There are probably many ways of producing curved sash, but I could find little information on building this type of window by hand. Avid old-tool enthusiasts, concerned with the function of the tools in their collections, seem to be doing most of the research. As a rule, most craftsmen rely on large millwork companies to produce on the assembly line the items they need, although cabinetmakers sometimes stumble into sash work when making such pieces as corner cupboards or secretaries.

Drawing the semi-ellipse—This is the first step in making curvilinear sash, and will determine the proportions of the window. I learned how to draw an ellipse from Asher Benjamin's

Curvilinear windows were popular features in houses built during the Federal and Greek Revival period. Even today curved sash must be made largely by hand, using traditional techniques of fine joinery.

The American Builder's Companion, an 1806 guide to neo-classical detail and proportions, reprinted by Dover Publications (180 Varick St., New York, N.Y. 10014) in 1969. The method I find most valuable for making a full-scale template is illustrated in figure 1. Using this technique, one can draw concentric ellipses, a necessity for making templates. Note, however, that the resulting curve merely approximates an ellipse, since the points d and d' on the major axis are not foci in the usual sense.

The distance between a and b will be the maximum length of the bottom rail (in my window, 3 ft., with a height of 13½ in.), but you must draw

the entire ellipse to be able to locate the compass points from which the top of the curve is described. The window I made (photo left) is rather elongated in comparison to many old windows I have seen. You can change the proportion of your ellipse simply by experimenting with the compass points d and d'. Moving them closer to the center point in equal increments will elevate the semi-ellipse by the amount of one increment. With a little experimentation, you'll eventually reach a pleasant proportion.

Making the templates—After determining the proportions of your semi-ellipse, begin to construct the templates. For these I use poster board, available in art-supply stores. Try to get board as large as the full-scale sash to avoid taped seams on the template. You'll also need a piece for exterior and interior casing templates.

To construct the templates, you'll need a compass capable of scribing large-diameter circles. For smaller sash, a pair of large dividers will suffice, though large dividers are difficult to locate. You can make dividers of wood, using a nail and a pencil as points and a wing-nutted bolt to secure the joint—a little crude, but functional. My beam compass, an old set of wooden trammel points fastened to a wooden beam, is shown in figure 2. One point is permanently attached; the other, which slides along the beam, can be set in place with a wooden wedge.

Begin by compassing your exterior sash dimensions on the poster board. Then determine the width of the rail (mine generally measure

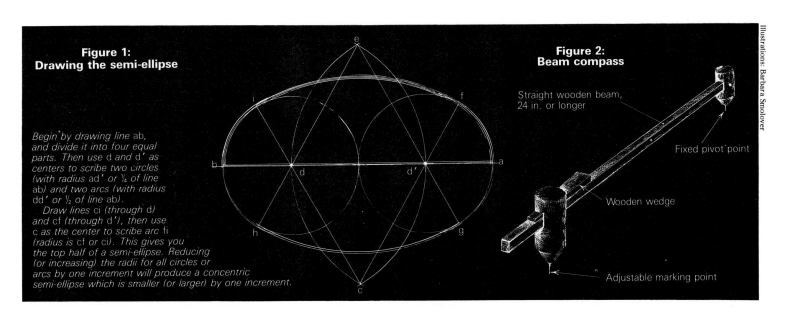

Illustrations: Barbara Smolover

Figure 1:
Drawing the semi-ellipse

Begin by drawing line ab, and divide it into four equal parts. Then use d and d' as centers to scribe two circles (with radius ad' or ¼ of line ab) and two arcs (with radius dd' or ½ of line ab).
Draw lines ci (through d) and cf (through d'), then use c as the center to scribe arc fi (radius is cf or ci). This gives you the top half of a semi-ellipse. Reducing (or increasing) the radii for all circles or arcs by one increment will produce a concentric semi-ellipse which is smaller (or larger) by one increment.

Figure 2:
Beam compass

Straight wooden beam, 24 in. or longer

Fixed pivot point

Wooden wedge

Adjustable marking point

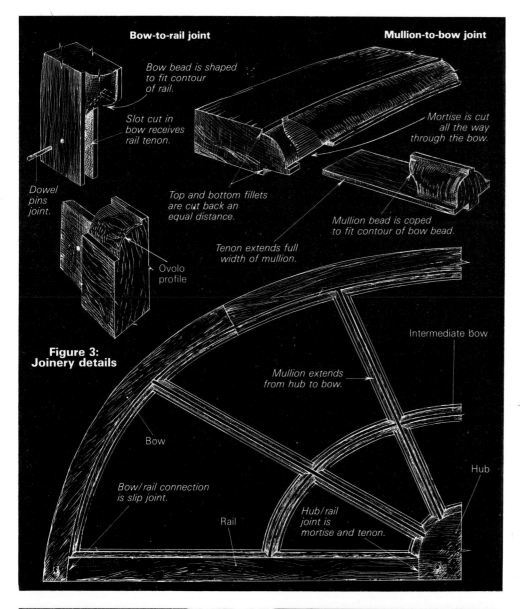

Bow-to-rail joint

Bow bead is shaped to fit contour of rail.

Slot cut in bow receives rail tenon.

Dowel pins joint.

Top and bottom fillets are cut back an equal distance.

Ovolo profile

Mullion-to-bow joint

Mortise is cut all the way through the bow.

Mullion bead is coped to fit contour of bow bead.

Tenon extends full width of mullion.

Intermediate bow

Mullion extends from hub to bow.

**Figure 3:
Joinery details**

Bow

Bow/rail connection is slip joint.

Rail

Hub/rail joint is mortise and tenon.

Hub

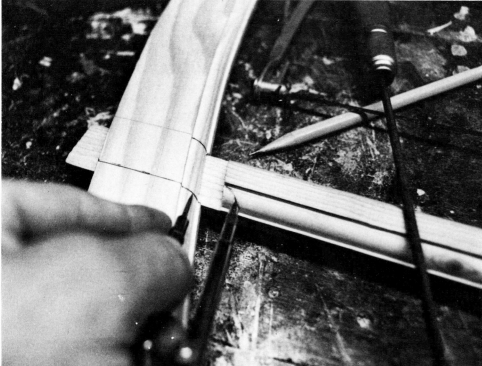

Making the mullion-to-bow joint: Once the intersection angle has been marked out on both mullion and bow, the mortises can be drilled out and squared with a chisel. Then the tenon is cut in the mullion and test-fitted in the mortise so the curve can be scribed in the bead, as shown. Use a coping saw to cut the curve and a rattail file or an in-cannel gouge to shape the joint for a snug fit.

$1\frac{1}{2}$ in. to $1\frac{3}{4}$ in.) and the width of the bow. Draw the line representing the top of the bottom rail parallel to the line ab. To draw the inner contour of the bow, you must reduce the radii of the circles with the centers d and d' by the exact width of the bow.

To avoid the disadvantages of short grain, it is best to make the bow out of three pieces of wood instead of one, orienting each to maximize long grain. Draw the lines ci and cf on the template to indicate the position and angle of the joints on the bow. Later, when you're cutting up the template, you can divide the bow into three segments. The sash shown in figure 3 has an intermediate bow, something you may want to avoid in early attempts at making curvilinear sash. If you decide to use an intermediate bow a short distance out from the hub, make it concentric with the curve of the primary bow. An intermediate bow will have one more segment than the number of mullions.

You need not cut templates for the mullions, as they are of straight stock and will be cut to proper length later. But you'll need to decide how many mullions you want in your sash, and to determine what angles they form with bow and hub sections. At this stage you may want to consult old work—the joinery details you are able to uncover in a traditionally made window will prove helpful from this stage onward. In terms of workability and historical accuracy, mullion width should be between $\frac{5}{8}$ in. and $\frac{3}{4}$ in. This will equal twice the width of the bead plus the width of the fillets.

The hub, where the mullions will converge, can be either semicircular or semi-elliptical, but make it large enough to keep the rays from coming into contact with each other, thus keeping the joinery less complicated. The hub must be cut separately, mortised to the rail, and then mortised to receive the mullions.

I usually make my templates $\frac{1}{8}$ in. longer than their finished dimensions. You need this extra length when you cut and fit the joints, and any excess can be trimmed off during final assembly. When you have completed the sash templates, make templates for the exterior and interior casings. My exterior casing, which overlaps the sash $\frac{1}{2}$ in. to keep it from falling out, is in two sections, divided by a keystone. Cut out the templates with a stencil knife. I use a surgical scalpel with sharp blades to keep the edges crisp.

Cutting the parts—Your cardboard templates will serve as patterns for laying out all the parts to be cut. I use clear 5/4 eastern pine stock, although mahogany, clear spruce or basswood would serve equally well. Curved cuts for bow sections and the hub can be accomplished with frame saw, bandsaw or saber saw, depending on your equipment and inclination. Remember to cut the mullions a bit long, since some trimming is inevitable with the type of joinery you'll be doing. To join the bow segments, I use a simple butt joint reinforced with dowels—two to each joint—and yellow wood glue.

During the Federal and Greek Revival periods, a complete set of highly specialized tools available from tool-makers and woodworking suppliers provided the joiner with everything needed to make all types of sash. While many of these tools are still around today, most are useful in dealing only with square or rectangular sash. One elusive tool is the sash shave, a cross between a spokeshave and a plane; without it curved sash is practically impossible to make by hand. I could not locate an old sash shave to match any of my planes, so I began to make one from scratch. I quickly realized it was going to take more time and research than I had anticipated (I still haven't finished it), so instead I used a router to cut the molding and the rabbets. You can also use a shaper. The only bits you'll need are a simple ¼-in. bead (with integral stop) and a ¼-in. rabbet (also with integral stop). The bead cutter gives an ovolo molding profile (figure 3), the easiest shape to work with when coped mortise-and-tenon joints have to be cut.

Mortise-and-tenon work—This is where the craftsmanship comes in. The two bow-to-rail joints are the first ones to make. Here you have the option of either mortising a bow tenon into the rail, or cutting a slip joint. I chose the latter because part of the slot and tenon-cutting work could be done on my radial arm saw. The mortise is more secure in terms of joint movement. The procedure for making all sash joints is the same: Lay out the joint first by aligning the parts, mark the stock, and then cut the tenons and mortises. I cut my mortises all the way through, since this makes squaring the cavity easier (you can chisel in from both sides). Check and adjust these for fit, mark where shoulder areas must be coped, and then cope the contoured part of the joint. This is by far the toughest part of the joinery work, and must also be a "check and adjust" operation. I use an in-cannel gouge (a curved gouge with its bevel cut on the inside) and a rat-tail file to shape the curved shoulders of the joint. You may want to use a coping saw to cut the cope into the mullions, but only a gouge can be used on the hub and bow-to-rail joints. Final shaping will consume the extra ⅛ in. or so added to finished length when the parts were first cut.

Once the bow/rail joints are secure, join the hub to the rail (temporarily) and lay out the mullions on the bow/rail/hub assembly to mark where mortises have to be cut. Fashioning the mortises and tenons for mullion-to-rail and mullion-to-hub joints can be made easier if you remember that the tenon extends the full width of the mullion and is made by cutting away the bead or contoured part of the mullion and the bottom fillet. To make the mortise, drill and chisel out the square central portion of the bow or curb between the bead and the rabbet.

Once the mortises and tenons have been cut and fit smoothly, you have to cope the curve in the mullion bead which completes the joint. First seat the mullion tenon in its bow (or curb) mortise, then mark the shoulder cut on the mullion bead, as shown in the photo on the facing page. This has to be done by eye. You can expect some trim work with gouge or file (after you make the cut with a coping saw) to achieve a flush joint, but it's surprising how tight a cope you can cut with a little practice.

When all the mullions have been cut and fitted, a trial assembly is in order. Slip the tenons of the mullions into the bow, and snug up the rail to the bow as you're fitting mullion tenons into the hub. With luck, everything will fit as you planned, but don't be alarmed if a little further trimming is necessary. When you are satisfied with the results, glue up all the joints and reassemble the sash. A peg glued in the two bow-to-rail joints will provide extra strength.

Making an elliptical sash without an intermediate bow, as described above, is the easier and more advisable route if you've never built a curved sash before. If you're incorporating an intermediate bow, as I did, then these bow sections would be cut and mortised into the mullions before final glue-up of the sash.

The jamb—Since all of the old curvilinear window units I observed were in place, I was not able to examine the way in which the jamb units were constructed and framed into the walls. I had to work out my own system, shown in figure 4; you may wish to do likewise. Total wall thickness, sheathing material and stud spacing will determine the width of the jamb and the framing details.

Before you can make the jamb unit, first make the window stool to support the casings. I use 8/4 native white pine, and bevel the stool 10° toward the outside to shed water. (You also have to bevel the bottom of the rail so the rail-to-stool joint is sound.) The stool should be longer than the rail if you want ears at each end. If ears are not part of the design, trim the stool flush once you've attached the jamb.

Although you can make the stool any width, extending it too far increases the amount of weather it will have to endure. Drip kerfs (grooves on the underside of the exterior edge of the stool) allow water to drip free from the building's face. A groove cut to accept the clapboarding under the stool is an aid to both fitting the siding and weatherproofing the unit.

The sides of the jamb unit can be made from two pieces of exterior-grade plywood—I used ¾-in. thick pine stock, although ½-in. plywood will be strong enough if you use more blocking. The elliptical opening in each panel should be a full ³⁄₁₆ in. greater than the final dimensions of the sash on all sides. Once these are cut, attach the blocking that connects both panels. Now you've got a single unit to work on. Invert it to attach the ³⁄₁₆-in. thick jamb sections. They have to be bent to conform to the curve in the plywood frame. I used clear pine and soaked the wood to make it more flexible before fastening it to the ply edges with glue and small brads.

The stool can be attached next, but test-fit the completed sash first—fine adjustments at sash, jamb and stool contact points are more difficult to make once the stool is in place. I used glue and dowel pegs to join the stool to the plywood, but screws will work equally well. Once this is done the exterior casing can be fastened to the exterior side of the frame. It should overlap the bow section of the sash by about ⅛ in. Several small stops, tacked to the jamb on the interior side of the bow, force it securely against the casing.

Allow ½ in. to ¾ in. on both the width and height of the wall opening to shim the jamb unit plumb and level. Fasten the unit by nailing through the exterior casing into the sheathing and frame of the building. Now the siding must be cut to fit the curvature of the casing, but don't despair. You still have the templates, from which you can fit the siding and the interior trim. ☐

Norm Vandal builds and restores traditional houses. He also makes period furniture.

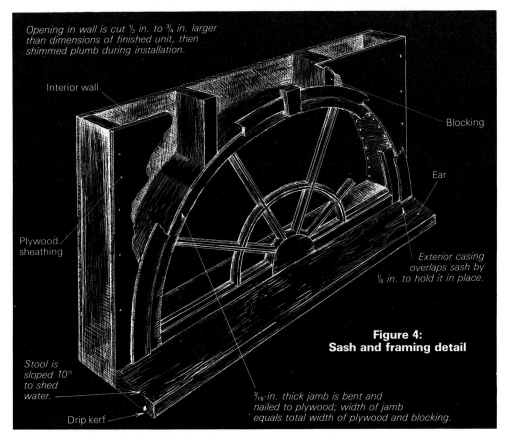

Opening in wall is cut ½ in. to ¾ in. larger than dimensions of finished unit, then shimmed plumb during installation.

Interior wall

Blocking

Ear

Plywood sheathing

Exterior casing overlaps sash by ⅛ in. to hold it in place.

Figure 4: Sash and framing detail

Stool is sloped 10° to shed water.

Drip kerf

³⁄₁₆-in. thick jamb is bent and nailed to plywood; width of jamb equals total width of plywood and blocking.

Classical Style in a Porch Addition

Tips from a restoration expert on deck construction, column building and weatherproof design

by Ted Ewen

Those who undertake exterior renovations on a Greek Revival house are faced with a touchy and challenging venture. Changes or additions must be compatible with the original design; they must also fit in with modern day-to-day activities. Even a small porch addition like the one I built recently must fulfill these two requirements. Of course another problem with porch construction is the weather: Porches won't last long if they are built without regard for weatherproofing. Fortunately I've been able to draw on quite a bit of boatbuilding experience, using materials and design details that have stood up to marine conditions.

The new porch was to replace a large laundry room that had been built onto the south side of a Greek Revival mansion in 1928. The main house, built between 1830 and 1840, overlooks the Hudson River from a high vantage point in Scarborough, New York. The old laundry room, like the kitchen it faced, was built at a time when cooking and cleaning were done by servants on sizable estates like this one. When the new owners of the house contacted me early in 1980, we decided to tear down the old laundry room. The new porch would expand the formerly viewless kitchen with its open deck and pleasant view. It would also give the kitchen a bit more formality as a main entry into the house. My job was to build something compatible with the main house. A repeated column and arch relief on the laundry room provided some important size and scale details. Having agreed with the owners on the design and drawn up plans, I set to work undoing the work of 50 years before.

A new deck—Tearing down the laundry room was a time-consuming job. It had been extreme-ly well built—knit together by carpenters who loved nails and knew how and where to drive them. The foundation was 12-in. thick reinforced concrete; the interior walls and ceiling were 1¼-in. concrete and plaster over galvanized wire lath. Over the years I've dismantled quite a few buildings, but this one really took the cake, especially considering its diminutive size (about 12 ft. by 16 ft.).

Before taking the roof off the laundry room, I removed the wood floor and the old floor joists to expose the crawl space underneath. Enlisting the help of the owner and a friend, I regraded the crawl space, installed 6-in. diameter PVC pipe to drain it outside the walls, and then floated 2 in. of concrete over galvanized lath laid on the crawl-space floor. We also poured two new concrete piers at the center of the foundation to support the new floor joists. The old roof

Having torn down an old laundry room and built a new deck on the south side of a Greek Revival mansion, the author prepares to extend the existing girder. Once in place, this girder will carry the ceiling joists for the new porch.

was left in place to protect the foundation from rain until the concrete cured.

With its beautiful view of Haverstraw Bay and the Hudson River Valley, the new porch and its columns would be exposed to heavy weather. Since the exposed section of the old deck had suffered considerably from water damage, I knew I would have to make the new construction and surface finish as durable as possible. I used pressure-treated 2x6s for all floor joists and spaced them 12 in. on center. Hot-dipped galvanized nails and screws, liberal applications of Woodlife preservative to all lumber (especially to the end grain) and a 2½-in. pitch in the deck's 24-ft. length were some of the measures I took to improve the weather resistance of the new construction. I also selected my lumber board by board whenever I could. Not all lumberyards will allow you to pick and choose, but using only straight, clear-grained stock was an extra assurance that the porch columns and their framework would last a long time.

For the deck I used 4-in. wide, 5/4 tongue-and-groove fir, toenailing directly into the 2x6 joists with 10d common nails. I took the time to drill pilot holes in both planks and joists to avoid splitting the wood. After nailing down every three courses of planks, I allowed a space of nearly 1/16 in. between the third and fourth courses. This gives the deck room to expand when moisture swells the boards. If you've ever seen a beautifully laid plank floor buckle in humid weather, you can appreciate the need for expansion joints. The total expansion space for the width of this deck is 1½ in., and I ran a seam of butyl rubber in each joint, to keep dirt out.

Making the columns—To duplicate the relief on the square columns of the main house, I measured them and then scaled the new columns accordingly. Proportions and detailing had to be kept the same to ensure compatibility with the rest of the house. The columns on the west side of the deck would be built in three parts: a large square main column flanked by the two smaller columns on which the arches would rest (see the drawing, below right). As an aid in laying out the columns, I drew full-scale cross sections on a piece of plywood and protected these working drawings with a wash coat of varnish. This way, they can be used on another job or as a guide for replacing a damaged column at some future date.

To build the columns, I used clear, No. 2 white pine boards, 3/4 in. thick. As shown in the drawing at right, I built the main columns with an inner and outer shell for load-bearing rigidity and general durability. The edges of the inner planks clear each other and the outer shell by about 1/8 in. The gap will permit the outer shell to shrink—as it will over the years—without having its joints forced apart by bearing on the inner planks, which will shrink much less. I fastened the inner pieces to the outer shell before putting the outer shell together, using 1¼-in. and 1½-in. #10 flathead screws.

To build up the corners and ends of each column to create a relieved panel in the center, I used a combination of ¼-in. lattice (strips of pine in various widths), marine-grade mahogany plywood (¼ in. thick) and stock moldings. The lines created by these raised strips and panels give the otherwise flat columns an appearance of lightness and fine detail. It's easy to understand why they were incorporated into the column motif on the original house.

The trim was fastened to the columns with aluminum nails and Phenoseal adhesive caulking (made by Gloucester Co., Box 428, Franklin, Mass. 02038). The adhesive caulking is a new material I discovered through my boatbuilding contacts. The waterproof bond is flexible and mildew resistant, two useful qualities as the wood ages and is exposed to moisture.

After planing and sanding the edges of the trim, I treated each column with Woodlife, waited several days for the preservative to penetrate and dry, and then brushed on a coat of Benjamin Moore white exterior alkyd primer. Then the columns were ready to go up, and construction could begin in earnest.

Column raising—Rather than rest the columns directly on the deck planks, I cut plywood pads conforming to the plan section of each column and screwed them to the deck where the columns would stand. The pads even out the downward pressure of the columns. More important, they elevate the vulnerable end grain of the column planks above the surface of the deck. Each

A full-scale template, drawn on ¼-in. plywood, serves as a guide during column construction. Here the arch post is test-fit on the half-column that will go against the side of the house.

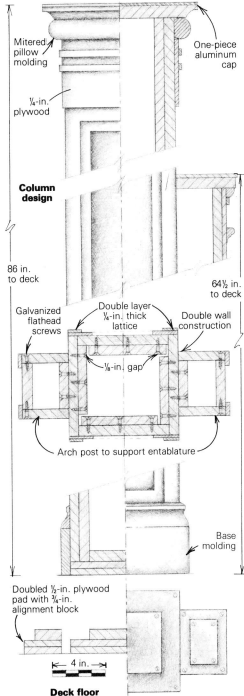

Mitered pillow molding

¼-in. plywood

Column design

86 in. to deck

Galvanized flathead screws

Double layer ¼-in. thick lattice

⅛-in. gap

One-piece aluminum cap

64½ in. to deck

Double wall construction

Arch post to support entablature

Base molding

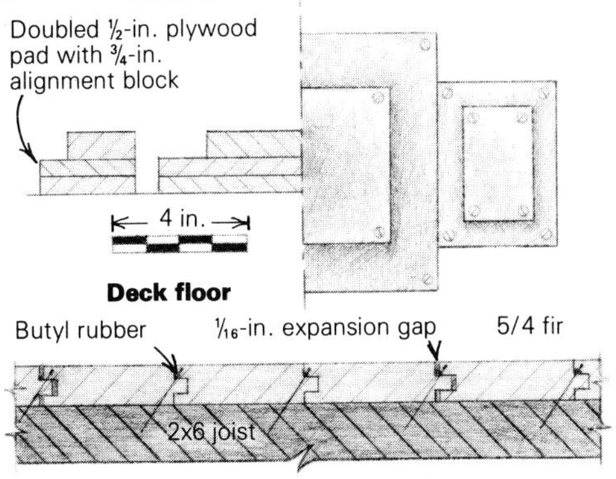

Doubled ½-in. plywood pad with ¾-in. alignment block

Deck floor

Butyl rubber   1/16-in. expansion gap   5/4 fir

2x6 joist

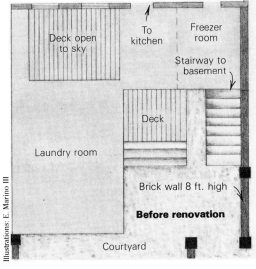

Deck open to sky

To kitchen

Freezer room

Stairway to basement

Deck

Laundry room

Brick wall 8 ft. high

Before renovation

Courtyard

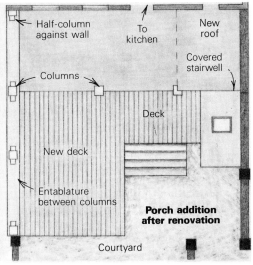

Half-column against wall

To kitchen

New roof

Columns

Covered stairwell

Deck

New deck

Entablature between columns

Porch addition after renovation

Courtyard

Test-fitting the soffit in the new roof. Joints between old and new soffit, fascia and frieze boards are staggered for lateral strength and to make the addition as inconspicuous as possible. One-by-three spacers build out the fascia to the full width of the column.

pad consists of two stacked pieces of ½-in. A/C plywood. Atop each pad I screwed a piece of ¾-in. plywood cut to match the interior dimensions of the column. This aligned the columns and also gave them extra resistance to horizontal pressures. Because of the pitch built into the deck, my square-bottomed columns would not rest plumb when raised into position. I had planned for this and constructed each column a bit longer than necessary. Using a large adjustable bevel, I transferred the pitch to the bottom of the column and then cut accordingly. The pad would be concealed by a base molding.

The procedure for column raising was as follows: I lifted the column into position on its pad, braced it temporarily with diagonal struts, installed an aluminum cap on the top of the column to prevent water infiltration, and then connected the top of the column to its neighbor with a new girder, soffit and fascia board. I had to run the girder (a 2x10 and 2x8 spiked together) all the way back to the corner of the freezer room because the existing beam had decayed badly. Then I nailed short lengths of 2x3s vertically

along the outside face of the girder to extend the width of the soffit that would cover these spacers. As for the soffit, fascia and frieze board, the joints where these new boards met their old counterparts were staggered to increase the lateral strength of the new roof section and to make the meeting of new and old less conspicuous (photo above). I vented the soffit by drilling ¾-in. dia. holes and covering them with small squares of galvanized hardware cloth.

One detail of the frieze bears mentioning. At 13½ in. from top to bottom, the frieze board had to be made from two planks, joined along one edge. Rather than use a 90° edge joint, which would allow moisture to make its way in behind the board, I cut a beveled edge joint (the detail can be seen in the drawing on the facing page). The top edge of the bottom board slants upwards, discouraging water penetration. I covered the joint with a taenia molding, completing the frieze and further increasing its resistance to the weather.

The roof itself posed no problems—I simply had to continue the line and pitch of the existing

roof that covered the freezer room. After lag-bolting a 2x10 header to the south wall, I anchored 2x6 rafters to the header with joist hangers. Each rafter had to be notched to fit the girder connecting the two columns (photo facing page, top right). Blocking (2x4s) followed between rafters for extra stability. Since the owners wanted a flat ceiling, the slanted rafters couldn't serve doubly as ceiling joists. I fastened 2x4 joists to the header and ran them level all the way across to the girder. Tapering the ends of the 2x4s allowed me to spike them directly to the rafters. I reinforced each joist-to-rafter connection by nailing plywood plates to both members. Then I nailed ⅝-in. plywood to the rafters and installed a triple-ply felt roof, with emulsion between layers and an aluminum/asphalt top-coat. The aluminum finish will reflect sunlight, thus preventing the heat build-up which shortens the life of dark-colored felt roofs.

The entablature—An arch taken from the old laundry room served as the model for the new arched entablature that would complete the classical detailing on the porch. I traced the arch on a 4x8 sheet of ¼-in. exterior-grade mahogany plywood, cut out the curve with a saber saw (photo below, center) and then cut the plywood to fit the space between columns. As shown in the drawing below, the entablature consists of two sheets of plywood held together by interior braces and a curved soffit (¼-in. plywood bent to the radius of the arch and glued to the interior braces). Once I had cut and trimmed the first

piece to fit between the two columns, I used it as a template for cutting the next piece on the west side of the porch. Rather than assemble the entablature completely before installing it between the columns, I reinforced each sheet and then lifted it into position against nailing strips fastened to the large columns. With this arch glued and nailed in place, I installed its mate in the same manner (photo bottom left). Gluing and tacking the curved soffit between the twin

arches was the next step. With the entablature in place, the porch was structurally complete. The arch and column motif will be repeated twice along the edge of the deck (photo, below right), creating a colonnade. The new addition is open and informal, yet very much part of the Greek Revival tradition of the rest of the house. ☐

Ted Ewen, of Scarsdale, N.Y., specializes in restoring and renovating historic buildings.

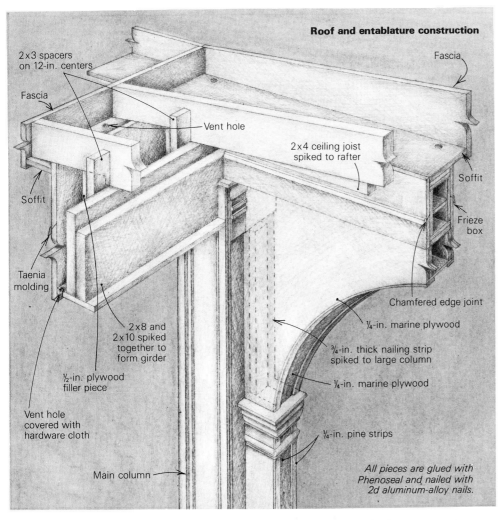

Roof and entablature construction

2x3 spacers on 12-in. centers

Fascia

Fascia

Vent hole

2x4 ceiling joist spiked to rafter

Soffit

Soffit

Frieze box

Taenia molding

Chamfered edge joint

2x8 and 2x10 spiked together to form girder

¼-in. marine plywood

¾-in. thick nailing strip spiked to large column

½-in. plywood filler piece

¼-in. marine plywood

Vent hole covered with hardware cloth

¼-in. pine strips

Main column

All pieces are glued with Phenoseal and nailed with 2d aluminum-alloy nails.

Rafters for the new roof are supported on one side by a 2x10 header bolted to the house, on the other by a girder between columns. Blocking between rafters adds rigidity.

Using an arch detail from the old laundry room as a model, the author cuts out the arch that will be fit between columns.

Nailing strips are fastened to the ceiling and new column sides (left) in preparation for attaching one plywood face of the entablature. Right, two identical plywood arches, fastened to nailing strips and joined with a curved soffit, complete the entablature between columns. Repeating this motif will connect the two additional columns on the finished porch, making a colonnade.

Door-trimming jig

The hinge side of a door must be trimmed to width at 90°. The latch side is usually trimmed at 5° off vertical for a beveled edge. Making these long cuts straight and smooth is easy with this two-sided jig.

First, buy a fine-toothed blade for your circular saw. I prefer a carbide-tipped plywood blade. Mount it on the saw arbor, and then use a combination square to make sure that the saw's base is 90° to the blade.

Build the jig with ¼-in. Masonite, 8 ft. long and about 12 in. wide. Accurately place a

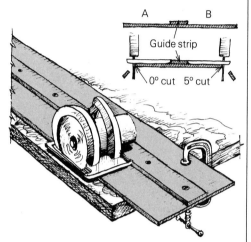

straight Masonite guide strip, 2 in. wide, down the middle, gluing and tacking it firmly into position. Leave dimensions A and B large at first. When the glue is dry, trim the sides with your saw—one side with the blade set at 0°, the other side with the blade set at 5°. Mark the respective sides clearly. When the jig is clamped to the work, the appropriate edge will register the line of cut. For smooth operation, apply furniture wax or car wax to the surfaces on which the saw's baseplate slides.

For crossgrain cuts, first run a knife along the cutline to sever the surface fibers of the wood. This will eliminate splintering, which can be a problem with veneered doors.

—Philip Zimmerman, Berkeley, Calif.

Preventing splits

Most people are vaguely aware that the points of nails are diamond-shaped. However, few people know that split boards can often be eliminated by proper orientation of a nail before driving it. This trick, shown to me by an elderly shipwright, has saved me a lot of pre-drilling and split moldings.

When driving a nail close to the end of a board, keep the rows of marks left by the nail-forming machine perpendicular to the grain of the wood. This causes the nail point to cut the wood fibers rather than wedging them apart, thus preventing a split from developing. It's easy to feel these ridges between your thumb and forefinger and then roll the nail to its proper orientation. This can be done so rapidly that you won't lose any time, even when framing. Brads and small finish nails inserted in a brad driver can be aligned by eye, since the diamond shape of their points is even more pronounced than on a box or common nail. —Steve Harman, Okanogan, Wash.

Crown molding, another way

For cutting crown molding I use a fixture made from an ordinary wooden miter box and two 1x stops. First, find the horizontal dimension (X in the drawing below) of the crown

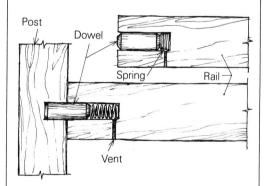

Crown molding in position for cutting an inside miter

molding, then screw the bottom stop in place so that its rear edge is that distance from the back of the miter box, as shown. Cut the other 1x wide enough to support the upper edge of the molding. Screw it in place, making sure that the screws are out of the saw's path.

Cut outside or inside mitered corners by setting the molding against the stops, wall edge down. Unlike Harry Stangel (next column), I think it's more accurate to cope inside corners to the molding's actual profile than to a penciled line, so after I cut the inside 45°, I cope that profile as shown.

—Peter Sacks, Somerville, Mass.

Spring-loaded dowels

Here's a simple way to dowel a rail in place between two fixed posts. Drill and dowel one end of the rail as you normally would. On the other end, drill the hole deeper and insert a spring before the dowel. Make the depth of

the hole equal to the length of the dowel plus the length of the compressed spring.

Once it's aligned with the hole in the post, the spring-loaded dowel will push its way home. I butter the dowel with hide glue before inserting it into the rail. I also drill a small vent hole in the bottom of the rail to prevent suction problems.

—Thomas Ehlers, Austin, Tex.

Crown-molding fixture

The fixture shown in the drawing below makes it easy for me to mark and cut crown molding for a coped inside corner. I use framing scraps to make the fixture—a 4x10 for a weighty base, ¾-in. plywood for the back, and 1x4s for the braces. These braces should be

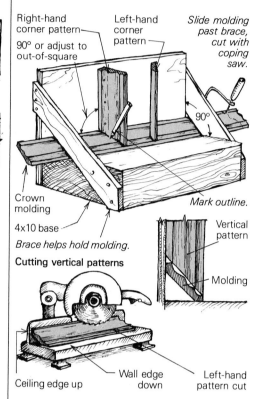

set so that the molding, with its face up, just passes underneath the bottom edge.

For the vertical pattern pieces, cut miters on two short lengths of crown molding, one right-handed, the other, left. These should be cut from stock of the same mill-run that you'll be installing. Nail these two pattern pieces so that they just touch the horizontal molding that is in the fixture for marking. The vertical pattern piece should be nailed at 90° to the table of the fixture, although they can be adjusted out-of-square if necessary.

To use the fixture, just mark the molding using the right- or left-hand patterns, and slide the scribed end out beyond the fixture for cutting with a coping saw. For marking and cutting, the molding is held securely by the pattern pieces and end braces. A generous undercut and a half-round file for touch-ups will yield a joint good enough to write home about.

—Harry W. Stangel, Palo Alto, Calif.

Index

A

Acrivue A, 135
Acrylic glazing, 132-35
Acrylite, 135
Air Changer Co., The, 19
Air Nail. *See* Hitachi.
Air/Vapor Barrier:
  in chimneys, 19
  form-fitting, 101
  importance of, 16
  locating, 16
  with windows, 17-18
Alexander, Geoff, on power planes, 128-31
Allen, Ed and Mary, brick work by, 63-65
All-Weather Wood Foundation, The: Why, What, and How, 144
American Builder's Companion, The (Benjamin), 229
American Colloid Co., 15
American Hydrotech, 15
Andrews, Bruce, on tablesaw molding, 204-5
Angle Square, 29
Asphalt. *See* Bituminous coatings.
Auger bits:
  for timber framing, 157
  sharpening, 159
Awls, 127
AWWF. *See* Wood foundations.
Axes, technique with, 170-75

B

Backfill. *See* Earth-sheltered houses, Foundations.
Bailey's, 184
Barden, Albie, 71
  Maine Wood Heat Company of, 73
  on masonry stoves, 73
Barge rafter, 33, 37
Barking spud, 170-71
Barnacle Parp's Chain Saw Guide (Hall), 184
Barthelmess, William C., 59
Bathtubs, wooden skirt for, 101
Batter boards, 11
BeA America Corp., 125
Beam compass, 229
Beams, rounded ends on, 185
Bed molding, 201, 202
Bender, Tom, 100
Bending moment, 160, 161
Bending strength, 161, 163-64
Benjamin, Asher, *The American Builder's Companion,* 229

Benjamin Moore exterior alkyd primer, 233
Benson, Tedd, on timber framing, 150-54
Bentonite, 14-15
Berglund, Magnus, on rammed earth, 82-87
B. F. Goodrich, 15
Bird's mouths, 26, 31, 34, 36
Bituminous coatings, 13-15
Black & Decker Workmate Bench, 126
Block foundations, 88-96
  layout for, 12
  See also Block work: surface-bonded; Foundations.
Block work:
  plaster-veneered, 120
  stone-faced, 97-99
  surface-bonded, 88-91
  surface-bonding mix for, 89
  See also Foundations.
Blocking, 108, 109, 110
Blue board, *See* Plaster veneer.
Board-and-batten, 108
Bollinger, Don, on flooring, 210-17
Borden, John, 104
Boring machine, 157
Bostitch, 125
  nail guns, 123
Brickwork, 62-69
  floors, 67-69
  with salvaged brick, 80
  silicone sealants for, 79
  See also Repointing, Russian fireplaces.
Broadaxes, 172, 173
Building terms, 31, 109
Bunce, Donald L., 221
Butyl:
  with acrylic, 135
  sheets, below-ground use of, 13
  sources for, 15

C

Cabinets. *See* Kitchen cabinets.
Calculators, for roof problems, 29
Cant strip, 43, 56
Cappelletti, Mick, 101
Carlisle Tire and Rubber Co., 15
Carpenter Handy Square, 29
Carson, George, 104
Catches, for cabinet doors, 193-94
Caulking:
  for acrylic, 135
  adhesive, 199, 202, 233
  for skylights, 52

Cement coatings, 13
Chainsaw mills, 182, 184
Chainsaws:
  for log building, 172, 175, 176
  rip chains for, 183
  supplies for, 184
Chalk box, 126
Chalk lines, 156
  joined, 101
  in log building, 172-75
Chemtron Metaseal, 16
Chimneys:
  and air/vapor barriers, 19
  block, 100
  crickets for, 56, 57
  flashing of, 56, 57
  prefabricated metal, 100
Chisels, 127
  corner, 156
  framing, 151, 156
  truing, 158
  See also Slicks.
Chromated copper arsenate, 181
Circular saw, portable:
  bevel cuts with, 111
  braces with, 110
  choosing, 109
  cut-off fixture for, 185
  door-trimming jig for, 236
  fit cuts with, 109-11
  notches with, 111
  plastic-laminate cutting with, 195
  plywood-cutting jig for, 104, 105
  pocket cuts with, 110
  protractor with, 147
  ripping with, 111
  safety with, 111
  stack cutting with, 109
  stair-gauge fixtures as fence for, 185
  for timber framing, 157
  trim with, 111
Clark, Jackson, 101
Clark, Sam:
  on-site shop of, 104-5
  Designing and Building Your Own House Your Own Way, 105, 108
Cole, John, and Charles Wing, *From the Ground Up,* 108
Collar ties, 31, 36
Columns, 233-34, 235
Combination squares, 152, 155
  truing, 158
Common difference, 31
Common rafters, 30, 31, 33, 34-36
Compasses, 127
Compression, 161
Concrete, 136-43
  cold joints, sealant for, 14
  gypboard form for, 100
  for log chinking, 181
  nails for, 123
  plaster-veneered, 120

Concrete blocks. *See* Block foundations, Block work: surface-bonded.
Conproco, 91
Consoweld Corp., 197
Contech PL 400 Adhesive, 199
Cornices:
  returns of, building, 201, 202-3
  types of, 200-1,
  water table for, 203
  See also Doorways.
Countertops, plastic-laminated, 195-97
Crayons, 127
Creosote, 181
Cripple-jack rafter, 31
Cripples, 109, 110
Crown molding, 201, 202-3, 236
  See also Moldings.
CYRO:
  Acrylite, 135
  Exolite, 135

D

Davis, Ron, 185
Dayton Speed Air Compressors, 124
Decks, 56, 57, 232-33
Deflection, 160-63
Designing and Building Your Own House Your Own Way (Clark), 105, 108
Dodge, John, 73
Doors:
  batten, making, 206-7
  18th-century reproduction of, 223
  flashing, 44-45, 55
  hinge-adjustment for, 101
  insulated, making, 208-9
  kitchen-cabinet, paneled, 193-94
  power-plane with, 130
  trimming jig for, 236
  vapor barriers with, 17, 18
Doorways, 218-23
Dornick, Ralph, 147
Dover Publications, 229
Dovetails, engineering of, 164
Drawers, 191-93
Drawknives, 156
Drills:
  plaster mixing with, 119
  for timber framing, 157
Drywall, 112-17
  nail gun for, 121
Dunkley, Don, on hip and gable roofs, 33-37
Duo-Fast Corp., 125
DuPont Co., Lucite, 135

E

Ear board, 201-2
Earth-sheltered houses, 13-15
Easton, David, Rammed-Earth Works houses of, 83-87
Eave, 31
Effective Building Products, 15
Ehlers, Thomas, 236
Eklind Allen wrenches, 127
Elastomers, 13, 14, 15
Electrical cable, 127
Ellipses, drawing, 229
Emglow Compressors, 123
Enercon Industries Ltd., 19
Entablature, constructing, 221-22, 234
Entrances. *See* Doors, Doorways.
EPDM:
  acrylic glazing with, 134, 135
  sheets, 13
  sources for, 15
Ewen, Ted, on porch addition, 232-35
Exolite, 134

F

Farmline Thermoply, 199
Farrell, Susan Caust, 101
Fascia, 31
Fiberboard Surface Bonding Cement, 91
Files, 127
Finish work, nail guns for, 121, 122, 124, 125.
  See also Trim.
Finnish fireplaces, 73
Firestone Tire and Rubber Co., 15
Flashing, 44-45, 47, 58
  for gutter ends, 203
  making, 52-53, 54, 55, 56, 101
  for skylights, 48-49, 50, 51, 52, 53
Flooring, 210-17
  nail guns for, 121, 124, 125
  nails for, 218-20
  sanding, 216-17
  sub, cutting in place, 110
Fly rafter, 31, 37
Ford, David K., on log building, 182-84
Formica. *See* Countertops, plastic laminated.
Formica Corp., 197
Foundations:
  backfilling to, 96
  concrete, disadvantages of, 88
  digging, 100
  drain tile for, 96
  insulating, 96

for log buildings, 180, 183
post, 20-21, 22
waterproofing, protection
for, 15
See also Block
foundations, Block work,
Earth-sheltered houses,
Wood foundations.
Framing:
nail guns for, 121, 122,
125
power plane for, 130
scissors lever for, 101
Framing squares, 152, 155
stair-gauge fixtures for, 29
truing, 158
French, James B., 101
Frieze, 201-2
doorway, 220-21
of joined wood, 234
Frieze blocks, 31, 36, 108
Frog Tool Co., 151
From the Ground Up (Wing
and Cole), 108
Fuge, Paul, on buying green
lumber, 106
Full Length Roof Framer,
The (Riechers), 28

G

Gable, 31, 33, 35, 37
Galvanic Action, 43, 58
Garrett Wade Co., 151
Gates Engineering Co., 15
Gin poles, 165, 166, 167
Glass Certification Council
Certified Product Directory,
52
Glass:
insulated, booklet on, 52
unputtying, 236
See also Caulking, Glazing,
Skylights.
Glazing, tapes for, 50, 51
See also Acrylic glazing.
Gloucester Co., Phenoseal
adhesive caulking of, 202,
233
Gomez, Chuck, 100
Gordon, Bruce, on batten
doors, 206-207
Gordon, J. E., *Structures, or*
Why Things Don't Fall
Down, 164
Gott, Peter:
house of, 171
log-building techniques
of, 170-75
Granado, Jesus, on plastic-
laminate, 195-96
Granberg Industries, 184
Grant Hardware Co., 193
Greenhouses, acrylic
glazing for, 134, 135
Greenhouse shutters, 198-
99
Greenlee Chisels, 151
Green wood, 106-8
Green Wood House, The
(Hackenberg), 108

Griskivich, Stan, on
greenhouse shutters, 198-
99
Gross, Gred, 185
Grouting, gun for, 100
Gutters:
and cornices, 201-3
flashing, 55
Gypboard. *See* Drywall.
Gypsum board. *See*
Drywall.

H

Hackenberg, Larry, *The*
Green Wood House, 108
Hacksaws, 127
Haddon Tool Lumber-
Maker Chainsaw Mill,
182, 184
Half-lap joints, engineering
of, 164
Hall, Walter, *Barnacle Parp's*
Chain Saw Guide, 184
Hammers:
choosing, 127
drywall, 114
Hanke, Paul:
on green wood, 106-8
on surface-bonded block,
88-91
Hasson, Will, on kitchen
cabinets, 188-94
Hawks, 114, 119
Header, 31
Heat exchangers, 19
Heller, Robert, and Mario
Salvadori, *Structure in*
Architecture, 164
Hewett, C. A., 152
High-Pressure Decorative
Laminates, 197
Hilley, John, 67, 69
Hilti Fastening Systems,
125
Hinges:
for batten doors, 206
mounting, 193-94
for overhead bifold
shutters, 199
for thick doors, 208
Hip rafters, 31, 33-36, 59
Hip roofs, 31, 33
Boston, 42, 43
See also Shingling, roof.
Hitachi:
Air Nail, 125
power plane, 128-29
Holland, Elizabeth:
on acrylic glazing, 132-35
on brick arches, 63-65
Hughes, John R., on
superinsulated houses, 16
Hydrocide, 96
Hydrozo Coatings Co.,
brick sealant, 69

I

Imperial Plaster. *See*
Plaster veneer.

Industrial Stapling and
Nailing Technical
Association, 123
Insulation:
of floors, 22, 23
of foundations, 14, 15
super-, 16-19
in truss frames, 22, 23
Interior Finish (Syvanen),
112
International Staple &
Machine Co., 125
Ireton, Kevin, 147
Isobutylene isoprene. *See*
Butyl.

J

Jaakkola, Sam, 73
Jack rafters, 30, 31, 34-36
Johns-Manville fiberglass
insulation, 23
Johnson Level and Tool
Mfg. Co., 29
Jointers, power planes as,
131
Joists, cutting to fit, 109
Jonsereds chainsaw, 176

K

Kalcoat. *See* Plaster veneer.
Karp, Lawrence, 168
Katzenbach, Chuck, on
acrylic glazing, 133, 135
Kaynor, Chapin and Donna,
88-89, 91
Kern, Ken, 75
The Owner-Builder's Guide
to Stone Masonry, 76
The Owner Built Home,
108
Kickers, 36
King common rafter, 31
Kitchel, Joseph, on
staircase renovation, 224-
28
Kitchen cabinets, 188-97
nail guns for, 121
See also Countertops,
plastic-laminated.
Klein screwdrivers, 127
Knives, 156
Koppers Co., Inc., 15
Kreh, Richard T.,
on block foundations, 92-
96
on repointing, 79
Kujawa, Andrew, 59

L

Ladder (roof), 31, 37
Lamin-Art, 197
Lane, Charles A., on
waterproofing houses, 13-
15
Lang, Paul, on Russian
fireplaces, 70-73

Langsner, Drew:
on ax techniques, 170-75
A Logbuilder's Handbook,
175
Lasar, Stephen, on making
skylights, 50-52
Lathing hatchet. *See*
Shingling hatchet.
Law, Tom:
on roof framing, 30
on site layout, 10-12
Layout tees, 29-31, 33-34
Layout, tools for, 126-27
Lee Valley Tools Ltd., 151
Lepuschenko, Basilio, 71
Level cut, 31
Levels, 126
Levin, Edward M.:
on timber framing, 155-59
on timber-framing
engineering, 160-64
Lindsey, Larry, on acrylic
glazing, 132, 133, 135
Linseed oil, for log
buildings, 181
Load, 161
Lo-Cal house, 16
Logbuilder's Handbook, A
(Langsner), 175
Log building, 171-84
Low-Cost Green Lumber
Construction (Seddon),
108
Loy, Trey:
on concrete, 136-40
on moving heavy timber,
165-66
Lucite, 135
Lumber. *See* Green wood,
Timber framing.

M

MacMath, Richard, on
stone walls, 74-78
Magers, Steve, Ken Kern
and Lon Penfield, *Owner-*
Builder's Guide to Stone
Masonry, The, 76
Maine Wood Heat
Company, 73
Makita:
circular saws, 157
power planes, 128-29,
131, 151, 156
Mallets, for timber framing,
157-58
Margolis, Angelo, on nail
guns, 121-23
Marking gauges, 156
Marples chisels, 151
Masonry. *See* Brickwork,
Stone walls.
Masonry Stove Guild
Newsletter, 73
Masonry stoves. *See*
Russian fireplaces.
Mathias, George, 182, 184
Mayes Bros. Tool Mfg. Co.,
28

McDaniel, Malcolm, 59, 101
McGaw, Sidney, 100
Measuring tape, 126
Metal brake, 54, 101
Miller, Amelia F., 221
Mills, Chas., 101
Milwaukee hacksaws, 127
Modulus of elasticity, 161
Moldings:
coping jig for, 236
miter box for, 236
with multiplane, 219
vocabulary of, 220
See also Doorways.
Moldings, tablesaw, 204-5
Moment of inertia, 161, 162
Montmorillonite clay. *See*
Bentonite.
Moose Creek Restorations
Ltd., 204
Morrell, C. Leigh, 208
Muriatic acid, 65, 69

N

Nailers, power, 213
Nail guns, 123-25
Nailing, without splits, 236
Nails:
for floors, 210, 212
for green wood, 107-8
for nail guns, 123
for shingling, 38, 41, 46
for staircases, 227
threaded, 107-8
Nailsets, 127
National Design
Specifications for Wood
Construction (USDA/
FPL), 164
National Electrical Mfgs.
Assn., *High-Pressure*
Decorative Laminates, 197
Neoprene, with acrylic, 135
Neufeld, Larry, stone-faced
block walls of, 97-99
Nevamar Corp., 197
Nicro Fico pulleys, 199

O

Oakland Carbide
Engineering Co., router
bit of, 168, 169
O'Hagan, Daniel, 170
Outriggers (roof), 33, 37
Owens-Corning fiberglass
insulation, 23
Owner-Builder's Guide to
Stone Masonry, The
(Kern, Magers and
Penfield), 76
Owner Built Home, The
(Kern), 108

P

Paint:
below-ground, 13
can drainage for, 147
for log building, 181
Paintbrushes, 147
Parker, H. F., *Simplified Engineering for Architects and Builders,* 164
Partridge, Lt. Samuel, 221
Paslode Co., 125
nail guns, 121, 123
Payne, George M., 147
Peake, Jud:
on portable circular-saw use, 109-11
on rafter square, 24-29
Peavey, 165
PEG. *See* Princeton Energy Group.
Pencils, 127
Penfield, Lon, Ken Kern and Steve Magers, *Owner-Builder's Guide to Stone Masonry, The,* 76
Pentachlorophenol, 181
Phenoseal, 202, 233
Picton, Jim:
on flashing, 52-53
on installing skylights, 47-49
Pioneer Plastics, 197
Pitch. *See* Bituminous coatings.
Pitch board, 29
Pitch line, 31
Pitch (roof), 31
Planes:
block, 127
compass, 222-23
jack, 156
metal, restoring, 158-59
multi-, 219
rabbet, 156
router, 156
smooth, 156
See also Power planes.
Plaster veneer, 118, 119, 120
Plastic laminate. *See* Countertops, plastic-laminated.
Plates:
cutting, 109, 110
roof, 31-33
Plexiglas, 135
Pliers, 127
Plumb bob, 126
extension for, 147
stabilizing, 100
Plumb cut, 31
Plumbing:
resoldering, 101
in truss frames, 23
Plumbing stacks, and air/vapor barriers, 19
Plywood:
cutting jig for, 104, 105
foundations of, 144-46
for subfloors, 210-11
Pneumatic tools, suppliers of, 125
See also Nail guns.
Polyethylene, below-ground use of, 13
Polyshim tape, 51
Polyvinyl chloride. *See* PVC.

Porch additions, 232-35
Porta-Tools, 213
Porter-Cable:
Power Block Plane, 128
Versa Plane, 128, 129
See also Rockwell.
Portland cement, fill coat, below-ground, 13
See also Concrete.
Post, Irwin L., on insulated doors, 208-209
Post, Irwin L. and Diane L., on wood foundations, 144-46
Power Nailer, 213
Power planes, 128-31
for timber framing, 156
PRC sealant, 16
Preshim tape, 50
Princeton Energy Group, on acrylic glazing, 132-35
Pulleys, 165, 199
Punches, self-centering screw, 127
Purlins, 31, 35, 36
Putty knives, 127
PVC sheets, below-ground use of, 13

Q

Q-Bond Corp. of America, 91
Quick Wall, 91
Quikcrete, 91

R

Rabek, Norman, 59
Rafter pattern, 31
Rafters. *See* Roofs.
Rafter square, 24-29
Rails, spring-fitted, 236
Rake, 31
Rammed-earth houses, 83-87
Ramps, 16
Record planes, 156
Red Cedar Shingle and Handsplit Shake Bureau, 44
Red Devil scrapers, 127
Rekdahl, Eric K., 185
Renovation, tools for, 126-27
Repointing, 79-81
Ridgeboards, 31, 35-36.
See also Shingling, roof.
Ridge reduction, 31
Riechers, A. F., *The Full Length Roof Framer,* 28
Rise, 31
Roberts, Earl K., 185
Rockwell:
Porta-plane, 128, 131
Speedmaster circular saw, 157
Unisaw, 205
See also Porter-Cable.
Rohn and Haas, Plexiglas, 135
Roof stacks:
and air/vapor barrier, 19
flashing of, 58
Roofs:
for additions, 234-35
books on, 28, 34

calculator equations for, 29
collar ties for, 31, 36
flashing of, 42, 43, 56-58
framing of, 30
frieze blocks for, 31, 36
glossary for, 31
layout of, 33
layout tees for, 29-31, 33-34
nail guns for, 121, 125
pitch boards for, 29
purlins for, 31, 35, 36-37
rafter installation for, 30, 35-36
rafter length calculations for, 29, 30, 34
rafters of, gang cutting of, 34-35
ridge height of, 35
stair-gauge fixtures for, 29
superinsulated, 18-19
tools for, 24-29, 33
waterproofing, 15, 86
See also Cornices; Rafter square; Scaffolds; Shingling, roof; Skylights.
Rope, trucker's hitch in, 167
Roto Stella skylights, 53
Routers:
bearing-over cutter trim bit for, 168, 169
dadoing jig for, 105
flush-cutting trim bit for, 168, 169
hand, 156
jointing jig for, 105
plastic laminate trimming with, 197
tablesaw mounting of, 105
timber-framing templates for, 168
Rules, 126
Run, 31
Russian fireplaces, 71-73

S

Sacks, Peter, 236
Salvadori, Mario, and Robert Heller, *Structure in Architecture,* 164
Sanders, 216-17
for moldings, 205
Sandstrom, John H., 185
Sanford, Jim, 133, 134-35
Saskatchewan Conservation House, 16
Sawhorses, 147
Saws, 127
keyhole, 114
for timber framing, 157
Sawzall, 185
Scaffolds:
need for, 35
roof brackets for, 41
wall brackets for, 38-39
Schipa, Gregory, on reproduction doorway, 218-23
Scraper, 127
Screwdrivers, 127
Screws, 127
Scribers:
double-bubble level, 177
homemade, 185
large, 229
Sealants:
for air/vapor barriers, 16

for brick, 69, 79
for concrete cold joints, 14
for rammed earth, 85
See also Caulking.
Sears Craftsman Screwdrivers, 127
Section modulus, 161, 162
Seddon, Leigh, *Low-Cost Green Lumber Construction,* 108
Sellers & Co., 132
acrylic glazing of, 133, 134
Sellers, David, on acrylic glazing, 132-33, 135
Senco Products, Inc., 125
coated nails, 123
nail guns, 123
Sharpening, 159
Shear, 160, 161, 163, 164
Sheathing, 109
cutting in place, 110
nail guns for, 125
See also Shingling, roof.
Sheetrock. *See* Drywall.
Sheldon, George, 221
Shelves, rests for, 194
Shingling hatchet, 38, 39
Shingling, roof, 38-43, 56-59
circular-saw corner trimming of, 110
Shingling, sidewall, 44-46
circular-saw corner trimming of, 110
Side cutters, 127
Siding:
nail guns for, 121
nailing gauges for, 59
power plane with, 130
Simplex Products, 199
Simplified Engineering for Architects and Builders (Parker), 164
Site layout, 10, 11, 12
Skil 860 circular saw, 157
Skimcoat. *See* Plaster veneer.
Skylights, 47-53
Slicks, 151, 154, 156-57, 175
Sliding bevel, 126, 155
Slope (roof), 31
Snyder, Tim:
on sidewall shingling, 44-46
on stone-faced block work, 97-99
on veneer plaster, 118-20
Sockets, 127
Sodium montmorillonite. *See* Bentonite.
Soffit, 201, 202
Songey, Mark, on timber templates, 168-69
Sonneborn Building Products, 96
Sorby chisels, 151
Span, 31
Speed Square, 28-29
Sperber Tool Works, 184
Spotnails, 125
Spring, Paul, 59
on ordering concrete, 141
on plastic laminates, 197
Squangle, 28
Squares, 126
truing, 158
Stack & Bond, 91
Staircases, 224-28

Stair-gauge fixtures, 29, 185
Stangel, Harry W., 236
Stanley:
Jobmaster screwdrivers, 127
multiplane, using, 219
planes, 156
Powerlock II tape, 126
router, 156
saws, 127
self-centering screw punch, 127
Surform, 114, 127
Wonder bar, 127
Staplers. *See* Nail guns.
Stead, Craig F., on renovating tools, 126-27
Stilts, 118, 119
Stone Mountain Mfg. Co., 91
Stone walls, 74-78
as facing, 97-99
Story pole:
for block work, 94
for sidewall shingling, 46
trench depths with, 12
Straightedges, truing, 158
Structure in Architecture (Salvadori and Heller), 164
Structures, or Why Things Don't Fall Down (Gordon), 164
Sunrise Builders, 100
Superinsulated houses, 16-19
Surewall, 91
Swanson Tool Co., 28
Swedlow Inc., Acrivue A, 135
Sweet william, 165, 166
Syvanen, Bob:
on brick floors, 66-69
on concrete, 142-43
on cornice construction, 200-3
on dry wall, 112-17
on flashing, 54-58
Interior Finish, 112
on roof shingling, 38-43

T

Tablesaws:
crosscut jig for, 189-90
homemade, 104-5
moldings on, 204-6
plastic laminate on, 197
as router table, 105
Tail, 31
Tail cut, 31
Tension, 161
Theoretical length, 31
Thermax, 199
Thermoply, 198, 199
Thincoat. *See* Plaster veneer.
3-4-5 rule, 11-12
3M V-Seal weatherstrip, 199
Thyrring, Chris, 147
Timber Construction Manual (USGPO), 164
Timber framing, 150-69
curves with, 185

Timber tongs, 165, 166
Tinsnips, 127
Tintle, Richard, 222
T-nailers. *See* Nail guns.
Tools:
    catalogs of, 127
    for renovation, 126-27
    shopping for, 127
    sliding chest for, 147
    *See also specific tool or
    machine.*
Tool Works, The, 151
Tremco:
    butyl tape, 134
    Polyshim tape, 51
    Preshim spacer-rod tape,
    50
    sealant, 16
Trim, cutting, 111
    power plane with, 130
Trimmer rafter, 31
Trowels, 115
Trucker's hitch, 167
Trusses, Fink, 37
Truss-frame construction,
    20-23
T-squares, aluminum, 114

U

Underground houses. *See*
    Earth-sheltered houses.
United States Department
    of Agriculture Forest
    Products Laboratory:
    *National Design
        Specifications for Wood
        Construction,* 164
    solar kilns of, 107
    truss-frame house of, 20,
    21
    on wood foundations, 144
    *Wood Handbook: Wood as
        an Engineering Material,*
    144, 164, 208
    *Wood Structural Design
        Data,* 164
Unit rise, 31
Unit run, 31
University of Illinois Small
    Homes Building Council,
    16
Urethane, with acrylic, 135
Utility knives, 114, 119, 127

V

Valenzuela, Chris, 101
Valley rafters, 30, 31, 59
Valleys (roof), 31
    See also Flashing,
    Shingling, roof.
Valspar Corp. Polyurethane
    Varnish, 69
Vandal, Norman, on
    curvilinear sash, 229-31
VanEE Air Exchangers, 19
Varnish, for log buildings,
    181
Velux skylight, 47
Vinyl, with acrylic, 135
Vise grips, 127
Volclay, 14, 15

W

Wade, Alex, 135
Wallace, Alasdair G. B., on
    log building, 176-81
Walsh, Valerie, acrylic
    glazing of, 132-35
Walter, David, 100, 101
Water level, 11
Waterproofing: roofs, 15, 86
    See also Earth-sheltered
    houses.
Waterstop-Plus, 14
Water table, 203
Weather Hill Restoration
    Co., 223
Weather stripping, V-Seal,
    199
Westinghouse Electrical
    Corp., 197
Whiteley, Bob, 101
White, Mark, on truss-
    frames, 20-23
Williams, John, house of,
    218-21
Wilsonart, 197
Windows:
    flashing, 44-45, 55
    vapor barriers with, 117-
    18
Windows, curved, 229-31
Wing, Charles, and John
    Cole, *From the Ground Up,*
    108
Winooski Block, 204-5
Wire strippers, 127
Wiring:
    in old walls, 59
    taping, 185
    in truss frames, 23
Wood:
    books on, 164
    defects in, 107
    green, 106-108
    heavy, moving, 165-67
    structural glossary of, 161
Woodcraft Supply Corp.,
    151, 177
Wood foundations, 144-46
Wood Handbook (USGPO),
    144, 164, 208
Woodlife, 233
Woodline-The Japan
    Woodworker, 151
Wood preservatives, 181
*Wood Structural Design
    Data* (USDA/FPL), 164
Workbenches, 126
Workshops, on-site, 104-5
W. R. Bonsal Co., 91
Wrenches, 127

Z

Zimmerman, Philip, 236